high frequency circuit design

by james k. hardy

with illustrations

by patricia hardy

RESTON PUBLISHING COMPANY, INC.
A Prentice-Hall Company
RESTON, VIRGINIA

Library of Congress Cataloging in Publication Data

Hardy, James K   High frequency circuit design.

Includes index. 1. Telecommunication—Apparatus and supplies—Design and construction. 2. Electronic circuit design. 3. Amplifiers (Electronics) 4. Oscillators, Electric. 5. Electric filters. I. Title.
TK5103.H38   621.3815'3   78-27778
ISBN 0-8359-2824-1© 1979 by Preston Publishing Company, Inc.
A Prentice-Hall Company
Reston, Virginia 22090

10 9 8 7 6 5 4 3

Printed in the United States of America

$ 21.95

contents

dedicated to

patricia, shawna, and sandra.

preface

High-frequency circuit design is written for those interested in a practical approach to the design of high-frequency amplifiers, oscillators, and filters. It will be valuable to technology and engineering students, practicing designers in the communications industry, and to experimenters and radio amateurs. The material complements a number of first-level communications circuit texts that are available and so concentrates on topics not normally covered by them in sufficient depth for design work.

This book is the outgrowth of a sixth semester course taught to technology students at Humber College. For maximum benefit, the reader should therefore have the equivalent of a course in ac circuit theory and the associate vector algebra and should also have completed a first-level course in communication circuits and modulation and demodulation. The mathematics within this book stays at the algebra level and involves no calculus.

Chapter 1 describes the circuitry problems that must be considered during the design process. These include internally generated noise and non-linear amplitude and phase characteristics, all of which can distort a signal as it passes through a communication system.

Chapter 2 discusses the practical nature of passive components and how their characteristics change with frequency, temperature, and in some cases, with the signal amplitude. The proper selection of components is necessary to make the final product operate as well as the initial paper work design.

The circuitry design begins with Chapter 3 where procedures for designing narrow passband filters to meet given bandwidth and skirt requirements are explained. A description of crystal and ceramic filters is included, not so much because the reader is likely to be designing them, but more for an explanation of the principles involved that may be useful in other filter designs.

Modern filter design, in Chapter 4 is a topic normally found only in high-level texts buried in a lot of mathematics. The general properties of filters are presented first, and then a specific description of input impedance, attenuation and delay characteristics are given for a group of standard designs. With the graphs, tables, and formulas provided the reader can design a filter to custom fit his exact requirements and with very good predictability.

For the student interested in a bit of the mathematical background to modern filters, Appendix 4A describes the *s*-plane and its relation to filter characteristics and also demonstrates how filter component values can be extracted from polynomial equations that describe the desired response.

To preserve precious signal levels, both large and small, communication circuits often require impedance-matching networks that will transfer maximum power from a source to a load. Often these are required to have bandwidths of varying amounts and provide specific attenuation outside of the passband. Designs to meet all requirements are given and are handled both mathematically and graphically—on the Smith chart.

As operating frequencies climb, the characteristics of transistors, be they bipolar or field effect, change. Their gains drop, input and output impedances decrease and become reactive, and stability decreases. Chapter 6 then serves as background information for the remaining chapters—all dealing with active circuits. The first of the active circuits are the small linear amplifiers commonly used in receivers. Chapter 7 concentrates on the design of such amplifiers and shows how to control internal noise, provide automatic gain adjustment, improve stability, and use alternate configurations to best ability.

Chapter 8 discusses oscillator circuits; what should be considered and how to design and measure the results. The traditional Colpitts and Hartley designs are presented and then the more modern technique of frequency synthesis is described. To fully appreciate this section the student may wish to brush up on the operation of logic gates first.

Chapter 9 is a collection of circuits common to transmitters and begins with a description of the specialized RF power transistor, often as complex as any integrated circuit. The chapter continues with the design of power amplifiers and how to amplitude modulate them.

Chapter 10 is a collection of receiver circuits, in particular the more critical areas which include the mixer and the detector sections. AM and FM IF amplifiers are also described.

It is a rare author who works completely by himself without outside help. In this case the author is very thankful for all the help received both from business associates and personal friends; without them there would be no book.

In particular I wish to heartily thank Linda Vince and Christine Adam for their expert typing of the manuscript, Lou Hale of Garrett Manufacturing for his assistance in providing information on components, and Bob Vince for his photographic advice. I also wish to thank my wife Patricia for her typing, drawing, and moral support throughout the project. A special thanks also to that special group of friends who provided both technical and moral support.

I hope that the readers enjoy the use of this material as much as the author did in writing it.

Jim Hardy
Georgetown, Ontario
July, 1978

chapter 1

signal distortion

The purpose of any communication system is to carry information from one place to another. However, for various reasons, the received signal will always be of poorer quality than the transmitted signal. Some of the frequency components may be of the wrong amplitude or missing altogether, new frequencies may appear, the relative time relation of the signals may be altered, and random noise may contaminate everything. Careful control of these problems during the circuit design will result in a more useful product.

1-1 THERMAL (JOHNSON) NOISE

Whenever resistance appears in a circuit, whether actual resistors for biasing or just losses in components, noise will be generated. *Thermal noise* is caused by the random movements of free electrons in any conductor. If the conductor has any resistance, a noise voltage will be generated. Since the electron movement increases with temperature, the noise voltage will, too. The average amplitude of this noise will be constant at all frequencies, and so the noise is *white*. (*Pink* noise has an amplitude that decreases with frequency.) The appearance of white noise on an oscilloscope is shown in Figure 1-1.

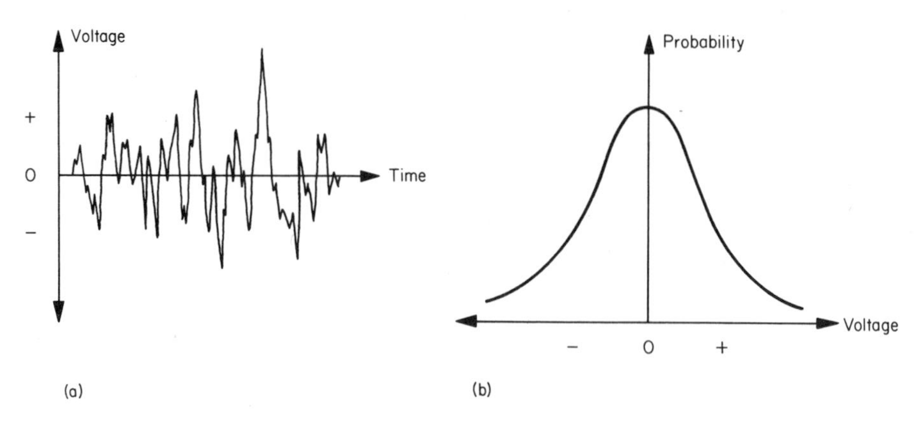

Figure 1-1 Typical white-noise voltage (a) and the probability of any voltage occurring (b).

At any instant, the amplitude of the noise voltage could be anything from very small to very large values. As shown by the probability curve, it is most likely that the amplitudes will be small; but, theoretically, spikes of infinite amplitude could occur. Noise cannot, therefore, be measured as peak voltages, and so the rms value must always be used. The open-circuit noise voltage across the terminals of a resistor is given by,

$$e_n = \sqrt{4KTBR} \qquad \textit{volts (rms)} \qquad (1\text{-}1)$$

where: K = Boltzmann's constant
$= 1.38 \times 10^{-23}$ J/°K
T = absolute temperature (°C + 273)
R = resistance (Ω)
B = bandwidth (Hz)

EXAMPLE 1-1

What noise voltage is generated in a 50-Ω resistor at room temperature if the measuring instrument has a 1.0-MHz effective bandwidth?

Solution

Room temperature will be taken as +20°C or 293 °K.

$$\begin{aligned} e_n &= \sqrt{4KTBR} \\ &= \sqrt{4 \times 1.38 \times 10^{-23} \times 293 \times 1.0 \times 10^6 \times 50} \\ &= 0.899\ \mu\text{V rms} \end{aligned}$$

If the 50-Ω resistor used in this example was actually the source resistance of a signal generator, the normal output signal would be contaminated with noise as shown in Figure 1-2. Therefore, when any signal generator is used for testing, this noise must be taken into account.

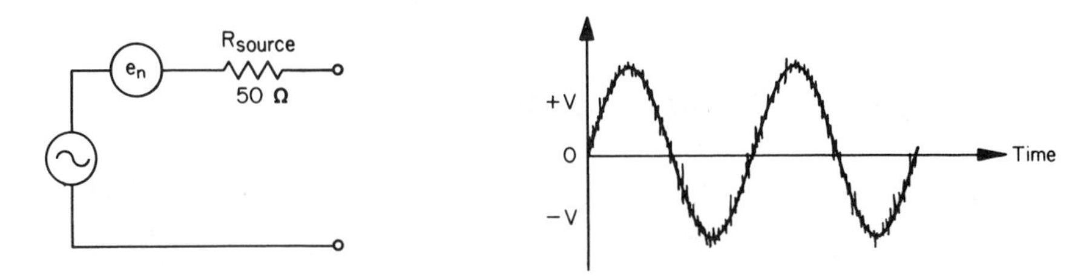

Figure 1-2 Output of a signal generator will contain some thermal noise produced in its source resistance.

The ratio of signal voltage to noise voltage from the generator is

$$\frac{S}{N}\text{ ratio} = \frac{\text{signal voltage (rms)}}{\text{noise voltage (rms)}} \qquad (1\text{-}2)$$

or

$$\frac{S}{N}\text{ ratio (dB)} = 20\log_{10}\left(\frac{\text{signal voltage}}{\text{noise voltage}}\right) \qquad (1\text{-}3)$$

It is normally assumed that all the generator noise is thermally produced in the source resistance and that the source resistance is room temperature (290°K). However, it would be easy to make a poor-quality oscillator that has a noise level far above the thermal level; fortunately, it would not sell very well.

1-2 SHOT (SCHOTTKY) NOISE

The second major source of noise in communication circuits is produced when current flows in any active device, be it bipolar or field-effect transistor, diode or vacuum tube. This noise is also white and has exactly the same appearance and probability distribution, as shown in Figure 1-1. Again, rms values must be used for its measurement, but in this case, it is usually expressed as a current. For a semiconductor diode,

$$I_n = \sqrt{2qIB} \qquad \text{amperes (rms)} \qquad (1\text{-}4)$$

where: q = charge on an electron
$= 1.60 \times 10^{-19}$ C
I = bias current (A)
B = bandwidth (Hz)

For bipolar transistors, the equation differs somewhat, since two currents and two junctions are involved. The noise produced in a transistor will be discussed in Chapter 6.

1-3 AMPLIFIER NOISE FIGURE

Because of thermal and shot noise produced in an amplifier, the signal at the output will have a higher percentage of noise than the input signal did. The amount of noise added will then determine how small a signal the amplifier can detect. A useful measure of this performance is the noise factor and corresponding noise figure:

$$\text{noise factor} = \frac{S_i/N_i}{S_o/N_o} \tag{1-5}$$

$$\text{noise figure} = 20\log_{10}\left(\frac{S_i/N_i}{S_o/N_o}\right)\ \text{dB} \tag{1-6}$$

where: S_i = input signal voltage
N_i = thermal input noise voltage from the source resistance
S_o = output signal voltage equal to $S_i \times$ amplifier gain
N_o = total output noise made up of the amplified input plus any added noise

Typical noise figure for several types of low-noise amplifiers are shown in Figure 1-3. The problem of obtaining low-noise figures at the higher frequencies is mainly due to the difficulty of obtaining reasonable power gains at those frequencies.

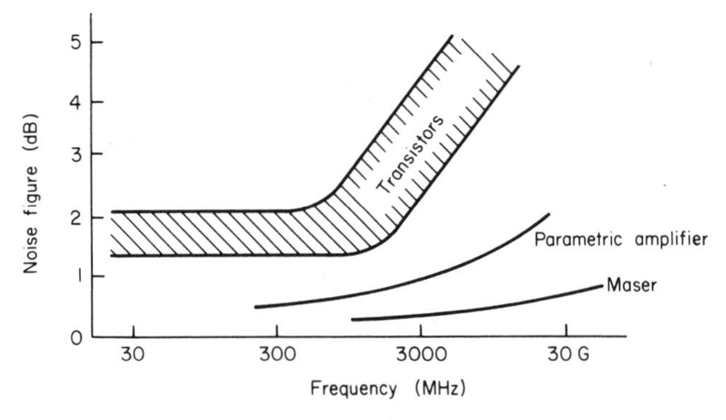

Figure 1-3 Noise figures that can be obtained with various types of low-noise amplifiers.

1-4 MEASUREMENT OF NOISE FIGURE

The *noise figure* (or *factor*) can be measured in one of two ways, depending on the type of generator used. The easiest measurement requires a calibrated noise source. This usually consists of a vacuum-tube diode whose dc plate current is controlled by changing the filament temperature (saturated operation). The shot noise from the tube is then added to the thermal noise of a 50-Ω resistor that acts as the source resistance of the noise generator. The excess shot noise can then be calibrated on a meter in terms of plate current.

To measure noise figure, the noise source is connected to the input of the receiver to be tested, and a true rms voltmeter is connected to the receiver output. (Since linearity is important, any automatic gain control must be defeated, so the voltmeter may have to be connected ahead of the receiver's detector.) With the noise source turned off, the voltmeter will read some level corresponding to the internally generated receiver noise plus the thermal noise from the generator's source resistance. The noise source is then turned on and adjusted so that the output noise level increases by 3 dB; the more excess noise that must be generated, the more noise the receiver is adding itself. The noise figure is then simply read off the meter of the noise generator. Figure 1-4 shows the general form of the noise generator and test setup. The big advantage of this measurement technique is that the receiver's gain and bandwidth characteristics do not have to be known, since the noise generator's output voltage is

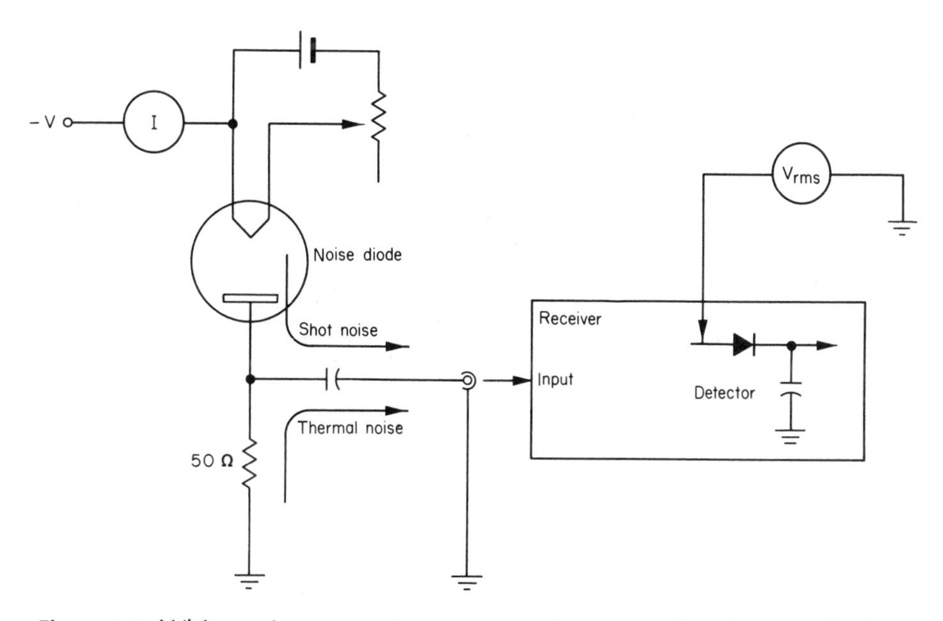

Figure 1-4 White-noise generator connected to a receiver for a noise-figure test. The rms voltmeter must be connected to a linear point in the receiver, which usually means ahead of the final detector.

a function of bandwidth anyway. Figure 1-5 shows a commercial noise-figure test set that includes both the source and the voltmeter. In this instrument, the noise source is turned on and off 1000 times per second, so the operation is completely automatic.

The second method of measuring noise figure is a bit messy. Without a white-noise source, the gain and bandwidth characteristics of the receiver must be known. A continuous-wave generator or sweep generator can be used to plot the amplitude characteristics of the amplifier. From this, the midfrequency gain and the effective noise bandwidth are determined. The effective noise bandwidth will be somewhat wider than the 3-dB bandwidth by an amount that depends on the skirt slope. For the ideal case of vertical skirts, the two bandwidths will be the same. For the typical response curve generated by a single tuned circuit, the noise bandwidth will be about 55% wider than the 3-dB bandwidth. Other filters will lie somewhere between these two limits.

The total output noise voltage from the amplifier is measured with a true rms voltmeter, and then the output noise power is calculated using this noise voltage and the value of resistive impedance across which the voltage was measured. This noise will be due to the amplified thermal source resistance noise (the amplifier input must be loaded with the appropriate value source resistance) plus the added noise from the amplifier itself. The amplifier noise figure is then

$$\mathrm{NF_{db}} = 10\log\left(\frac{N_o}{GKTB}\right) \qquad (1\text{-}7)$$

where: N_o = noise power output (W)
G = amplifier power gain (not in dB)
K = Boltzmann's constant
T = 290 °K (room temperature)
B = effective noise bandwidth (Hz)

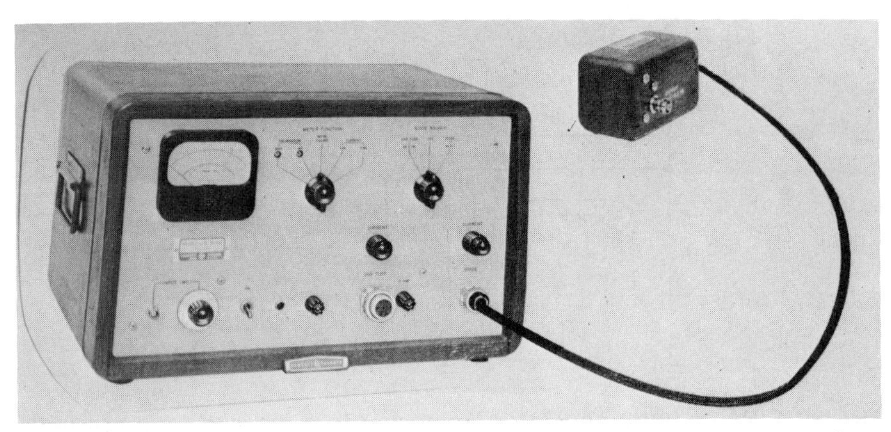

Figure 1-5 Commercial noise-figure test set. (Courtesy of Hewlett-Packard.)

For wide-band amplifiers, where noise figure must be measured at several different frequencies, an external narrow-band filter is placed at the output to set the noise bandwidth at each test frequency. Either method of measurement can then be used to measure the noise figure.

1-5 AMPLITUDE DISTORTION

When an amplifier has a perfectly linear amplitude characteristic, there will be a constant relationship between the output and the input voltage; for example, $e_{out} = 10 \times e_{in}$ represents a linear voltage gain of 10. The output wave will have exactly the same shape as the input and no new frequencies will appear. For low-frequency circuits using plenty of negative feedback, the gain can be kept linear to better than 0.1%, as long as the output signal stays within the power-supply limits. For high-frequency circuits, where less gain is available, amplifiers are often designed with little or no feedback. The inherently nonlinear characteristics of transistors can then create numerous problems: the output wave from a nonlinear amplifier will not be the same shape as the input wave, new frequency components will be created, and mysterious modulation may "appear" from signals that are completely outside the passband of the amplifier. The problems that are created depend on the "types" of amplitude nonlinearities present.

1-5.1 Square-Law Gain

If the gain of an amplifier can be described by an equation similar to

$$e_{out} = 10e_{in} + 3e_{in}^2$$

certain types of distortion will occur. The transfer characteristic from this amplifier is shown in Figure 1-6. The curve is referred to as *square-law* or *second-order* because of the e_{in}^2 term in the transfer equation. The larger the coefficient of this term, the more curved the characteristic and the greater the distortion problem. The result of a single sine-wave input is shown in the time domain (b) and the frequency domain (c) in Figure 1-6. The changes that have occurred in the output are due to *second harmonic distortion*, since a new frequency has been created at twice the input frequency. (A dc component is also produced.)

The amplitude of this second harmonic component is worth taking a look at. The fundamental frequency component will always have an amplitude at the output that is 10 times the input signal. The second harmonic, however, has an amplitude that depends on the *square* of the input *and* also the coefficient 3. As a result, the second harmonic increases in amplitude at a faster rate than does the fundamental, as shown in Figure 1-7. The use of logarithmic scales

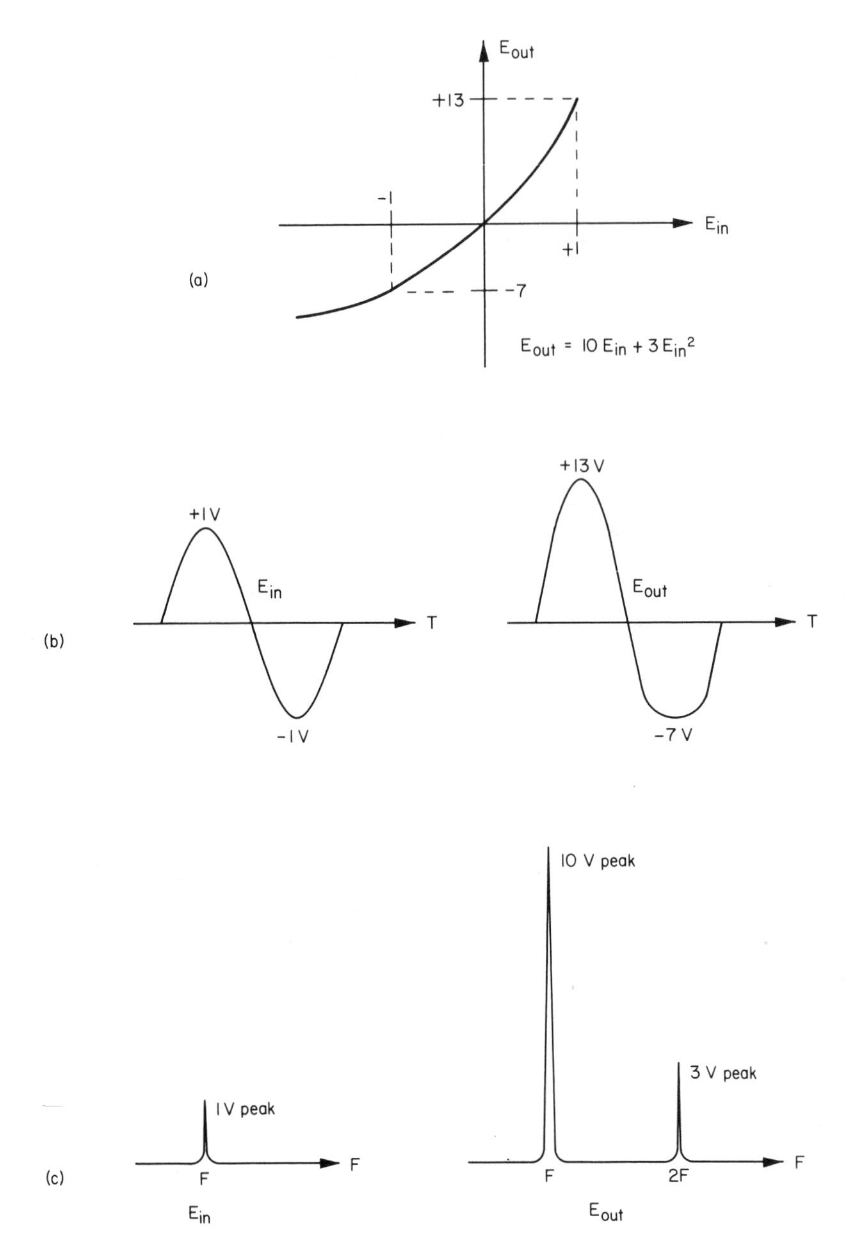

Figure 1-6 Typical transfer characteristic for an amplifier with second-order distortion (a). The resulting distortion of an input sine wave is shown in (b), and the corresponding frequency spectrums of the input and output waves are shown in (c).

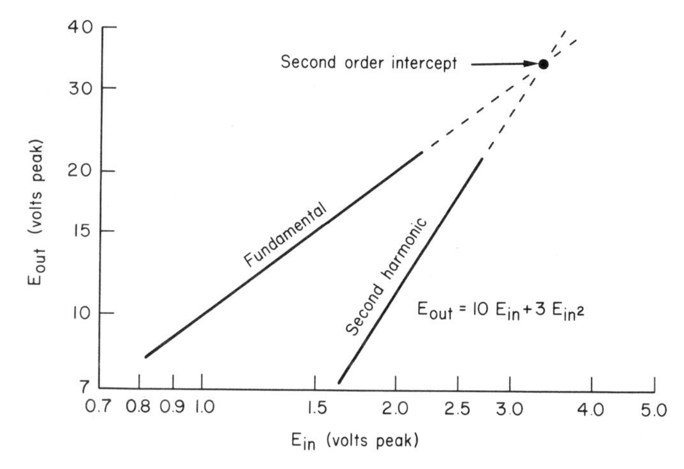

Figure 1-7 Relative size of fundamental and second harmonic outputs will depend on the size of the input signal.

conveniently results in straight lines. The top end of each line is shown dashed, since very high values of input and output voltages would not be possible in a practical amplifier with fixed power-supply voltages and the like. The amount of distortion could be expressed as *second harmonic amplitude* as a percentage of *fundamental amplitude*, but this keeps changing with the input signal size. The other possibility is to indicate the input level that produces an equal amount of fundamental and second harmonic output. This is called the *second-order intercept point* and for our example occurs when e_{in} is 3.33 V.

1-5.2 Two Input Signals

When two signals of different frequencies are applied to an amplifier with square-law characteristics, the results become more complex. Each signal will again have a second harmonic and additional sum and difference frequencies appear. Frequency spectrums of the two input signals and the resulting outputs are shown in Figure 1-8. All the new signals that appear in the output are called *second-order intermodulation products*, since they are created by the e_{in}^2 term. The sum of the terms for each frequency is always 2 (i.e., $1 \times f_2 - 1 \times f_1$, $2 \times f_2$, $1 \times f_1 + 1 \times f_2$). The amplitude of each product is proportional to either the input voltage squared (for $2f_1$ and $2f_2$ products) or to the product of the two input amplitudes (for $f_2 \pm f_1$).

While any type of distortion is not particularly desirable in an amplifier, other types can be worse than these. For example, if the two signals are applied to a narrow-bandwidth amplifier (up to 1-octave bandwidth), all the intermodulation products will likely fall outside the passband and so can be ignored. If

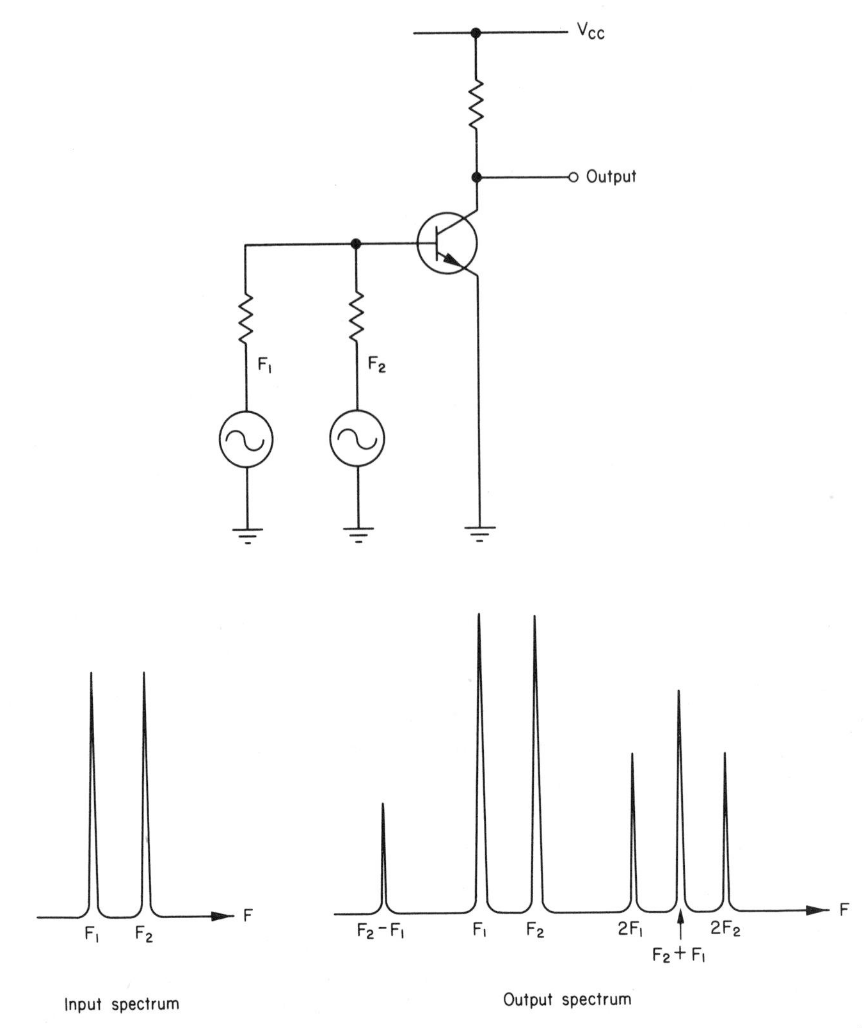

Figure 1-8 Second-order intermodulation products that are added to the output of a square-law amplifier with two input signals.

automatic gain control is required in a narrow-band transistor amplifier, the square-law characteristic is mandatory. Different gains are obtained by changing the dc bias to provide operation on different parts of the curve. As long as the square-law characteristic is maintained, the amplitude relation between input and fundamental output, over a 1-octave bandwidth, will be perfectly linear and no compression will occur.

Another useful application of the square-law characteristic is in mixer circuits for superheterodyne receivers, where one input is a local oscillator and the other the incoming RF signal. A tuned circuit is then used to select the difference frequency (down-converter) or the sum frequency (up-converter) for further amplification in the intermediate frequency amplifier.

1-5.3 Third-Order Gain

A different type of distortion problem occurs if the amplifier has a third-order gain characteristic. The transfer equation might be of the form

$$e_{\text{out}} = 10e_{\text{in}} - 2e_{\text{in}}^3$$

This would result in the transfer curve shown in Figure 1-9. The distortion of a single-input sine wave is shown in both the time and frequency domain. Notice that the distortion of the output wave is symmetrical above and below the horizontal axis. A third harmonic now appears along with the original input signals, hence the name *third harmonic distortion*. (No dc component exists for third-order distortion.)

The amplitude of both components is worth looking at. The third harmonic, as you would expect, has an amplitude dependent on the *cube* of the input voltage *and* also on the coefficient 2. But, unlike the second-order characteristic, there is no longer a linear relation between the input signal and the fundamental output. One of the *third-order products* is $2f_1 - f_1$ (the sum of the coefficients is 3), and this of course occurs at the fundamental frequency and so partially cancels[1] the fundamental input that is 10 times the input. This loss of linearity is an important problem, as we shall see shortly.

The amplitude relationship between input and output signals of an amplifier with third-order distortion is shown in Figure 1-10. The third harmonic amplitude rises very steeply and the fundamental amplitude rises proportional to input initially and then slows down and decreases. The amount of distortion can again be described in terms of an intercept point—in this case the *third-order intercept*. As before, these signal amplitudes may be beyond the

---

[1]When e_{in} is a single sine wave with a peak amplitude E_m so that

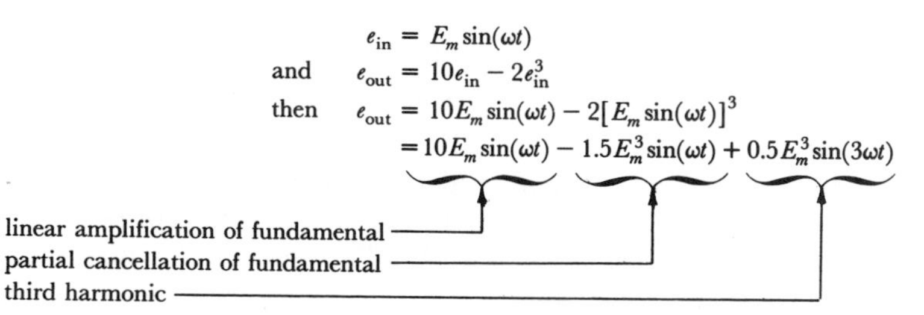

$$e_{\text{in}} = E_m \sin(\omega t)$$

and $$e_{\text{out}} = 10e_{\text{in}} - 2e_{\text{in}}^3$$

then $$e_{\text{out}} = 10E_m \sin(\omega t) - 2[E_m \sin(\omega t)]^3$$
$$= 10E_m \sin(\omega t) - 1.5E_m^3 \sin(\omega t) + 0.5E_m^3 \sin(3\omega t)$$

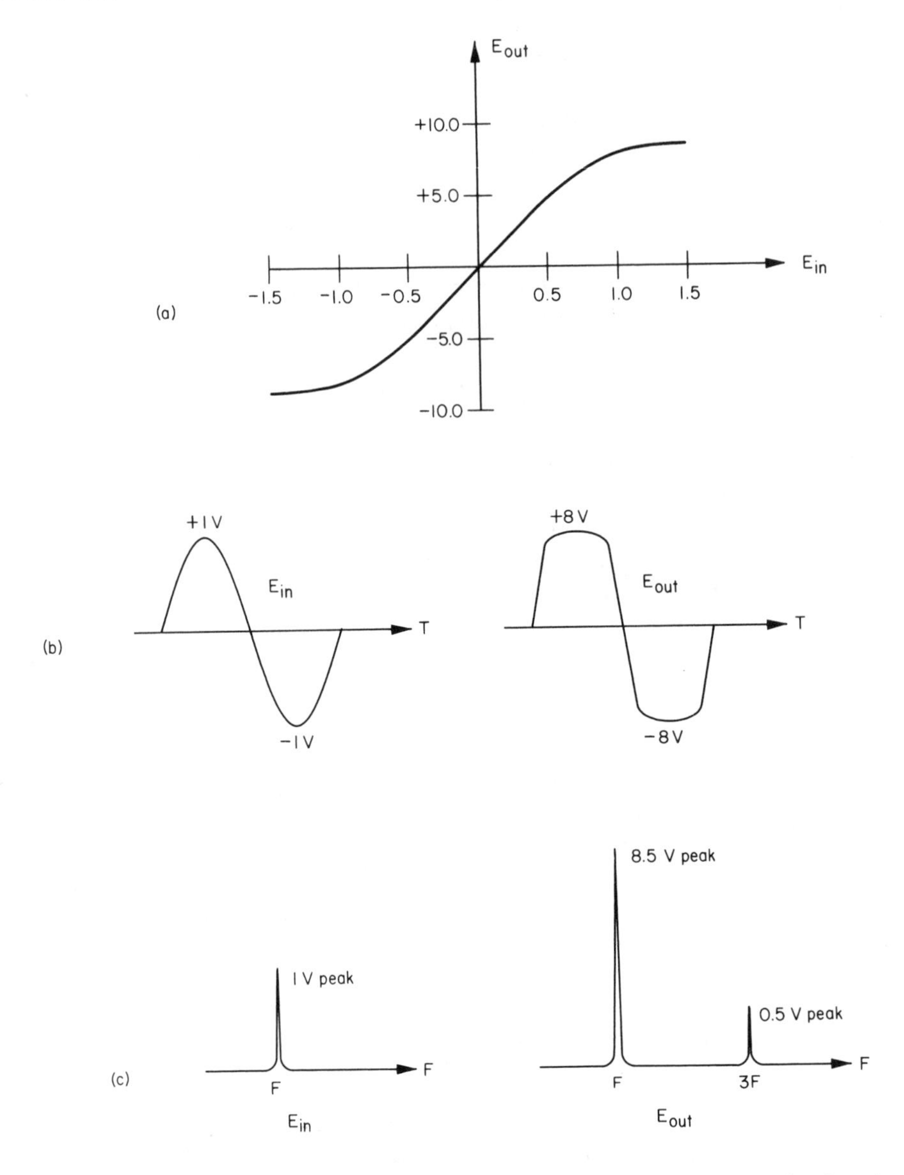

Figure 1-9 Typical transfer characteristic for an amplifier with third-order distortion (a). The distortion of a single-input sine wave is shown in (b) and the spectrum of the frequency components is shown in (c).

saturation cutoff limits of the transistors in the amplifier, and so the lines may have to be projected (dashed lines) to find this theoretical intercept point. For our example, this occurs when e_{in} is 4.47 V. The smaller the third-order distortion, the higher the intercept point.

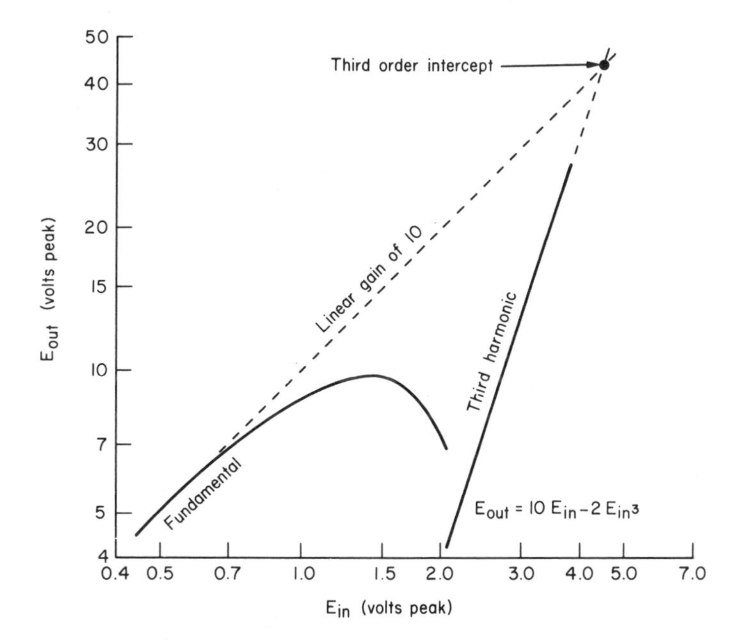

Figure 1-10 Amplitudes of the output fundamental and third harmonic from an amplifier with third-order distortion. The fundamental amplitude is no longer proportional to the input.

1-5.4 Two Input Signals

When two frequencies are applied to the input of an amplifier with any order distortion, new frequencies will be generated at

$$f = n \cdot f_1 \pm m \cdot f_2 \qquad (1\text{-}8)$$

where:

n and m can be any whole number $0, 1, 2, 3, \ldots$
$n + m$ is the order of the distortion

For two input signals to an amplifier with third-order distortion, we will find new outputs at

$$f = 3f_1, \quad 2f_1 \pm f_2, \quad 1f_1 \pm 2f_2, \quad 3f_2 \qquad (1\text{-}9)$$

These, the *third-order intermodulation products*, are shown on the spectrum of Figure 1-11. As with the single signal input, the two original signals f_1 and f_2 that appear in the output will be smaller than the $10 \times e_{in}$ that would be expected due to partial cancellation by additional third-order products. This effect is called *compression*. The most serious problem associated with third-order distortion is shown in Figure 1-11. Even for a narrow-band amplifier, some of the

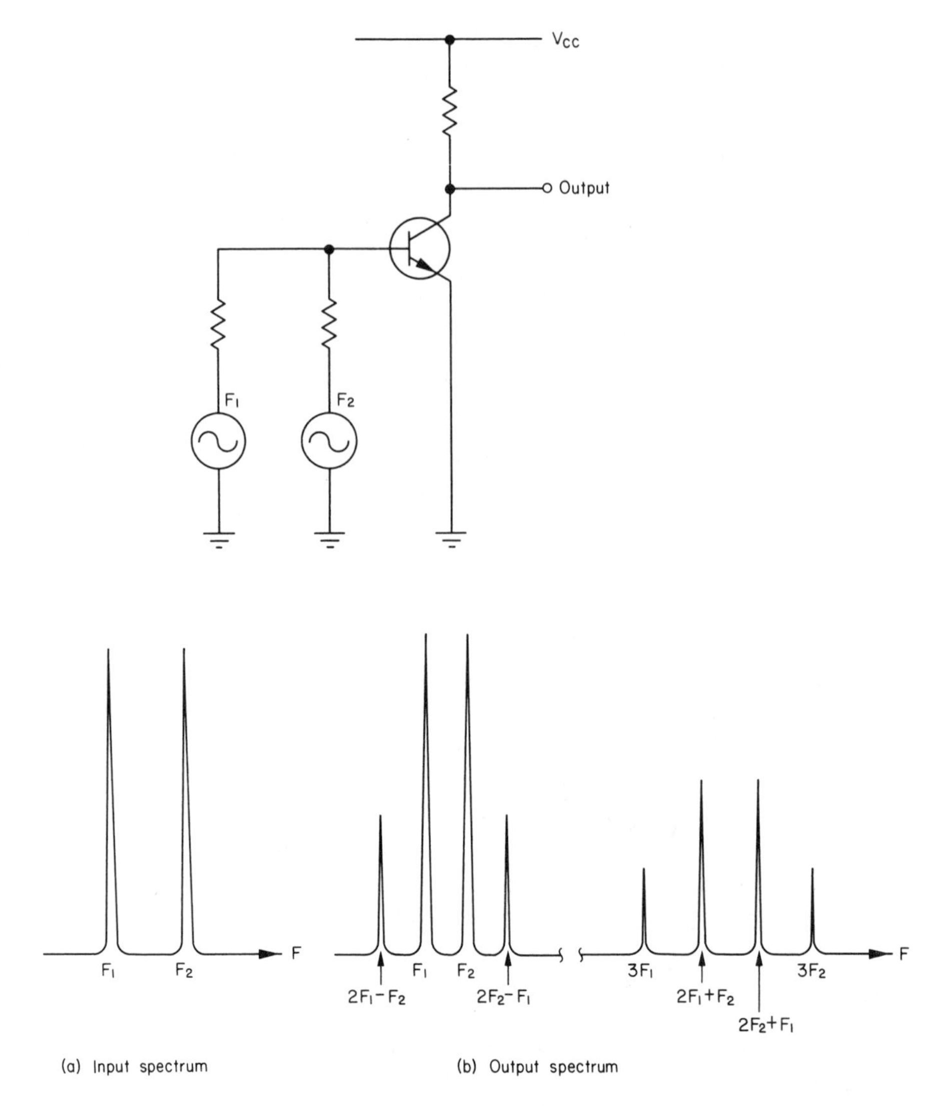

Figure 1-11 Input (a) and output spectrum (b) of an amplifier with third-order distortion. The new signals appearing in the output spectrum are the third-order intermodulation products.

distortion products will fall within the passband, distorting the desired wave-shape. Another problem is associated with the compression of the desired signal. Since the amount of compression is proportional to the amplitude of the other inputs, it is possible for one undesired signal that is amplitude-modulated to produce a corresponding amplitude change in a desired signal; in other words, new modulation "mysteriously" appears. This problem is called *cross-modulation*.

Notice that no simple sum and difference frequencies exist, so third-order characteristics cannot be used for superheterodyne mixer circuits. In fact, this characteristic has very little going for it and is avoided whenever possible.

1-6 PARASITIC SIGNALS

Some spurious signals that appear at the output of an amplifier have no particular frequency relationship with the desired signals; they are not the result of predictable intermodulation products. Instead, they may appear and disappear at random. They are usually the result of poor construction techniques that allow an amplifier to produce a continuous oscillation or a short burst of oscillations at some point on the output wave of a desired signal. A spectrum analyzer connected to the output of an amplifier while it is being tested will usually disclose these and other distortions. A wide-band oscilloscope may also be used, in which case the wave of Figure 1-12 would appear.

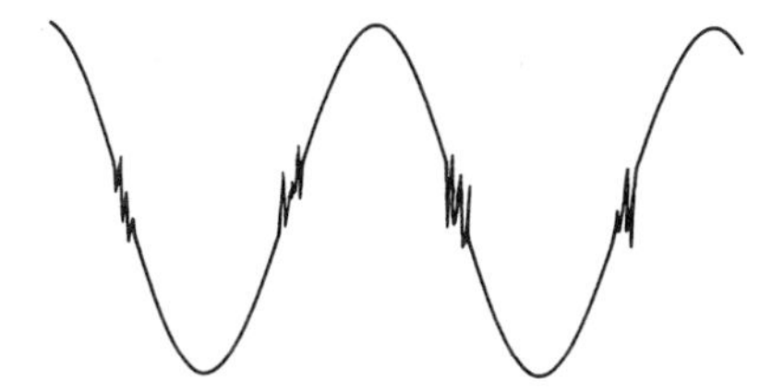

Figure 1-12 Oscilloscope presentation of the output wave of an amplifier with parasitic oscillations.

1-7 PHASE AND DELAY DISTORTION

One form of distortion still exists. Even though a circuit is carefully designed to minimize random noise and to pass all important frequencies with equal amplitude and have absolutely linear amplitude characteristics, a waveform may still pass through badly distorted. The problem is delay distortion; if all frequency components do not pass through a circuit with the same time delay, the output waveform will not look the same as the input. This is illustrated in Figure 1-13 for the first three frequency components of a 100-Hz square wave. The initial waveform is shown in (a) along with its three frequency components drawn in the proper time relationship. If this wave were to pass through a circuit that produced different time delays at different frequencies, the waveform at (b) might result.

Each sine-wave component has had its phase shifted by the reactances within the circuit. The relation between the phase shift and the time delay is

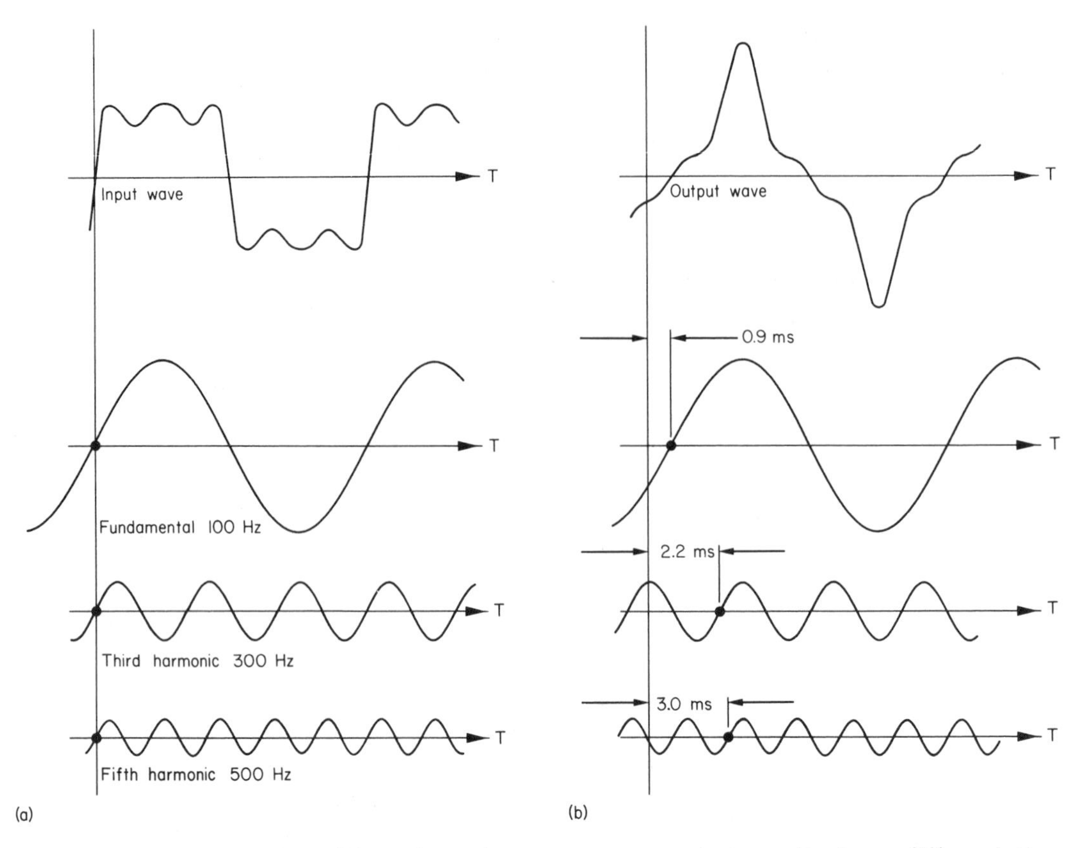

Figure 1-13 Distortion of a square wave that results from different time delays for each frequency component.

given by

$$\text{time delay (seconds)} = \frac{\text{phase shift (degrees)}}{360 \times \text{frequency (hertz)}} \tag{1-10}$$

This relation is shown in Figure 1-14 for two sine waves shifted in phase by 30°. The period for one cycle can be estimated as 5 ms. The frequency is therefore $1/(5 \times 10^{-3}) = 200$ Hz. Using Equation (1-10), the time delay between the two waves would be

$$\text{time delay} = \frac{30°}{360 \times 200} = 0.0004166 = 0.4166 \text{ ms}$$

This agrees with the time interval between identical points on the two waves. The time delay will be the same for all frequency components of a complex waveform if the phase shift increases proportional to frequency.

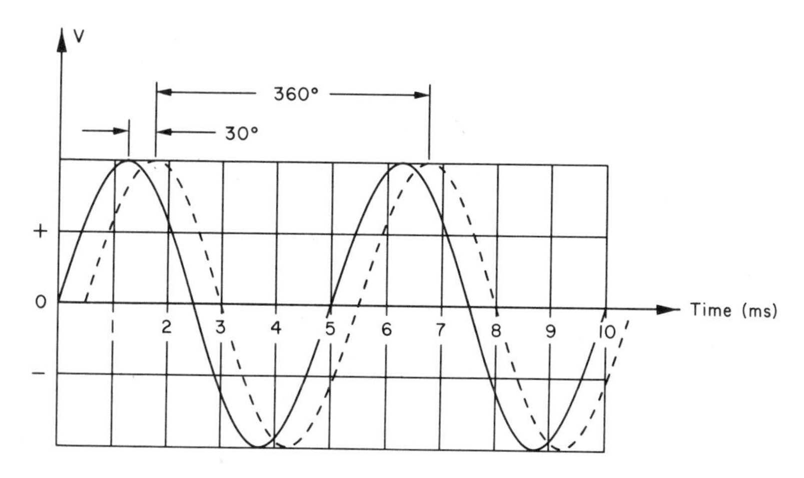

Figure 1-14 Two sine waves, showing a relative phase shift of 30°.

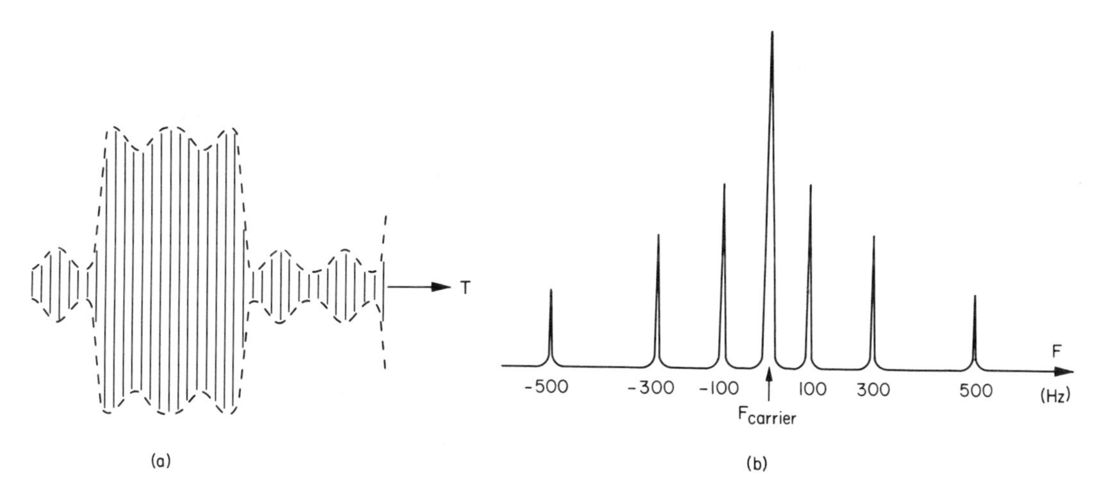

Figure 1-15 Three harmonically related frequencies forming a *square wave* are amplitude-modulating a high-frequency carrier. The time scale is shown in (a) and the frequency spectrum in (b).

If we take our *three-frequency square wave* and amplitude-modulate a high-frequency carrier with it, the waveform of Figure 1-15 will result. This signal consists of a carrier frequency and sidebands at 100, 300, and 500 Hz above and below the carrier. What phase relationship is required in this situation to eliminate delay distortion? This depends on what we are going to do with the wave. In most applications, the wave will be demodulated at some point, so all that is important is the *envelope* of the wave rather than the high-frequency structure. To preserve the envelope shape, the phase change between the carrier and each sideband must change linearly with sideband frequency. The actual phase shift at the carrier frequency is not important. *Envelope delay* is given by

$$\text{envelope delay } (T_e) = \frac{\theta_{\text{sideband}} - \theta_{\text{carrier}}}{360(F_{\text{sideband}} - F_{\text{carrier}})} \tag{1-11}$$

$$= \frac{\Delta\theta}{360 \times \Delta f} \quad \text{(seconds)}$$

For example, the linear change of phase angle with sideband frequency shown in Figure 1-16 will prevent envelope distortion that would result with unequal delay times.

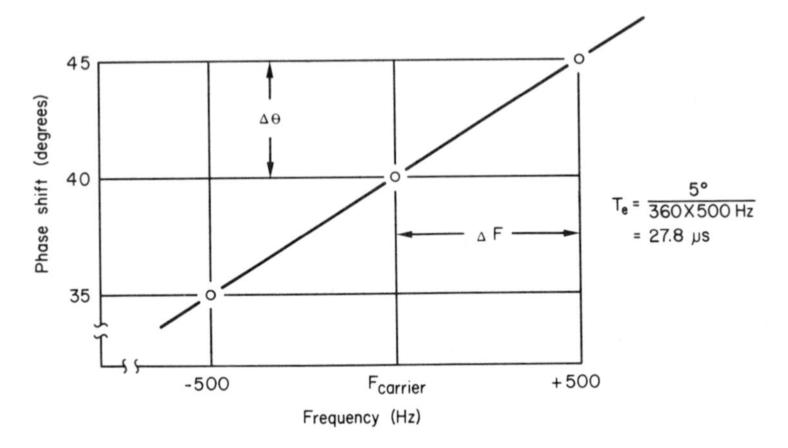

Figure 1-16 Linear phase change proportional to sideband frequency is necessary to prevent envelope delay distortion.

If all sideband components experience the same delay, the envelope shape is retained. The actual amount of delay between input and output is rarely of any great importance[2]; rather, it is the variation of this delay that causes distortion. In some cases, then, only the delay variation is plotted for a filter, and some statement is usually made that this is the delay variation relative to some average value.

Envelope delay can be measured on an oscilloscope by observing the time between identical points on the input and output waves of the circuit, as shown

[2]One exception is the chominance bandpass amplifier for a color TV receiver. The total delay in these amplifiers would result in the chroma information being horizontally shifted on the screen from the luminance information. This is corrected by placing a matching delay in the luminance amplifier.

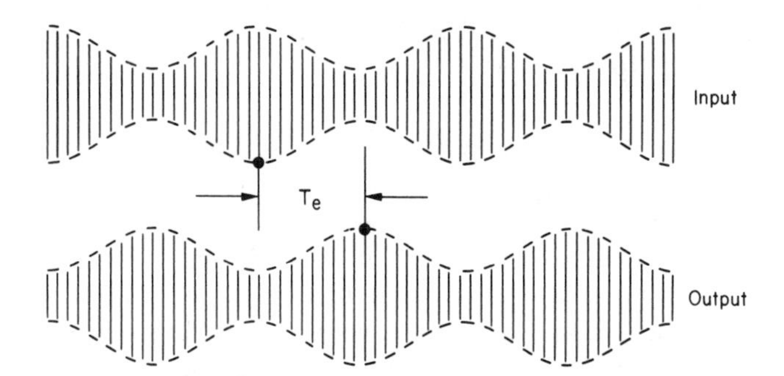

Figure 1-17 Measuring the envelope delay time required for a modulated sine wave to pass through a filter.

in Figure 1-17. The true delay could, of course, be greater than that measured by integer multiples of the modulating period.

1-8 GROUP DELAY

Group delay is a measurement of the delay caused by filters and amplifiers designed to carry many modulated channels. It was developed by the common carrier (telephone) companies for measurements on wide-band multiplex systems where many carriers exist at different frequencies. Group delay is the average delay time that a specified narrow range of frequencies experiences when passing through a circuit. It is measured in the same way as envelope delay, but, in this case, the amplitude-modulated test signal is swept across the passband of the circuit under test rather than remaining at a central frequency. The assumption is made that, because of the wide bandwidths involved, delay will only vary slowly with frequency, and so is more apt to result in a delay of one modulated channel relative to the next, rather than cause distortion of the envelope of any one channel. The amplitude-modulating frequency for the swept carrier is chosen low enough to show up small delay variations without being so low as to limit accuracy. A typical modulating frequency is 20 kHz, so the delay measured would be the average of a 40-kHz group of frequencies when both upper and lower sidebands are considered.

Group delay is defined as

$$\text{group delay } (T_g) = \frac{1}{360} \times \frac{d\phi}{df} \qquad \text{seconds} \qquad (1\text{-}12)$$

In other words, group delay is proportional to the rate of phase shift at each frequency of interest.

1-9 COMPARISON OF ENVELOPE AND GROUP DELAY

The terms "envelope" and "group delay" are often used interchangeably, even though they refer to two separate measurements. The following example will demonstrate the difference between the two.

Consider a very simple bandpass filter consisting of one parallel tuned circuit as shown in Figure 1-18. The amplitude response is also shown; it has a perfectly symmetrical shape about the vertical axis located at the parallel resonant frequency, but this is only because the frequency axis is a logarithmic scale.

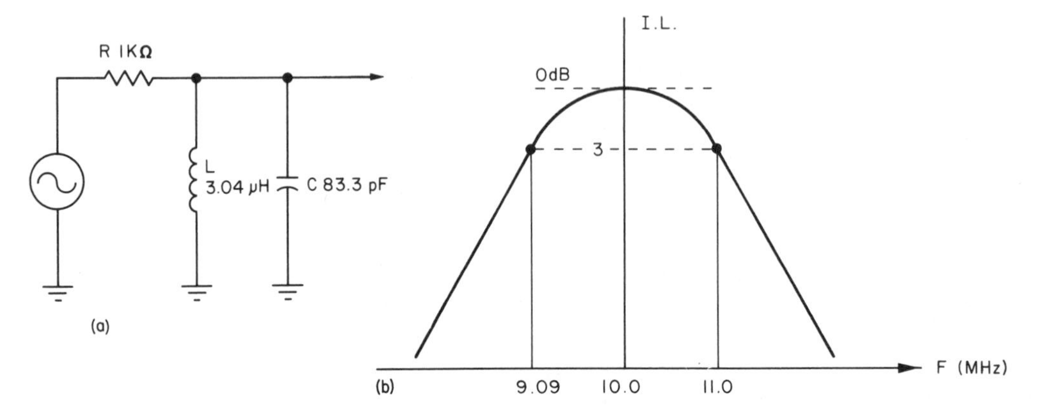

Figure 1-18 Single tuned-circuit bandpass filter (a) with its frequency response (b).

The phase-shift characteristics are plotted on a linear frequency scale in Figure 1-19. The linear scale is used to emphasize the nonlinear variation of the phase angle and also because sidebands of a modulated carrier within the passband would be linearly spaced with frequency.

Now to calculate the different delay times. First we will find the envelope delay for a carrier frequency of 10.045 MHz (which is the true center frequency of the filter). This delay will be the difference between the phase angle at each frequency and the phase angle at 10.045 MHz ($-2.7°$) divided by the frequency difference. This is given in Equation (1-11). The actual measurement on such a circuit could be made by observing the delay time of the envelope that would result from the two test frequencies, one fixed at 10.045 MHz and the other variable. This is actually a hard measurement to make, since the envelope frequency changes from zero up to 1.0 MHz as the variable frequency is moved across the band. The results are shown in Figure 1-20. The figure also includes the group-delay curve for the same filter. This is found by calculating the change in phase angle over a narrow range of frequencies (10

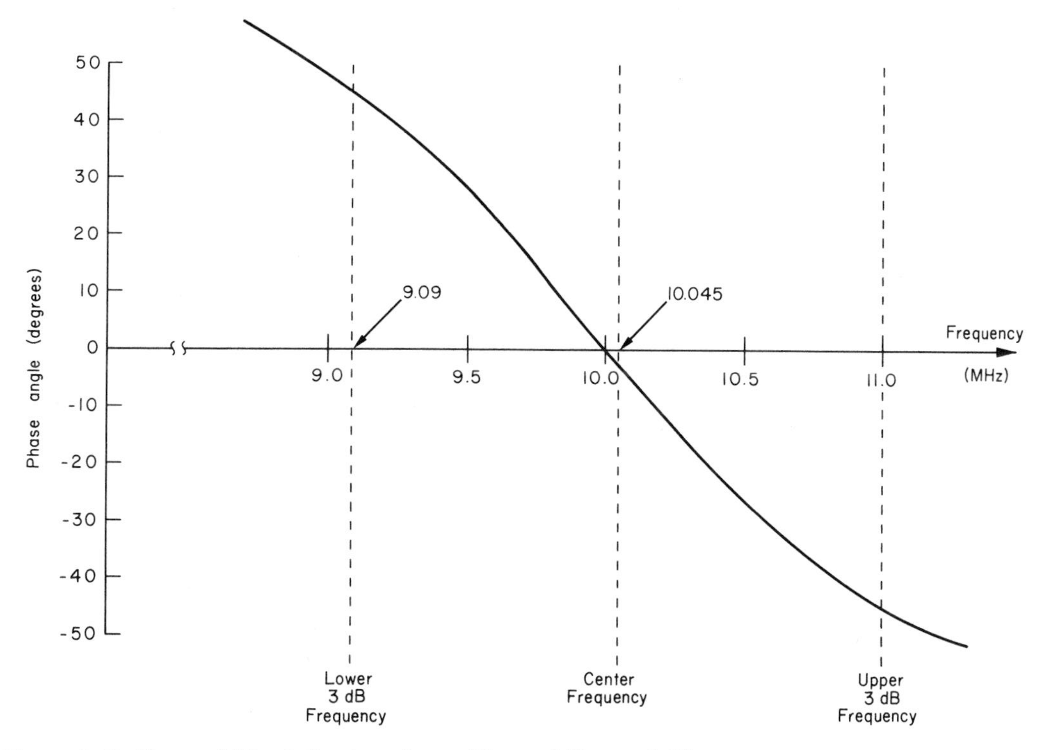

Figure 1-19 Phase shift of the bandpass filter of Figure 1-18.

kHz) at a number of different center frequencies, as indicated by Equation (1-12). The measurement is simpler since the envelope now has a constant frequency.

Looking at Figure 1-20 it is obvious that envelope delay and group delay, as we have defined them, are not the same thing. Group delay shows a wider variation across the passband, and for this reason is often measured when it is required to adjust a filter for as close to zero variation as possible. When flat delay is achieved, the group and envelope delays will be identical. Both delays are nonsymmetrical about the center frequency of the filter, owing to the logarithmic rate of change of phase with frequency. A better choice for the center frequency would be about 9.9 MHz, but then the amplitude response would be more asymmetrical. The final choice would depend on the application. Amplitude modulation is more sensitive to distortion from amplitude versus frequency distortion, whereas frequency modulation is more sensitive to delay distortion.

The actual delay times associated with envelope delay measurements are more meaningful. The values represent the delay time for each sideband frequency envelope. Group delay, on the other hand, is an average value for a

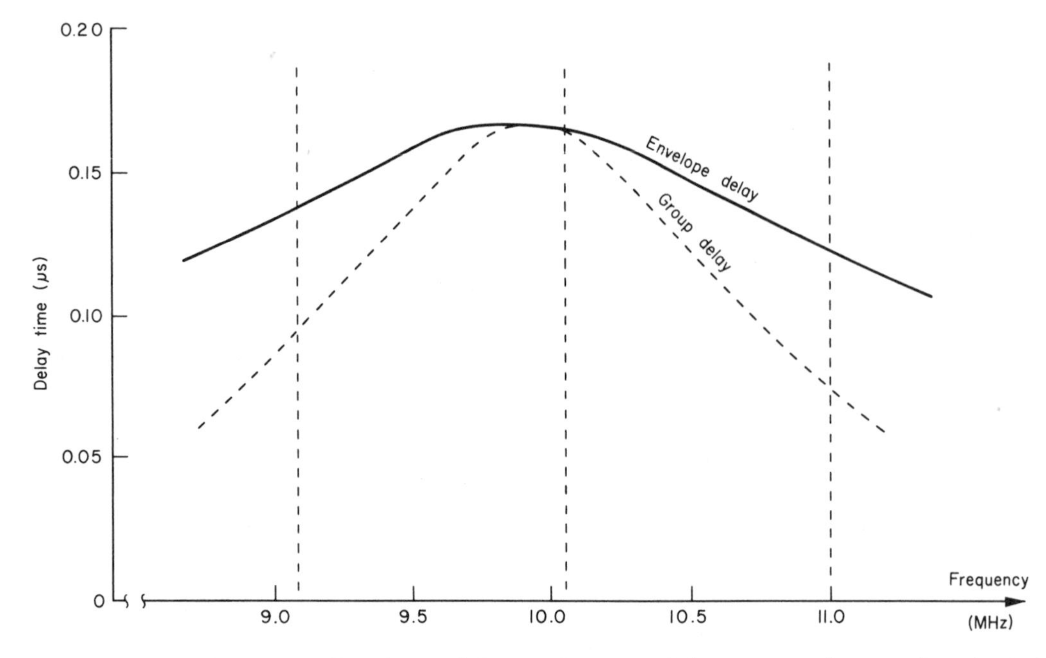

Figure 1-20 Envelope delay and group delay across the passband of the filter of Figure 1-18.

group of frequencies and shows no variation over bandwidths less than the modulating frequency used for the measurement. In this text, then, all delay times will be true sine-wave delay for low-pass filters and sideband envelope delay for bandpass filters.

QUESTIONS

1 Calculate the rms noise voltage developed across an 800-Ω resistor at 20° C if the measuring bandwidth extends from 30 Hz to 4.0 MHz.

2 If the 800-Ω resistor from Question 1 is the internal resistance of a signal generator that produces a 50-μV rms sine wave, find the S/N ratio in dB at the generator's output.

3 A video amplifier with the bandwidth specified in Question 1 is used to amplify the output of the signal generator. The amplifier's voltage gain is 18 and an extra 200 μV of noise is added by the amplifier to the output signal. What is the amplifier's noise figure in dB?

4 For an amplifier with the nonlinear characteristic $e_{out} = 20\ e_{in} + 1.5 \times 10^3\ e_{in}^2$, plot a graph of the amplitudes of the fundamental and second harmonic output for inputs from 0.5 mV up to 20 mV.

5 From the graph drawn in Question 4, determine the second-order intercept point and also the largest input signal that will keep the second harmonic amplitude 20 dB less than the fundamental at the output.

6 For an amplifier with the nonlinear characteristic $e_{out} = 20\ e_{in} - 3000\ e_{in}^3$, plot a graph of the amplitudes of the fundamental and third harmonic outputs for inputs from 10 mV up to 200 mV. Find the third-order intercept point.

7 A filter has a phase shift of 95° at 1.0 MHz, 105° at 1.01 MHz, and 85° at 0.99 MHz. What envelope delay would occur for a 1.0-MHz carrier amplitude modulated at 10 kHz?

8 Explain the difference between envelope delay and group delay and how each can be measured.

9 Write a computer program to calculate envelope delay (Equation 1-11) and group delay (Equation 1-12) for a filter similar to that of Figure 1-18. For component values use, $R = 10\ k\Omega$, $C = 820\ pF$, $L = 150\ \mu H$ and the frequency range 445 to 465 kHz. For the envelope delay calculation, use a carrier at the exact frequency of zero phase shift.

REFERENCES

Hewlett Packard Co. 1968. *Swept Frequency Group Delay Measurements*. Application Note 77-4. Palo Alto, Calif.

Hewlett Packard Co. 1973. *Spectrum Analyzer Series*. Application Note 150-4. Palo Alto, Calif.

Kennedy, G. 1970. *Electronic communication systems*. New York: McGraw Hill.

Simons, K. 1968. *Technical handbook for CATV systems*. Philadelphia: Jerrold Electronics Corporation.

chapter 2

components

High Frequency...
where capacitors become inductive,
and inductors capacitive.
where amplifiers oscillate,
and oscillators refuse to.

Much of the apparent mystery surrounding high-frequency circuits is often the fault of the components—one in particular—the inductor. When the paperwork design is nearly finished, it is time to start selecting components and, at the higher frequencies, this will require as much care as did the initial design. In fact, the final component characteristics may require some circuit redesign. Even after an appropriate component is selected, there is still room for error, as lead length, orientation, and spacing can still alter the characteristics of a circuit.

2-1 WIRE

Wire has a number of applications. As the connecting link from one point to another, it should not have any resistance, inductance, or capacitance. When used for inductors and transformers, its inductive properties become important. For transmission lines, both the inductive and capacitive properties are considered. Finally, for wirewound resistors, only the resistive properties are useful, all other properties causing undesirable effects.

2-1.1 Wire Types

For use in electronic circuits, most wire is soft-drawn copper; tin or silver coatings may be added to the surface for better soldering and conductivity, respectively (Figure 2-1). The wire can be either one solid conductor or a number of stranded conductors in contact with each other, resulting in greater flexibility. For limited coil-winding applications, a third type is available, consisting of strands of fine wire insulated from each other. This is called *Litzendraht* or *Litz wire*.

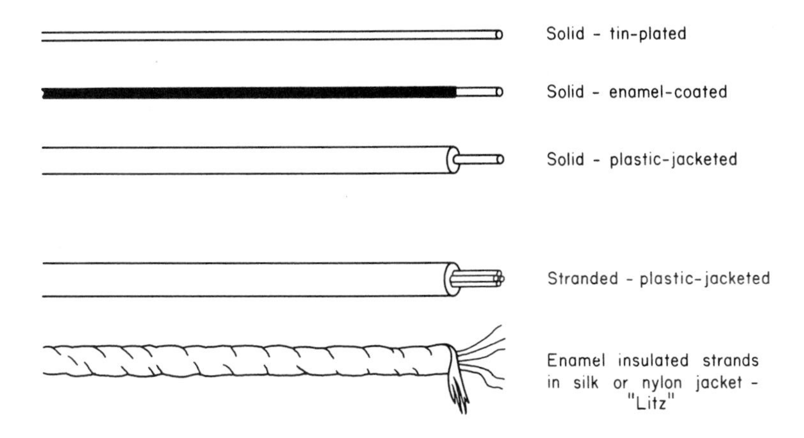

Figure 2-1 Five wire types used for electronic circuits.

A wide range of coatings may be provided to insulate the wire. Important characteristics of these coatings are their melting and flame behavior and their mechanical resistance to cutting and abrasion. Popular coatings are varnish, plastics, nylon, silk, Teflon, polyethylene, and polypropylene.

2-1.2 Wire Sizes

The diameter of electrical wire is expressed as the *American Wire Gage (AWG) number*. Table 2-1 shows the diameter for bare and enamel-coated wires, and dc resistance. Note that the gage number is inversely proportional to diameter and that there is a 2 : 1 diameter change for every jump of six gage sizes.

TABLE 2-1

WIRE DIAMETER AND RESISTANCE FOR WIRE GAGES 12–32 MEASURED AT 20°C. LITZ WIRE IS EXPRESSED AS NUMBER OF INSULATED STRANDS OF NO. 44 WIRE.

GAGE	*BARE DIAMETER*		*DOUBLE ENAMEL-COATED DIAMETER*		*RESISTANCE*	
(awg)	*thousand*	*mm*	*thousand*	*mm*	*per 1000 ft*	*per km*
12	80.81	2.052	83.8	2.13	1.67	5.488
14	64.08	1.628	67.4	1.71	2.614	8.576
16	50.82	1.291	53.8	1.37	4.646	15.24
18	40.30	1.024	43.1	1.10	6.693	21.96
20	31.96	0.812	34.6	0.879	10.46	34.30
22	25.35	0.644	27.6	0.701	18.58	60.97
24	20.10	0.511	22.2	0.564	26.77	87.82
26	15.94	0.405	17.8	0.452	40.81	133.9
28	12.64	0.321	14.4	0.366	64.89	212.9
30	10.02	0.255	11.6	0.295	103.2	338.5
32	7.95	0.202	9.5	0.241	164.1	538.3
5×44 Litz	nylon-wrapped		8.0	0.203	530.4	1740
6×44 Litz	nylon-wrapped		9.0	0.228	415.0	1361
7×44 Litz	nylon-wrapped		10.0	0.254	331.0	1086

2-1.3 High-Frequency Resistance of Wire

At dc, charge carriers are evenly distributed through the cross section of a wire. As the frequency increases, the carriers move away from the center, toward the edge of the wire and, as a result, the apparent resistance increases. This is the *skin effect*, which is a result of the stronger magnetic field near the center of the wire that increases the local reactance. The charge carriers then find it easier to move nearer the edge, and so the effective cross-sectional area of the wire decreases.

This tendency to travel nearer the edge is shown in Figure 2-2. Even though there is a gradual decrease in charge carriers closer to the center, an effective thickness of the charge-carrier region can be considered for calculation purposes. This is called the *skin depth*, and its values for several metals are shown in Figure 2-3 for a wide range of frequencies. Once the skin depth is known, the information in Table 2-1 can be used to calculate the ac resistance.

EXAMPLE 2-1

Find the ac resistance of AWG No. 22 copper wire at 10 MHz.

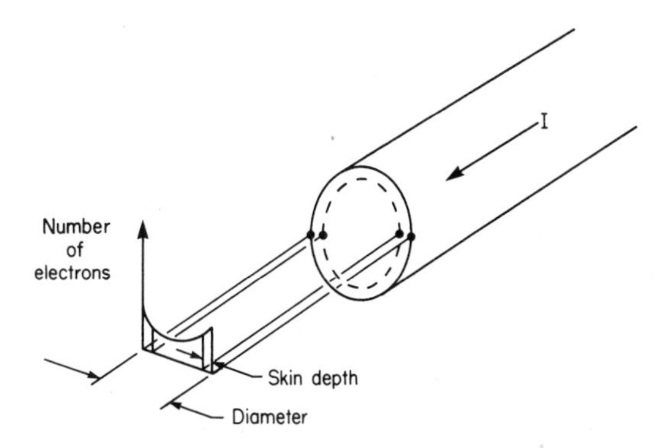

Figure 2-2 High-frequency skin effect causes charge carriers to stay near the circumference of a wire. The effective thickness of this layer is skin depth.

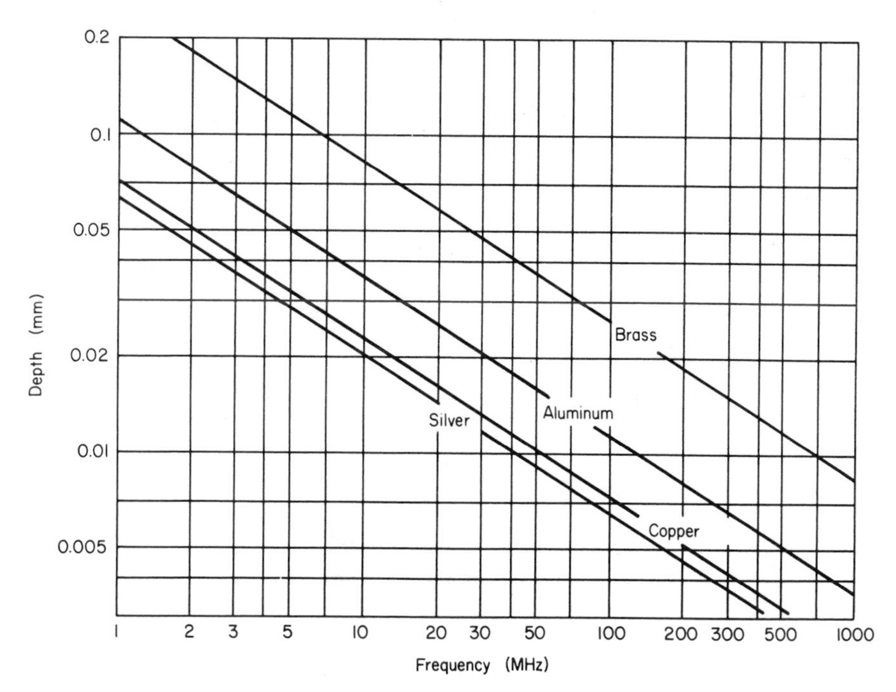

Figure 2-3 Skin depth of charge carriers decreases as the frequency increases.

Solution

From Table 2-1 we find the following for No. 22 wire:

$$\begin{aligned} \text{diameter} &= 0.644 \text{ mm} \\ \text{dc resistance} &= 60.97\ \Omega/\text{km} \end{aligned}$$

From Figure 2-2 we find that the skin depth for copper at 10 MHz is 0.0216 mm. The ratio of the ac resistance to the dc resistance will be the same as the ratio of the total cross-sectional area to the shaded area shown in Figure 2-4. The total

$$\text{dc area} = \pi r_1^2 = 3.14159 \times \left(\frac{0.644}{2}\right)^2$$
$$= 0.3257 \text{ mm}^2$$

The area of the shaded ring can be found as the difference between the total area and the area of the inner circle.

$$\text{ac area} = \text{total area} - \pi r_2^2$$
$$= 0.3257 - \pi\left(\frac{0.644}{2} - 0.0216\right)^2$$
$$= 0.3257 - 0.2835$$
$$= 0.0422 \text{ mm}^2$$

Since

$$\frac{\text{ac resistance}}{\text{dc resistance}} = \frac{\text{dc area}}{\text{ac area}}$$

then

$$\text{ac resistance} = \frac{\text{dc area}}{\text{ac area}} \times \text{dc resistance}$$
$$= \frac{0.3257}{0.0422} \times 60.97$$
$$= 7.72 \times 60.97$$
$$= 450.6\ \Omega/\text{km}$$

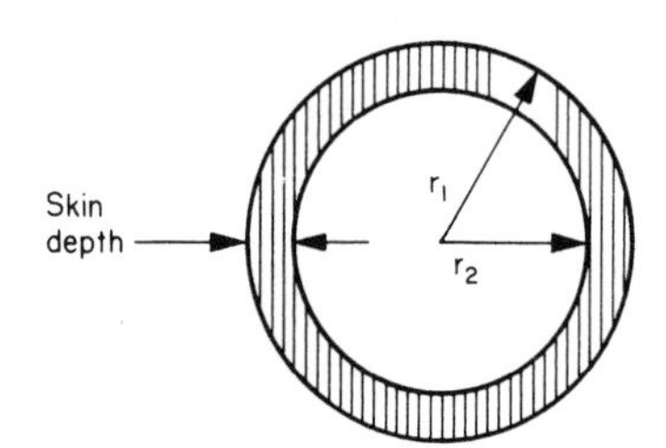

Figure 2-4

2-1.4 Inductance of Straight Wire

Because of the magnetic field that surrounds all current-carrying conductors, the conductors will have some self-inductance. At low frequencies, the reactance of short lengths of wire is insignificant; but, as frequency increases, even the 1-cm lead lengths of transistors and capacitors can be a problem.

The low-frequency inductance of a straight wire depends on its diameter and length and is given by the formula

$$L = 0.129l\left[2.303\log_{10}\left(\frac{4l}{d}\right)-0.75\right](\mu\text{H}) \tag{2-1}$$

where: l = wire length (mm)
d = diameter (mm)

The skin effect also causes a small decrease in the wire's inductance as frequency increases. This low-frequency inductance is also shown in Figure 2-5. For very high and ultra-high frequency (VHF and UHF) applications, hollow

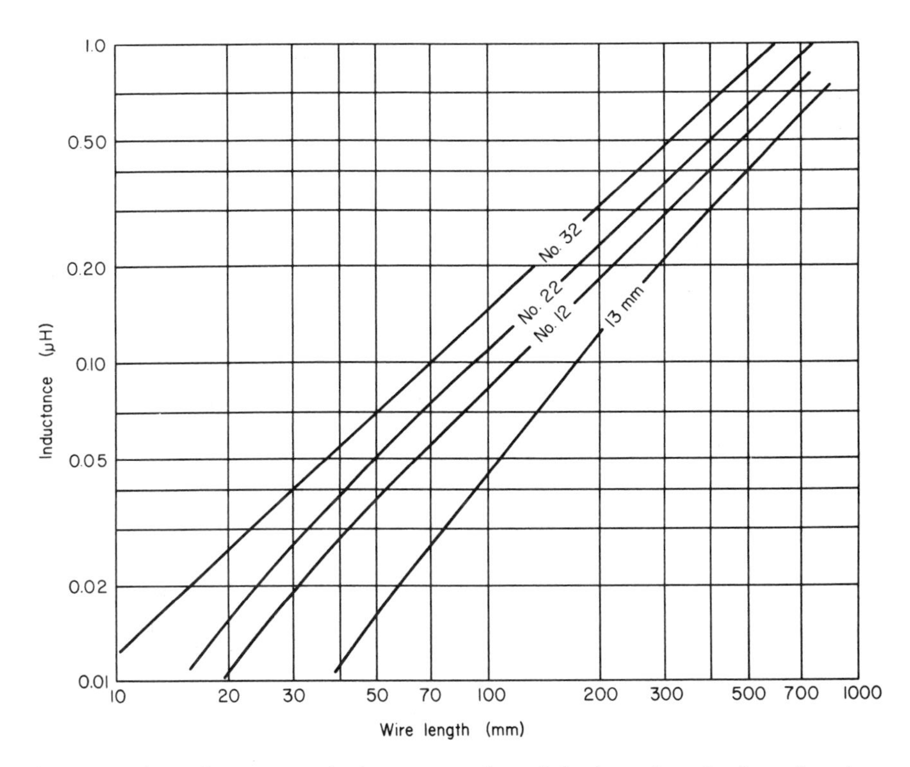

Figure 2-5 Low-frequency inductance of straight lengths of wire of various gages.

metal tubing or wide metal strips are often used in place of thinner wires to minimize inductance and resistance problems. A thin plating of silver is also used to reduce resistive losses in some critical areas.

2-2 INDUCTORS

An *inductor* consists of wire coiled to increase its inductance beyond what would be possible for the same length of straight wire. The increase occurs because the magnetic lines of force from each turn join up with that of the other turns to create a stronger field. Any technique that improves the concentration of the field will increase the inductance. The shape of the coil will therefore affect the inductance, as will the use of any material that changes the shape of the magnetic field.

2-2.1 Inductance Formula

For a single-layer coil with an air core (Figure 2-6), the low-frequency inductance is approximately

$$L = \frac{r^2 n^2}{22.9l + 25.4r} \quad (\mu H) \tag{2-2}$$

where: r = coil radius (cm)
l = coil length (cm)
n = number of turns

For best Q the length should be roughly the same as the diameter (i.e., $l \approx 2r$), but this is not critical.

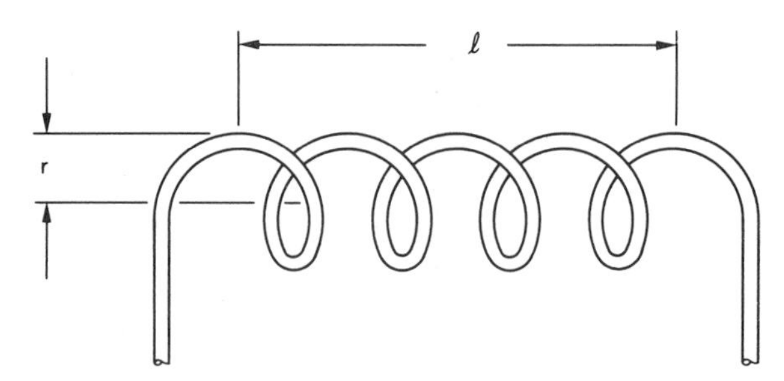

Figure 2-6

EXAMPLE 2-2

Design a coil with an inductance of 1.0 μH using No. 22 double-coated enamel wire.

Solution

From the wire table, the diameter of double-coated wire is 0.701 mm. If the wire is to be spaced out by an amount equal to its diameter (to lower winding capacitance), the length of the coil will be $l=0.14n$ (cm). We will also let the length be twice the radius, $l=2r$. Then

$$L=\frac{r^2n^2}{22.9l+25.4r}$$

$$1.0=\frac{(0.07n)^2n^2}{22.9(0.14n)+25.4(0.07n)}$$

$$n=10 \text{ turns}$$

$$\text{and diameter}=0.14n=1.4 \text{ cm}$$

For larger inductances it is more efficient to use multiple-layer coils, as shown in Figure 2-7, than to just increase the radius and number of turns of a single-layer coil. For multiple-layer coils that have a winding cross section that is more or less square, the low-frequency inductance is approximately

$$L=\frac{0.315a^2n^2}{6a+9b+10c} \quad (\mu\text{H}) \tag{2-3}$$

where: a = average radius of windings (cm)
b = length of winding (cm)
c = thickness of winding (cm)
n = number of turns

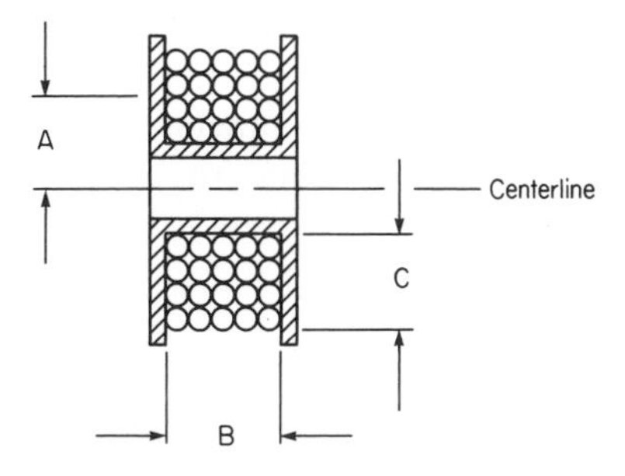

Figure 2-7 Rectangular cross section of multiple-layer coil.

2-2.2 Coil Capacitance

Because each turn of a coil is close to another turn and a voltage difference exists between turns, interwinding capacitance results. If a metal shield is used, additional coil-to-can capacitance appears. This all acts as an equivalent shunt capacitance across the coil, as shown in Figure 2-8. The reactance of all practical inductors will therefore be modified as shown. At medium frequencies

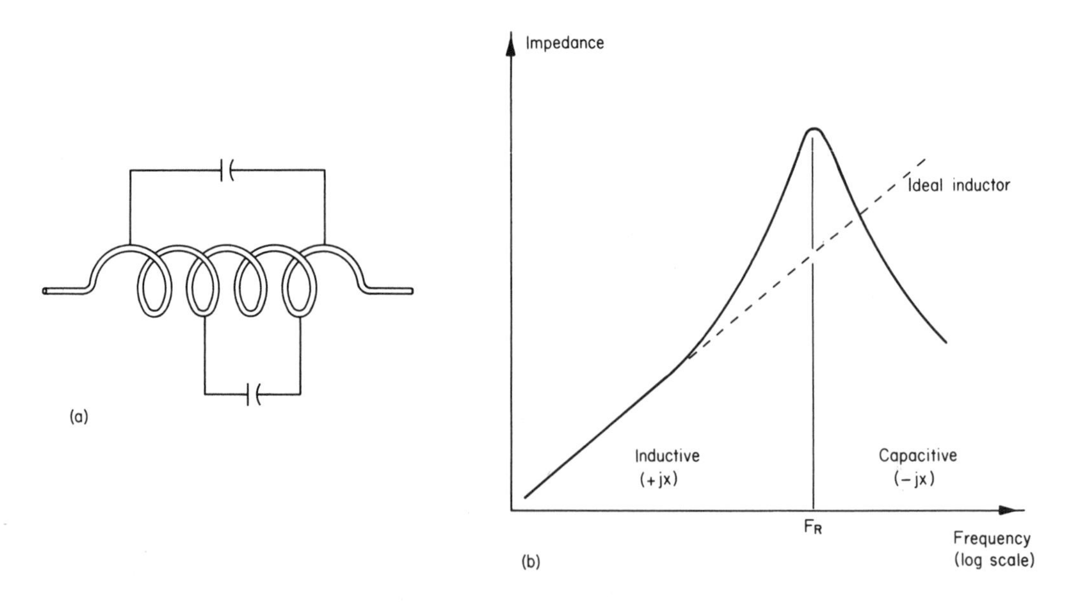

Figure 2-8 Stray capacitance between windings of an inductor (a) and the resulting changes that this causes on the component impedance (b).

the inductive reactance will climb faster than a theoretically perfect inductor. At some higher frequency, parallel resonance will occur, and above this the component will have a decreasing capacitive reactance. For most applications this capacitance has an undesirable effect, and so it must be minimized. This can be accomplished by different techniques used during the winding of the coil.

For single-layer coils, the overall diameter should be kept small and the turns should be spaced slightly to leave an air gap—air having a lower dielectric constant than any of the wire insulations used.

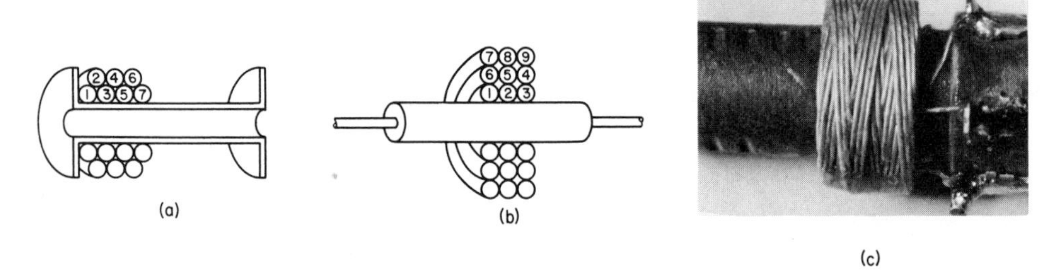

Figure 2-9 Winding variations for inductors used to reduce the total shunt capacitance.

For multiple-layer coils, the capacitance increases considerably. The windings should be arranged so that the voltage difference between adjacent turns is minimized. If the coil is to be long, a *bank winding* can be used. This is wound, as shown in Figure 2-9, by placing the first turn on the first layer and then the second turn on the second layer. The third turn, if only two layers are used, goes back on the first layer, and so forth. The greatest voltage difference is then along the longer dimension, and so the capacitance is minimized.

Making the windings short and tall, as shown in (b), will also reduce the capacitance as long as the voltage increases along the longer dimension—the radius, in this case. For larger inductance values, additional *pie sections* can be added with a large air gap between to reduce intersection capacitance. The *universal winding* of (c) takes this one step further. The turns of any layer are all parallel, but the turns of the next layer cross the windings of the first at some angle to distribute the capacitance around. The parallel resonant frequencies and construction methods of several air-core coils will be found in Table 2-3.

2-2.3 Coil Losses

Because coils are wound with wire and wire has some resistance, coils will have losses. Since the wire is often thin and its resistance increases with frequency, these losses are often very significant. The equivalent circuit for a practical inductor will show these losses as a series resistance whose value will have to change somewhat with frequency, as the wire resistance (and other losses) are frequency-sensitive (Figure 2-10). To indicate the relative size of the series resistance and the inductive reactance, the term *component Q* is used.

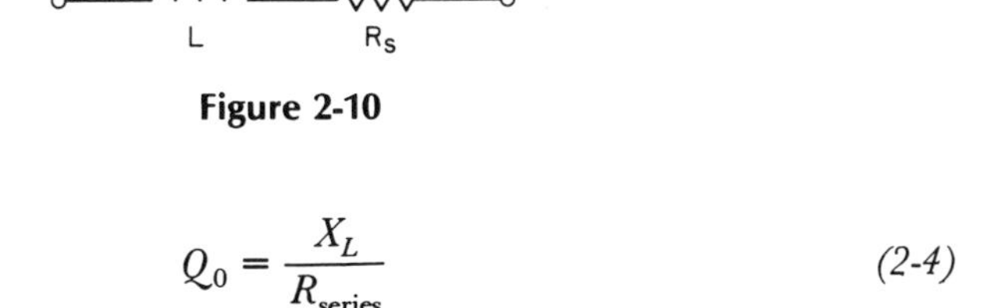

Figure 2-10

$$Q_0 = \frac{X_L}{R_{series}} \tag{2-4}$$

Both X_L and R_S will change with frequency, but not necessarily at the same rate. A practical inductor, therefore, could have a Q_0 that changes with frequency, as shown in Figure 2-11. At low frequencies, the Q_0 will increase as fast as the frequency does, as long as the wire resistance stays constant at the dc value. As soon as the skin effect causes the resistance to increase, the Q_0 curve slows down until it reaches the peak. Here the resistance and the reactance are increasing at the same rate. At higher frequencies, the wire resistance continues to increase, but now the shunt winding capacitance magnifies the losses. The Q_0 curve now decreases until it reaches zero at the parallel resonant frequency of the coil.

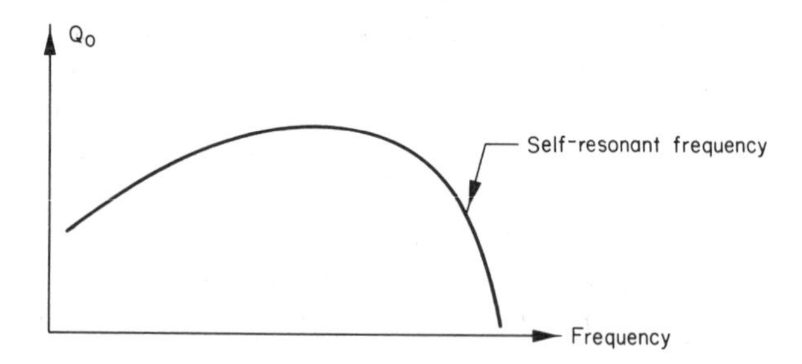

Figure 2-11 Component-Q_0 variations with frequency of several inductors.

2-2.4 High-Q Inductors

Increasing inductor Q means increasing X_L without significantly increasing R_S at any one frequency. There are a number of methods that will accomplish this:

1 Choose the best shape to maximize the inductance for a given length and size of wire while maintaining low winding capacitance; these have already been mentioned.

2 Pick a larger wire.

3 Plate the wire with silver to improve conductivity. This is often done above 150 MHz.

4 Use Litz wire to minimize the skin-effect increase of series resistance while keeping the wire diameter down. This is useful up to 2.0 MHz.

5 Use a magnetic core to increase the inductance so that fewer turns will be needed (the added core losses will be offset by a reduction in copper losses).

2-2.5 Measurement of Inductors

Any evaluation of an inductor will require the measurement of two components, the inductance and the series resistance. Since both are frequency-sensitive, the measurements should be made at the intended operating frequency of the component. For small inductors, the lead lengths should also be trimmed to the length to be used in the final application. Any metal cans used for shielding should be included and the shield connected to the "cold" end of the coil to complete the circuit for any coil-to-shield capacitance.

The actual readings obtained from an inductor measurement could be in any of the following forms:

1 Inductance (or reactance) and series resistance.

2 Inductance (or reactance) and parallel resistance.

3 Inductance and Q.

4 Impedance magnitude and phase angle.

An instrument that measures inductance and Q is shown in Figure 2-12 (see also Figure 2-13). Both measurements are obtained by placing the inductor in a series resonant circuit. Resonance is achieved by varying the capacitor until the voltmeter peaks. Calibration marks on the capacitor control will then indicate the inductance (the oscillator frequency must be set to predetermined points for an inductance measurement) and the voltmeter reading will show the Q.

When measuring inductors, several things should be considered. As already mentioned, lead length, especially of small inductors, should be cut to the final length. If the inductor is variable, two Q and inductance measurements should be made, one at each extreme of the core position. The operating point should be about halfway between these two measurements, thereby allowing equal adjustment up and down.

The coil form may affect the stability of the coil with changes in temperature and humidity. For air-core coils, the inductance will usually increase with temperature as the copper wire expands and as the wire resistance increases. The thermal expansion of the form may add to this. A more serious problem is the absorption of moisture into unprotected cardboard forms, causing them to swell.

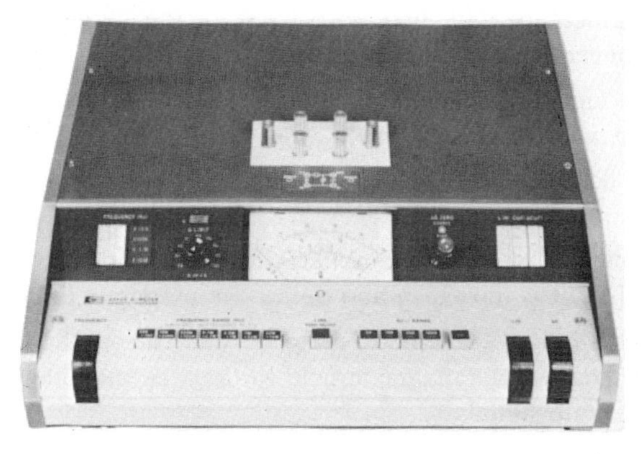

Figure 2-12 Hewlett-Packard 4342A Q meter used to measure inductance and Q_0 from 20 kHz to 70 MHz. (Courtesy of Hewlett-Packard.)

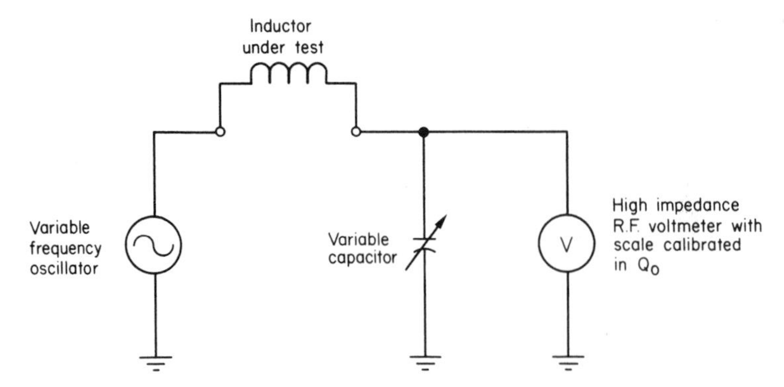

Figure 2-13 Simplified schematic of a *Q* meter similar to that of Figure 2-12.

When ferrite or powdered iron cores are used, the inductance will be current-sensitive. Measurements should therefore be made at the same ac and/or dc levels to be experienced in the final design. Ferrites and the like are also temperature-sensitive, so the final *Q* and inductance may have either a positive or negative temperature coefficient.

Even after an inductor has been properly measured, its operation in a finished circuit may change if other metal items and coils are mounted close enough to add stray capacitance or alter the magnetic field. Therefore, care should be taken to either mount coils properly or to simulate their final environment when measurements are made.

2-2.6 Magnetic Core Materials

Soft[1] magnetic materials are very often used as core material for inductors to obtain the following benefits:

1. The inductance will increase because of the higher-permeability core that concentrates the magnetic field.
2. The inductor *Q* could be improved because of the higher inductance without added wire length.
3. If two coupled windings are required, the core material will improve the coupling.
4. A small slug of magnetic material can be used to make an inductor variable if it is moved in and out of the magnetic field.
5. Certain shapes of magnetic material will surround the coil and completely contain the magnetic field so that the inductor neither radiates nor picks up signals.

---

[1]"Soft" magnetic materials retain very little residual magnetism and are used to increase inductance values, for example. "Hard" magnetic materials, such as nickel irons, act as permanent magnets or memory cores for computers. Their hysteresis curve encloses a larger area.

However, core materials do introduce some problems, and so their use should be carefully considered for each application:

1 The core material will have losses of its own (usually hysteresis losses) which may actually decrease the coil Q instead of increasing it.

2 The permeability of most materials drops off rapidly above some frequency, making them useful only below 2 or 5 or 20 MHz, depending on the actual material.

3 Some materials are extremely temperature-sensitive, with permeabilities that normally increase with temperature but may show negative coefficients near room temperature.

4 The permeability will also change with the applied signal. The average inductance may then decrease as the signal level increases, and harmonics of the signal may also be generated. Any dc current through the coil may also cause a change in inductance.

The materials most often used for radio-frequency applications are the various ferrites. These are complex ceramic compounds of metal oxides that have such a wide range of characteristics that it is absolutely essential that the material type be known and the manufacturers' literature be available. In addition to the many material types, each can be built in a wide range of shapes and sizes, as shown in Figure 2-14.

The most important characteristic of any magnetic material is its *magnetization curve*. This relates the applied magnetic field intensity (H) and the resulting magnetic flux density (B)[2], as shown in Figure 2-15. The permeability of the material (its ability to conduct magnetic flux) is found as

$$\text{permeability, } \mu = \frac{B}{H} \qquad \text{(Wb/A-turn)} \tag{2-5}$$

Its value will change depending on the amount of current and number of turns used for the winding. At some value (H_0) of applied field intensity, the core will *saturate* and the *incremental permeability* beyond that point will be that of air, a much lower value. For linear circuit operation, the peak input current must be kept safely below this saturating value. The permeability over this initial, linear portion of the curve is called the *initial permeability* (μ_i). (This is usually expressed as the ratio of the material permeability relative to that of air, so no units are involved.) Figure 2-16 shows the typical variation of this initial permeability with frequency and temperature for four different ferrites. In general, the higher the initial permeability at room temperature, the more it will change with temperature. For most radio-frequency applications, ferrites

[2]Flux density (B) can also be expressed with the SI unit *tesla*. which is the same as a *weber per square meter* ($\text{T} = \text{Wb/m}^2$).

Figure 2-14 Various ferrite core shapes and sizes.

with $\mu_i = 10-150$ are used, retaining their permeability up to 1000 MHz and 20 MHz, respectively.

For linear inductor operation, the inductance of a coil whose magnetic field is completely contained within a magnetic core is

$$L = \frac{0.4\pi N^2 \mu_i A \times 10^{-2}}{l} \qquad (\mu H) \tag{2-6}$$

where: N = number of turns
μ_i = initial permeability, relative to permeability of free space (no units)
A = cross-sectional area of core (cm^2)
l = average length of core (cm)

The formula is made simpler by some ferrite manufacturers, who give an *inductance index* for specific core shapes and materials. This includes the effective cross-sectional area and length of the core and its initial permeability:

$$L = N^2 A_L \times 10^{-3} \qquad (\mu H) \tag{2-7}$$

where: A_L = inductance index (μH/1000 turns)

For applications in transmitters where large signals are likely to be encountered, it is important to check that linear operation of the core is being maintained. When the rms voltage across the inductance is known, the peak

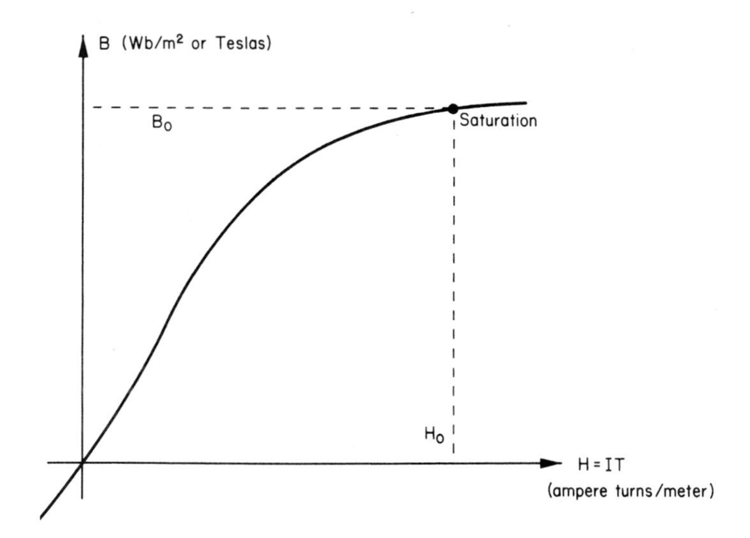

Figure 2-15 Typical magnetization curve showing the nonlinear permeability (μ) and the maximum flux density (B_0).

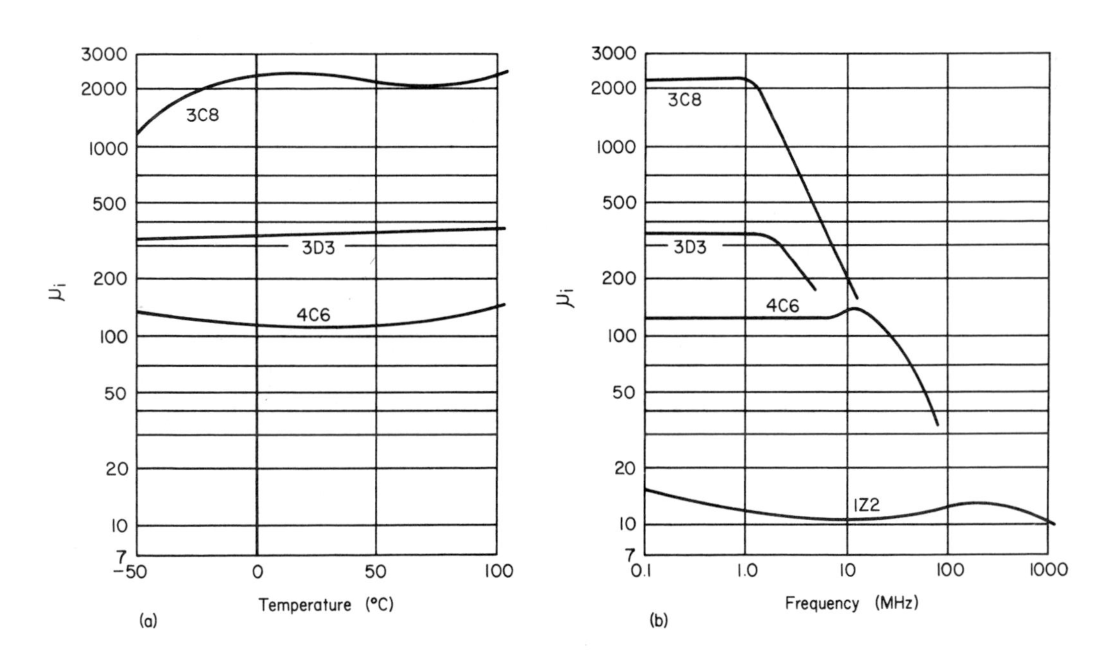

Figure 2-16 Variation of initial permeability with temperature (a) and frequency (b) for four different ferrite materials.

flux density will be

$$B_{peak} = \frac{E \times 10^8}{4.44f \times N \times A} \quad \text{(Teslas)} \tag{2-8}$$

where: E = rms voltage across the inductor
f = operating frequency (Hz)
N = number of turns
A = effective cross-sectional area of the core (cm^2)

If this value is safely less than the saturation flux density published by the manufacturer, linear operation will be maintained, the inductance will hold reasonably constant, and harmonic generation (mainly third) will be negligible. Any dc bias currents through the winding, incidentally, will increase this peak flux density and so must be considered.

2-2.7 Magnetic Core Losses

One of the major benefits of most ferrites over powdered irons is their high resistivity. Eddy-current losses in the material are therefore very low, and the main losses are due to hysteresis—the energy needed to realign the magnetic particles as the field keeps changing.

Knowing these losses is important for two reasons, the first being the obvious reduction of the inductor's Q below what it would otherwise be. The second reason is that the core loss and any other losses will heat up the core material. This changes its permeability (usually increases it) and lowers the saturation level, making the inductor less linear. At some temperature above 130°C, called the *curie point*, the permeability of the material drops very rapidly, and continued operation at this temperature may damage the core material. These losses depend on the applied signal amplitude and increase with frequency. The units are given in mW/cm^3 of material, so if the volume of a particular core is known, the total core loss can be calculated. Typical losses for a high-frequency ferrite are shown in Figure 2-17.

EXAMPLE 2-3

Design a toroidal inductor for operation at 2.0 MHz with an inductance of 0.25 mH.

Solution

For this design, the 4C6 ferrite material was selected since it is recommended by the manufacturer for use over the optimum frequency range

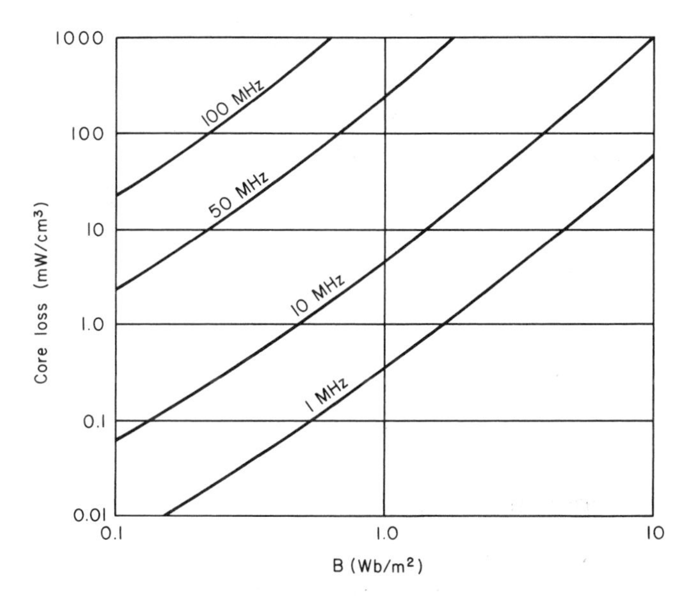

Figure 2-17 Energy losses in core of a typical ferrite used for high-frequency applications. The losses increase both with frequency and applied field intensity.

0.2–2.5 MHz. Its initial permeability (μ_i) is 120±20% at room temperature.

A core with a 14 mm outside diameter was selected, as it provided a large-enough opening for the number of turns required. The inductance index (A_L) for this size core is about 50.

Using Equation (2-7), we can find the number of turns:

$$L = N^2 A_L \times 10^{-3} \qquad (\mu\text{H}) \qquad (2\text{-}7)$$

$$N = \sqrt{\frac{L(\mu\text{H})}{A_L \times 10^{-3}}}$$

$$= \sqrt{\frac{250}{50 \times 10^{-3}}}$$

$$= 70.7 \text{ turns}$$

Seventy turns of No. 28 wire, enamel-coated, were wound on the core and an inductance of 0.265 mH was measured. This is a little above the value we set out to design, but the error is within the tolerance given for the initial permeability. Further details on the performance of this coil and its appearance are given in Table 2-3 and Figure 2-18.

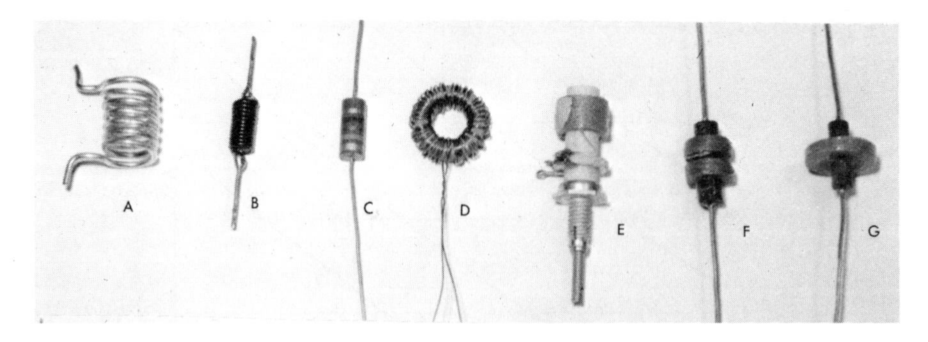

Figure 2-18 Seven inductors described in Table 2-3.

2-2.8 *Variable Inductors*

Variable inductors are often needed to compensate for tolerances that accumulate in any design work, mostly from capacitors and the reactive impedances of transistors. A small, threaded core placed inside an air-core coil will usually be sufficient. The turning range will depend on the size of core used relative to the coil size and also on the permeability of the core material. Tuning slugs can be ferrite, powdered iron, or metals such as copper, brass, or aluminum. The first two materials will increase both the inductance and the Q slightly as they are moved into the coil. The metals, on the other hand, act as a shorted turn in the magnetic field and so reduce both the inductance and the Q. To minimize the Q reduction with metal slugs, they are only used above approximately 50 MHz, where they are often better than the powdered irons.

Tuning slugs are often identified with a single color dot according to Table 2-2.

TABLE 2-2

APPROXIMATE RELATIVE PERMEABILITY AND RECOMMENDED FREQUENCY RANGE FOR COLOR-CODED, POWDERED IRON TUNING SLUGS.

COLOR DOT	*APPROXIMATE* μ_r	*RECOMMENDED RANGE (MHz)*
Yellow	30	0.2–1.5
Red	12	1–20
Purple	10	2–40
Green	10	20–50
Blue	5	40–300
White	3	30–200

TABLE 2-3

SEVEN INDUCTORS AND DETAILS OF THEIR CONSTRUCTION. THESE INDUCTORS ARE SHOWN IN FIGURE 2-18.

COMPONENT	*INDUCTANCE*	*CONSTRUCTION*
A	0.22 μH	7 turns of No. 18 tinned copper wire with an air core; diameter is 10 mm and length is 15 mm
B	0.22 μH	13 turns of No. 22 enamel-coated copper wire wound on a $\frac{1}{2}$-W 68-kΩ carbon resistor
C	10 μH	35 turns of No. 36 enamel-coated copper wire on 3-mm core and molded in plastic
D	0.265 mH	70 turns of No. 28 enamel-coated copper wire on a 14-mm toroidal form (design example)
E	0.4–0.9 mH (variable)	150 turns of 7×44 Litz wire universally wound on a 6-mm-diameter ceramic form with a yellow tuning slug
F	0.9 mH	Two universally wound pies on a 4-mm core with a total of 100 turns of 7×44 Litz wire
G	1.0 mH	One universally wound pie on a 4-mm core with 120 turns of No. 36 silk-covered solid copper wire

2-2.9 Sample Inductors

A range of inductors were measured to illustrate some of the differences in construction methods and the results in terms of inductance and component Q. Table 2-3 describes the construction; Figure 2-18 is a photograph of the inductors, and Figure 2-19 shows how the Q of each inductor varies with frequency.

It is interesting to compare the Q of the two inductors A and B as shown in Figure 2-19. They both have the same inductance, yet the one wound with the heavier wire and with the larger diameter has a much higher Q than the other. Also of interest are F and G, with approximately the same inductance and roughly the same type of winding. The inductor wound with the Litz wire has a much higher Q.

2-3 CAPACITORS

The *capacitor* is used for coupling and bypassing applications and, in conjunction with inductors, for filters and tuning elements. Unlike inductors, capacitors are rarely manufactured by the circuit designer, and so it is more important to

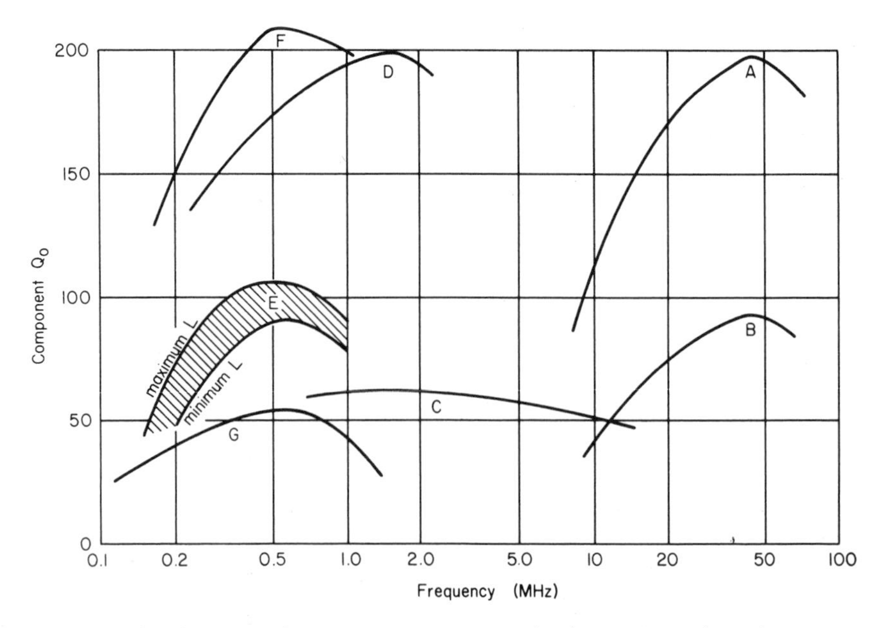

Figure 2-19 Variation of component Q with frequency for the seven inductors of Table 2-3.

know what characteristics to look for in manufacturers' catalogs than to have design formulas available. The important points to look for are:

1. Does the capacitance change with frequency? This could be a dielectric problem or a series resonance problem with body and lead inductance.
2. Does the capacitance change with applied voltage? Some types will show a 30% decrease in capacitance between zero and rated voltage.
3. How much does the capacitance change with temperature? Most capacitors have a positive temperature coefficient, some have a negative coefficient, and some are specially designed for no change over a limited range of temperature.
4. How serious are the losses in the capacitor? This is given either as Q or dissipation factor, where

$$Q = \frac{1}{\text{dissipation factor}}$$

The capacitor properties are determined mainly by the characteristics of the dielectric material. The following section describes some of the more important parameters of dielectrics used for high-frequency work.

2-3.1 Ceramic

This dielectric can be manufactured with relative dielectric constants (k) ranging from 5 to 10,000, and the capacitors using it will have a very wide range of characteristics that change with the composition dielectric constant.

Class I Ceramic.

These are made with lower dielectric-constant material (5–250) that has controlled positive or negative temperature coefficients (P-150 to N-5500 ppm/°C) or even zero temperature coefficient (NP0).

The capacitor Q can be quite high (250–5000), and the applied voltage has very little affect on capacitance. Typical capacitor applications would be tuned circuits, filters, and so on. The only disadvantages are the high cost and larger size compared to other ceramic types.

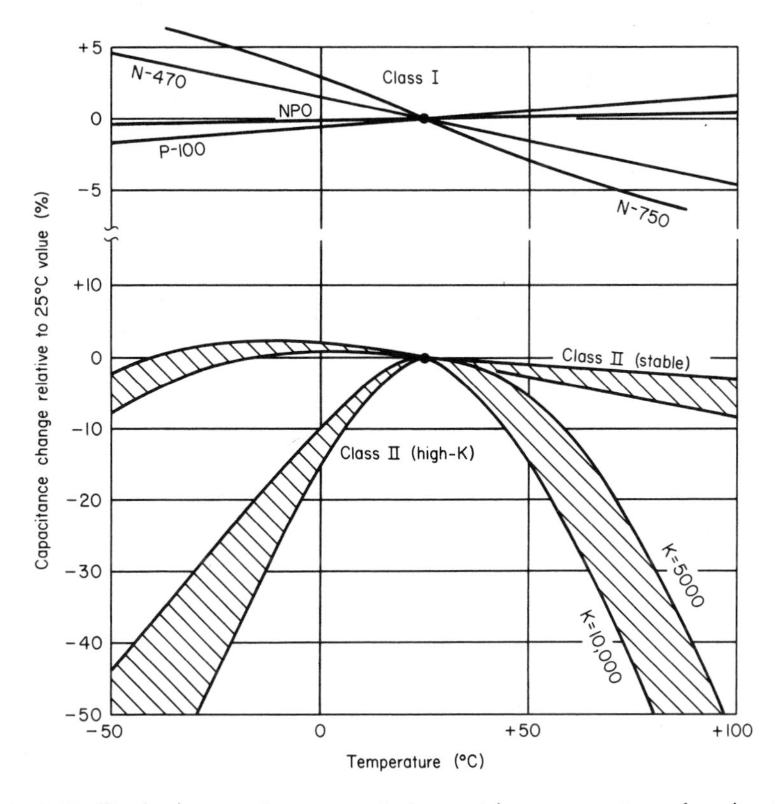

Figure 2-20 Typical capacitance variation with temperature for the three ceramic capacitor types.

Class II Ceramic (stable).

The higher dielectric constant (250–2500) produces a smaller capacitor with less stable characteristics. The temperature coefficient is nonlinear and the dissipation factor is higher, resulting in component Q's of 100–250. Capacitance values are slightly voltage-sensitive. Applications are limited to bypassing and coupling.

Class II Ceramic (high-k).

For greater volumetric efficiency and lower cost, dielectric constants in the range 250–10,000 are used. The high-temperature coefficient and greatly reduced Q make these capacitor types useful only for bypass applications. The capacitance is more sensitive to voltage, and some piezoelectric properties may be noted (i.e., sensitivity to mechanical vibration).

Class III Ceramic.

These are very similar to class II. Losses are higher and capacitance drops off rapidly at low temperatures. The voltage sensitivity is also higher than for class II. Used only for room-temperature bypass applications.

The capacitance changes with temperature and voltage for the three capacitor types are shown in Figures 2-20 and 2-21. The variation of Q with frequency is shown in Figure 2-22.

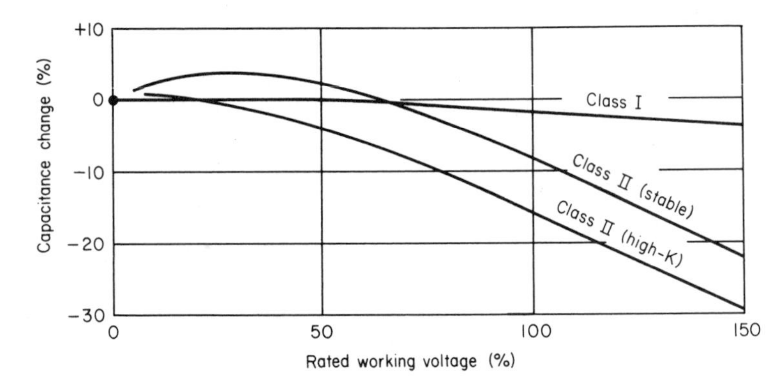

Figure 2-21 Typical capacitance change with dc voltage for the three ceramic capacitor types.

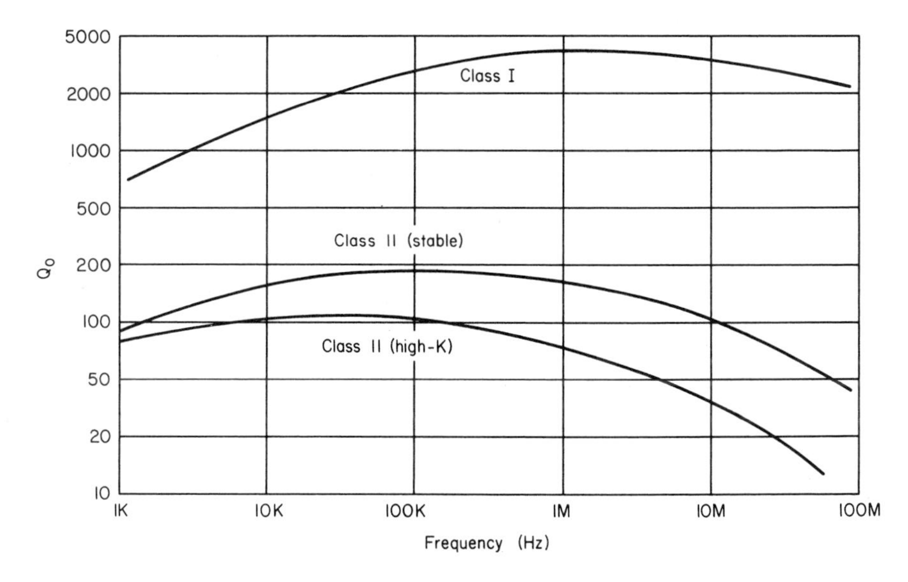

Figure 2-22 Typical component Q's with frequency for the three ceramic capacitor types.

2-3.2 Mica, Glass, Porcelain

All these materials have dielectric constants between 4 and 10 and provide high-quality, stable operation for high-frequency circuits. Their temperature coefficients are all less than ± 500 ppm/°C and Q's well above 1000. None of the materials are sensitive to applied voltage.

2-3.3 Practical Capacitors

A practical capacitor will have more to it than pure capacitance, unfortunately. All capacitors will have some extra inductance created by lead and body dimensions (see Table 2-4), and the various losses will appear as an equivalent series resistance. The equivalent circuit will then be as shown in Figure 2-23 and will result in the typical impedance curve shown. A large part of the inductance is caused by the attached leads, so a number of different lead configurations are available to suit different requirements, as shown in Figure 2-24.

TABLE 2-4:

APPROXIMATE RESONANT FREQUENCY OF SOME TYPICAL CAPACITORS WITH LEAD LENGTH AS STATED.

CAPACITOR (pF)	*WITH $\frac{1}{2}$-cm LEADS (MHz)*	*WITH $\frac{1}{4}$-cm LEADS (MHz)*
3000	25	35
1000	40	50
500	55	60
200	85	100
100	100	150

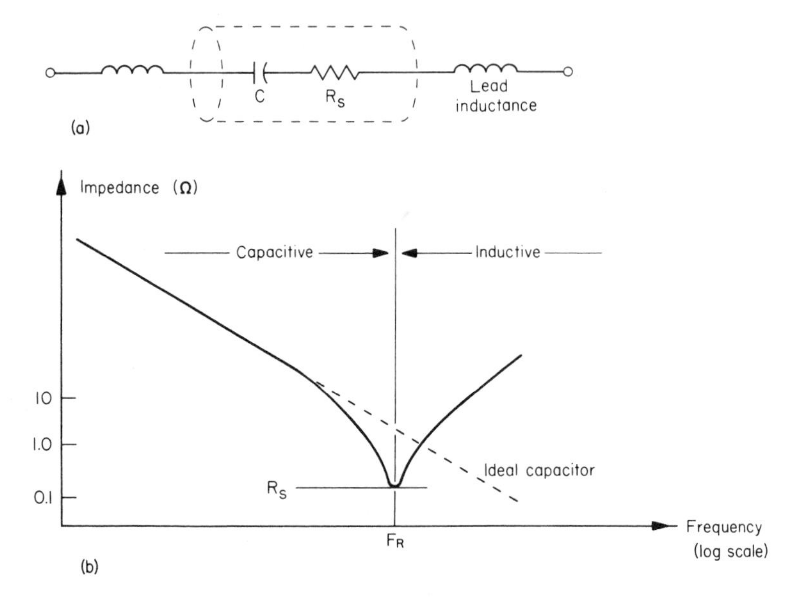

Figure 2-23 Equivalent circuit of a practical capacitor (a) and the effects of the added inductance and resistance on the total impedance of the component (b).

The temperature characteristics of capacitors are coded in one of two ways. If the change is carefully controlled and is fairly linear, then the temperature coefficient will be given as, for example, N-1400. This indicates a negative temperature coefficient of 1400 ppm for every °C change in temperature, the same as 0.14% change for each °C. A P-150 value would then indicate a positive temperature coefficient of 0.015%/°C.

For class II ceramics, where the change is nonlinear, a letter and number code are used, as shown in Table 2-5. A capacitor with an X5F characteristic would then change a maximum of ±7.5% from its capacitance at room temperature over the temperature range −55 to +85°C. A Z5U characteristic

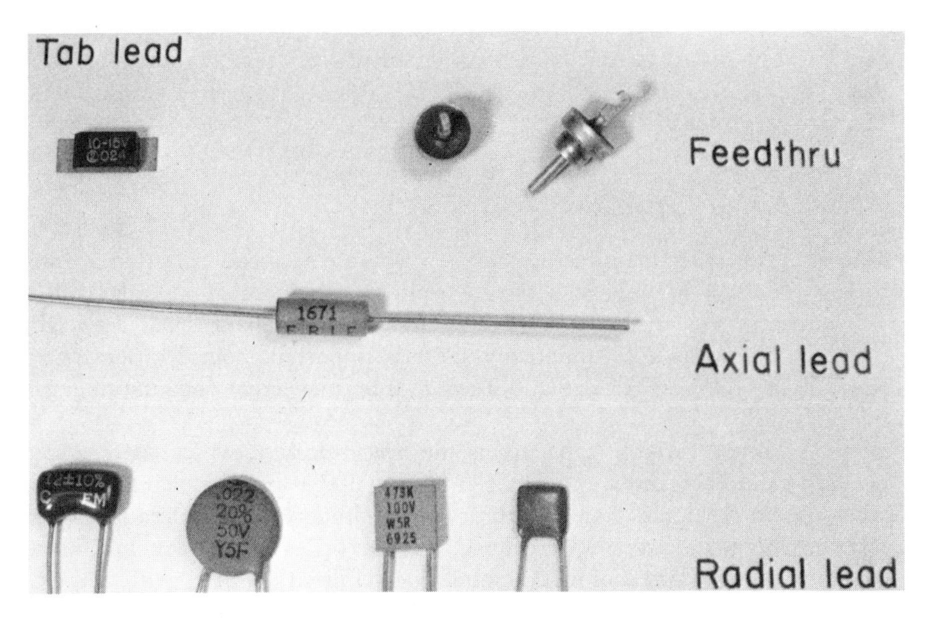

Figure 2-24 Variation of capacitor package styles.

TABLE 2-5

E.I.A. SYMBOLS USED FOR TEMPERATURE CHARACTERISTICS OF CLASS II CERAMIC CAPACITORS.

LOW TEMPERATURE (°C)	*LETTER SYMBOL*	*HIGH TEMPERATURE (°C)*	*NUMBER SYMBOL*	*MAXIMUM CAPACITANCE CHANGE (%)*	*LETTER SYMBOL*
+10	Z	+45	2	±1.0	A
−30	Y	+65	4	±1.5	B
−55	X	+85	5	±2.2	C
				±3.3	D
		W5 = −55 to +125°C		±4.7	E
		V5 = −55 to +150°C		±7.5	F
				±10.0	P
				±15.0	R
				±22.0	S
				+22 to −33	T
				+22 to −56	U
				+22 to −82	V
				+22 to −90	W

could be 22% higher or 56% lower than its stated room-temperature value over the temperature range +10 to +85°C, a rather sloppy characteristic with limited applications.

2-3.4 Variable Capacitors

Variable capacitors are manufactured for two separate applications, trimming and tuning. *Trimmers* are adjusted during equipment alignment, usually with a screwdriver or alignment tool, and then rarely touched again. Their design life may be as low as 250 complete rotations. Important considerations when selecting are the physical size and freedom from movement once the tuning is set.

The other variable capacitor is the *tuning capacitor*, whose shaft is connected to the knob mounted on the front panel of a receiver. Rotational life would be much higher than the trimmer type. Important considerations when selecting might be capacitance change characteristics, linear frequency, linear capacitance, smoothness of motion, and repeatability.

The variable capacitors, as the fixed values, are characterized by their dielectrics. Tuning capacitors are mainly air dielectric, although some vacuum variables are used in high-powered transmitters. For lower-cost tuning, thin metal plates can be used if insulating plastic sheets are placed between the opposite surfaces to maintain spacing.

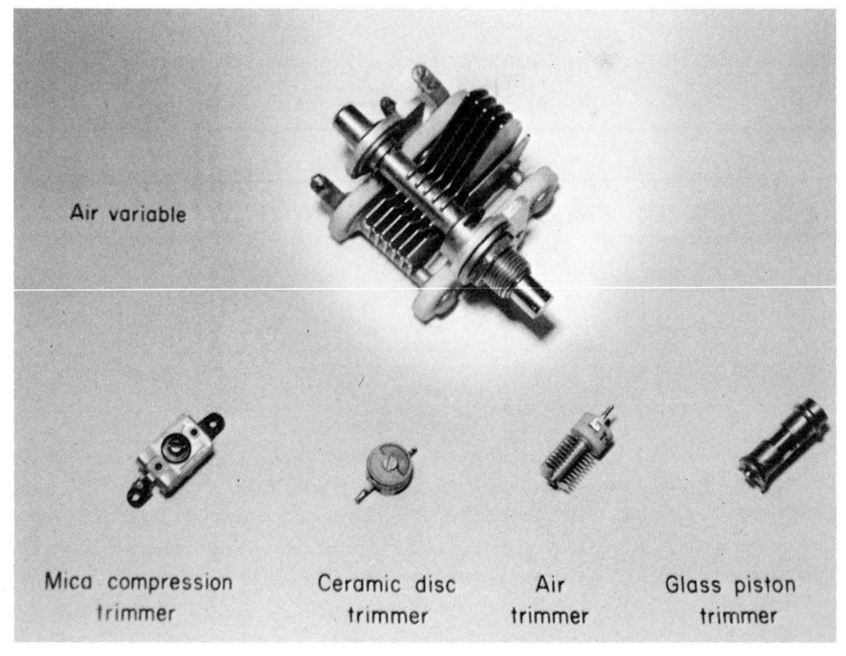

Figure 2-25 Variable capacitors used for trimming and tuning circuits.

Trimmer dielectrics can be air, polystyrene, ceramic, glass, or mica. Each dielectric is made in a number of different operating shapes—piston, disc, or compression, as shown in Figure 2-25. Trimmers can have Q's that range from as low as 200 to over 20,000. Each will also have a temperature coefficient that will depend on the material used.

2-4 RESISTORS

Resistors suffer from the normal component ailments, such as lead inductance and body capacitance between leads and end caps. The resulting equivalent circuit is shown in Figure 2-26. The total inductance of the leads and body of the resistor will be about the same as for an equivalent length of wire, about 0.035 μH for a total length of 3 cm. The capacitance across the body is somewhat less than 1 pF for a $\frac{1}{2}$-W size. The final effect that these two reactances will have depends on the value of the resistor. For high-resistance values the small amount of added series inductance will be insignificant; the shunt capacitance, on the other hand, will be more important. For low resistor values the series inductance plays the dominant role. Values around 100 Ω seem to be affected equally by capacitance and inductance and so show a flatter impedance line. Figure 2-27 shows the impedance variations that would be expected for metal film resistors.

The type of resistance element used will also affect the impedance variation. One obvious problem is the wirewound resistor, since the resistance wire is coiled around the core in the same manner as the windings of an inductor. Special *low-inductance windings* reduce the problem somewhat but still leave the resistors unusable above 2 MHz, particularly for lower resistance values. This assumes that the effective impedance at some frequency is vital; if the resistor is used just for biasing a 200-MHz amplifier, there is no problem.

Carbon resistors are usable to higher frequencies, since their inductance is mainly lead inductance. They do, however, have a higher body capacitance, which is probably due in part to capacitance between the carbon granules. As a result, higher resistance values will probably start to show an impedance decrease at about half the frequency shown in Figure 2-27.

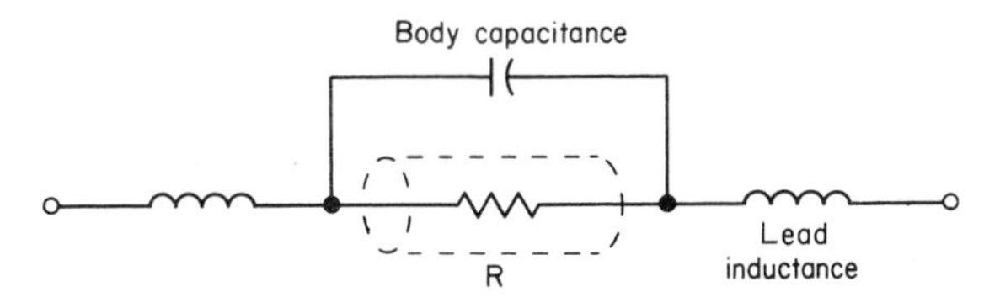

Figure 2-26 Equivalent circuit of a resistor.

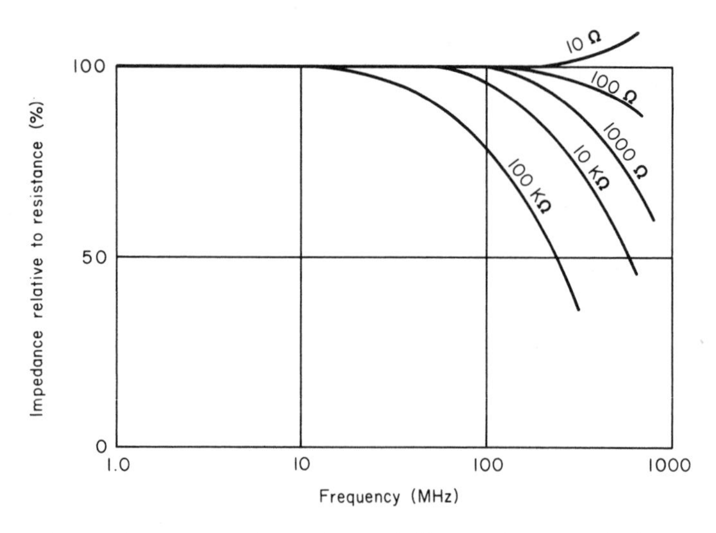

Figure 2-27 Impedance variation of metal film resistors with frequency.

2-4.1 Excess Noise

All resistors produce a certain amount of thermal noise (Chapter 1), but some resistor types produce additional noise over and above the thermal value. This excess is caused by the movement of current through the resistive material in a random path due to the local internal variations. Carbon resistors, particularly of the *slug* type, are bad for this because of the granular nature of the resistance element. Higher wattage ratings exhibit lower excess noise because of the larger cross section. Metal-film types also generate excess noise, but at a much lower level, because of the more uniform nature of the resistance element. Wirewound types generate only thermal noise. For critical applications, metal-film resistors are usually the best choice for radio-frequency work, where low noise and constant value are required.

QUESTIONS

1 What is the room-temperature dc resistance of an inductor using 25 metres of No. 32 AWG copper wire?

2 What is the ac resistance of this coil at 50 MHz?

3 Design a single-layer air-core coil with an inductance of 0.3 μH using No. 18 wire.

4 Design a multilayer air-core coil with an inductance of 1.0 mH using 7×44 Litz wire. Keep the winding cross section square ($b = c$ in Figure 2-7) and the maximum diameter less than 2.0 cm.

5 Design a toroidal inductor using 4C6 ferrite material for an inductance of 0.15 mH and using the largest wire size that will result in a single-layer winding. The core cross section is 4×4 mm and the overall diameter is 15 mm.

6 A tuned circuit uses a coil whose inductance increases 150 ppm/°C. Its value at 20°C is 1.5 μH. It is resonated with an N-750 characteristic ceramic capacitor whose 20°C value is 130 pF. What is the resonant frequency at 20°C and by how much will it change if the temperature climbs to 75°C?

7 Describe the temperature characteristics of X4V and Y5P ceramic capacitors.

REFERENCES

Harper, C.A. 1977. *Handbook of components for electronics*. New York: McGraw Hill.

Johnson, F.L. 1965. *Which capacitor*. A technical report of Marshall Industries, Capacitor Division.

Perna. *Taurus non est* (translated from the original Latin version). American Technical Ceramics Application Note.

Perna and Klein. November 5, 1973. To you it's a capacitor—but what does the circuit see? *Electronic Design News*.

Snelling, E.C. January 1972. Ferrites for linear applications. *IEEE Spectrum*.

Terman, F.E. 1943. *Radio engineers handbook*. New York: McGraw Hill.

chapter 3

tuned circuits

In this chapter, we will discuss the use of series- and parallel- resonant circuits as narrow-band filters. We will look at the effects of the source and load resistances and of any component losses on the filter response. The circuits looked at in this chapter would be used as RF and IF filters in receivers and in the frequency-generating circuits of transmitters.

The formal design of broad bandpass and high- and low-pass filters is left for a later chapter.

3-1 SPECIFICATIONS FOR BANDPASS FILTERS

A bandpass filter is used in applications where a narrow range of frequencies must pass with very little attenuation. All signals above and below the selected bandpass must be attenuated by increasingly large amounts. Figure 3-1 describes some important specifications for the amplitude response of a bandpass filter.

Although it would be nice to have ideal filters that passed a selected band of frequencies with no attenuation, with everything outside attenuated by an infinite amount, this is impossible, both in theory and practice. All filters will have some slope to the sides or skirts of the passband characteristic. There will

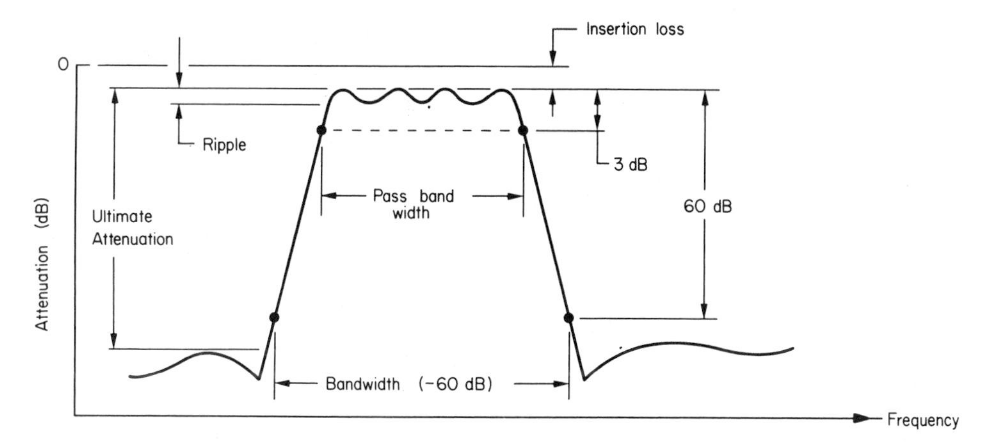

Figure 3-1 Amplitude-response specifications for a bandpass filter.

also be some loss of energy in the filter's passband caused by resistance in the reactive elements that make up the filter.

With reference to Figure 3-1, the specifications can be described as follows:

1 *Insertion Loss.* If the filter is removed from the circuit and the generator connected to the load, a certain load voltage will be obtained. If the filter is then returned to the circuit, the load voltage will normally be several decibels less than before, even at frequencies within the passband. (This assumes that the filter is not performing any impedance matching.) This loss produced by inserting the filter into the circuit is caused by the resistive losses in the reactive elements.

2 *Passband Ripple.* The insertion loss is not constant over the full width of the passband. There will be a gradual rolloff as the cutoff frequencies are approached, and there may be some additional fluctuations above and below this. Depending on who is doing the specifying, the ripple may refer to the total variation or just the minor fluctuations. Note that this specification does not include the absolute value of the Insertion loss but rather its variation over the passband.

3 *Passband Width.* The cutoff frequencies will usually be specified as being 3 dB below the attenuation at the center frequency. The passband width is then the frequency range between these two points.

4 *Shape Factor.* To further describe the shape of the attenuation curve, additional bandwidth measurements can be made at larger values of attenuation. The shape factor of the curve is then the ratio of two of these bandwidths. For example, if the bandwidth at 3 dB below the center frequency was 10 kHz and at 60 dB was 23 kHz, the shape

factor would be 23/10=2.3. A shape factor as low as 1.15 (60–3 dB) would represent a fairly sophisticated design.

5 *Ultimate Attenuation.* Because of the imperfect nature of all electronic components, a practical filter will not be able to provide extremely large values of attenuation (i.e., >100 dB). The attenuation may be greater than the ultimate value at a few frequencies but should never be less when used at frequencies outside the main peak.

The amplitude response of a filter is one important characteristic. But equally important, for many applications, are the phase-shift and time-delay characteristics. Phase shift and time delay are important when complex waveforms are passing through the filter. Ideally, all frequency components should pass through with the same time delay. All components would have the same relative phase at the output as they did at the input and the waveform would appear undistorted. One example is a television receiver, where uneven time delays can cause smearing of the picture and excessive phase shifts can change the color of the scene.

We can now add to our list of filter specifications:

6 *Time Delay.* The time required for any single frequency to pass through a filter circuit. Important where harmonically related waves are passed such as a television video amplifier.

7 *Envelope Delay.* The time required for the modulation envelope of a modulated carrier to pass through the filter. This is important in bandpass amplifiers where sidebands must undergo symmetrical phase shift to prevent distortion of the envelope.

3-2 INSERTION MEASUREMENTS

The amplitude and phase response of any system can be measured in two different ways, but only one of these will accurately describe the system. The difference depends on what point is taken as the input or reference point of the circuit.

All amplifiers and filters are operated between a signal source, which will have some resistance, and a load resistance as shown in Figure 3-2. If the ratio of V_3 to V_2 is plotted, the results, for the simple circuit shown, would be a straight line at the 0-dB level and there would be no phase shift at any frequency. This, of course, is what you should measure for this circuit, since the two meters used for the measurements are directly connected in parallel.

However, we know that this should be a bandpass filter and our measurements should show this. The problem was in using V_2 for the measurement. This effectively eliminated any effect that the input impedance of the filter had on the overall response, and in this circuit it was the total effect.

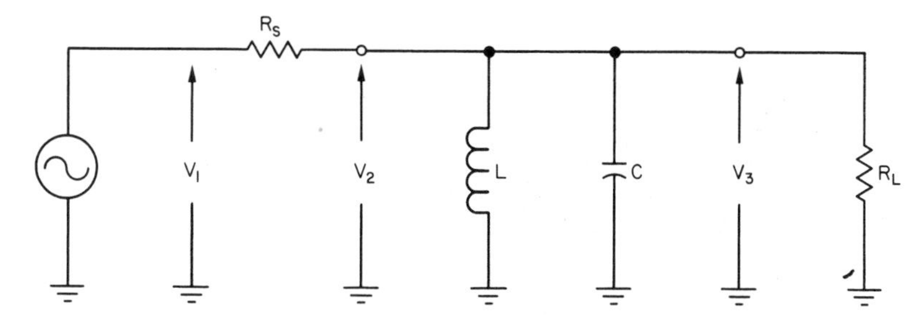

Figure 3-2 Simple tuned circuit with three points that can be measured.

The correct measurement should reflect the difference between the load voltage with and without the filter in place. Mathematically, this is the output voltage (V_3 as shown above) divided by one-half of the generator voltage (V_1) assuming that the load and source resistances are equal.

For a practical measurement, two separate tests are necessary. First, the generator is connected directly to the load without the filter and the frequency of the generator is varied over the range of interest. For a generator of any quality, the resulting load voltage should be constant within 1 dB or better. This level is then taken as the 0-dB reference. The filter is now put in place and the attenuation measured at each frequency. If the initial measurements with the generator and the load showed a large variation, then the true attenuation of the filter would be the difference between the two sets of readings. If the initial readings were less than 1 dB, no further correction is usually necessary.

Figure 3-3 is a photograph of a swept measurement of a filter's amplitude response. The system was initially calibrated by directly connecting the sweep generator output to the detector input with a short length of coaxial cable. By using the attenuator on the sweep generator, the output level was reduced in 10-dB steps and the horizontal lines set on the display. The coaxial cable was then removed and the filter inserted in its place. The display now shows the attenuation characteristics.

Phase or delay measurements are a little more difficult to properly make in a practical situation. This is still the phase relation between V_1 and V_3 as shown in Figure 3-2, but the problem involves actually getting at V_1, since it is some theoretical point inside the signal generator. The difficulty can be overcome by re-creating the reference point and the generator impedance outside the generator box. A directional coupler or an attenuator can be used as shown in Figure 3-4. The directional coupler is ideal since it provides a sample of the output signal from the generator that is not affected by any reactive portion of the input of the filter being tested (any reflection from this would end up in the reverse termination of the coupler).

The actual measurement of the overall phase/delay characteristics can be made in several ways. An oscilloscope or phase meter could be used to read the phase angle between input and output for a number of different frequencies

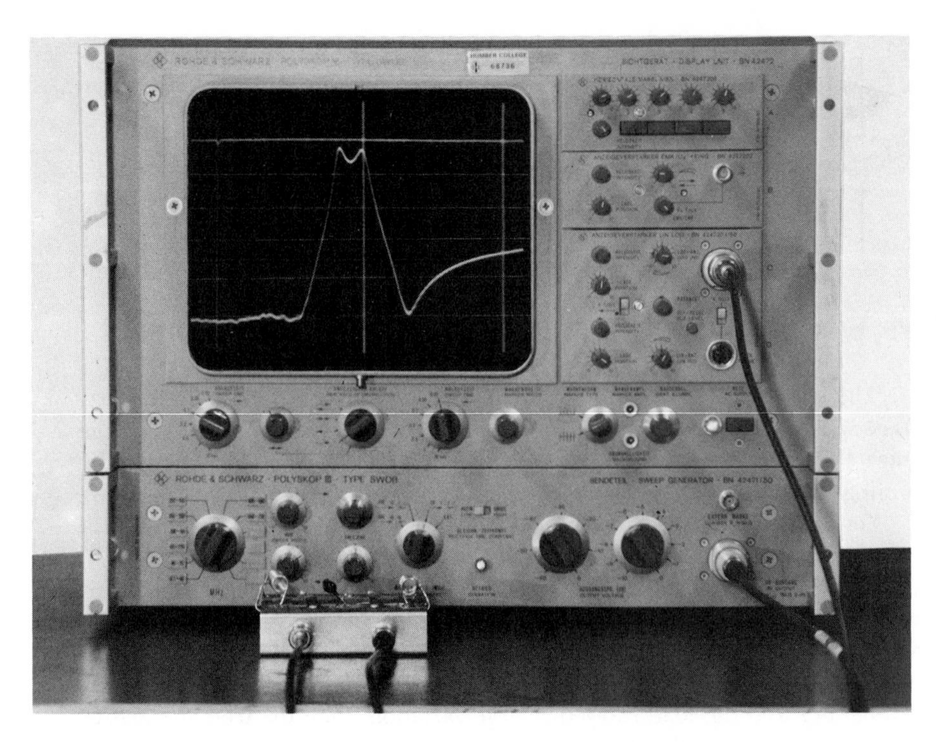

Figure 3-3 Swept amplitude response of a 10-MHz bandpass filter.

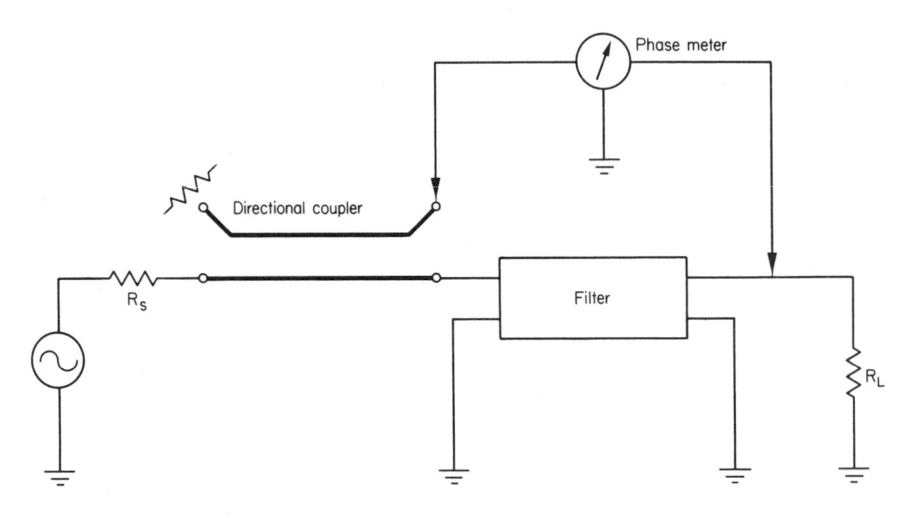

Figure 3-4 Directional coupler; can be used to obtain a proper reference signal for phase and delay measurements.

and the results plotted. The envelope delay times could then be obtained from the results. Phase measurements, however, can only be obtained from 0° to 360°, so some ambiguity will exist for phase angles greater than 360°. Calculated delay times could then be in error if the proper overall phase shift is not properly estimated.

Delay test sets are also available that use modulated waves to measure the group delay. Other test sets sweep the input signal at a precise rate. Any time delay in the system being swept will result in an output signal that is different from the input frequency, the difference being proportional to sweep rate and the time delay in the system.

Measurements of system phase and delay are occasionally made without using a proper reference point for the input signal. Instead of using the phase angle of the source itself, the phase of the signal at the input to the filter or amplifier is used. An error will be produced with this measurement technique since the phase shift caused by any changes in input impedance angle would be canceled; but when the amplifier or filter is placed into a system, this additional shift exists. If, however, the input impedance of the circuit stays perfectly resistive over the frequency range of interest, the input point could be used for reference. The phase angle would remain the same as that of the source.

3-3 BANDPASS FILTER DESIGN

The ideal bandpass filter will have a flat passband with no insertion loss and vertical sides producing infinite attenuation for all frequencies above and below the passband. As already stated, this is an impossible dream, and all practical designs are approximations of the ideal. The closeness of the approximations improves with the amount of time and money spent.

3-3.1 Parallel Tuned Circuit

The single *LC* parallel tuned circuit can be used as the first approximation of the ideal bandpass filter, or it can be used in greater numbers to improve the approximation. For these applications each *LC* circuit is referred to as a single pole (see Appendix 4A), so the simplest bandpass filter is referred to as a *single-pole bandpass filter*.

We will look first at the characteristics of the single resonant circuit when it is associated with only the source resistance. The load, for the moment, will consist only of a high-impedance ac voltmeter. Initially, we will use only perfect components, without losses and without stray reactances. Later we will consider the effects of these.

The circuit to be considered is shown in Figure 3-5. A switch is used so that the circuit response can be measured with capacitor only, inductor only, and then both together. There are two extreme situations that can occur.

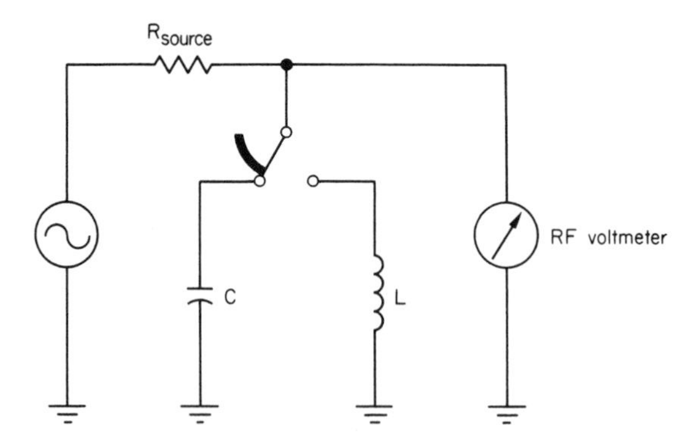

Figure 3-5 Parallel tuned circuit used as a single-pole bandpass filter.

The first situation arises when a small capacitor and a large inductor are used. The individual and combined amplitude response is shown in Figure 3-6. The "capacitor-only" response resembles that of a low-pass filter and the "inductor only" response is that of a high-pass filter. The two cutoff frequencies are widely separated, and so the overall result is a flat passband and an attenuation slope of 6 dB/octave above and below the passband. The 6-dB slope results when only one reactance is significant at any one frequency. For example, on the high-frequency slope, the inductor will have a much higher reactance than the capacitor, so, for the parallel combination, the capacitor has the dominant effect.

The second situation results from the use of a larger capacitor and a smaller inductor. The center frequency will still be the same, but the shape of the overall curve will be quite different, as shown in Figure 3-7. The two separate curves cross over far down their respective slopes. When both elements are in the circuit at the same time, their reactances will be equal and opposite at this crossover point and so will cancel. The output voltage at this frequency will be the same as the source voltage and no attenuation occurs. This crossover point is the *resonant frequency* of the tuned circuit. At slightly higher and lower

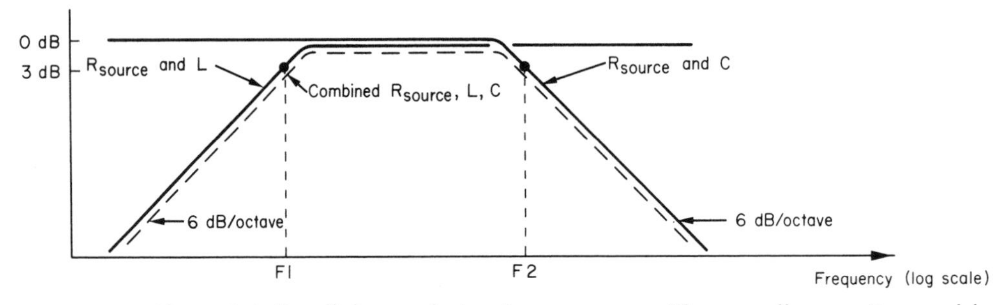

Figure 3-6 Parallel tuned-circuit response with a small capacitor and large inductor.

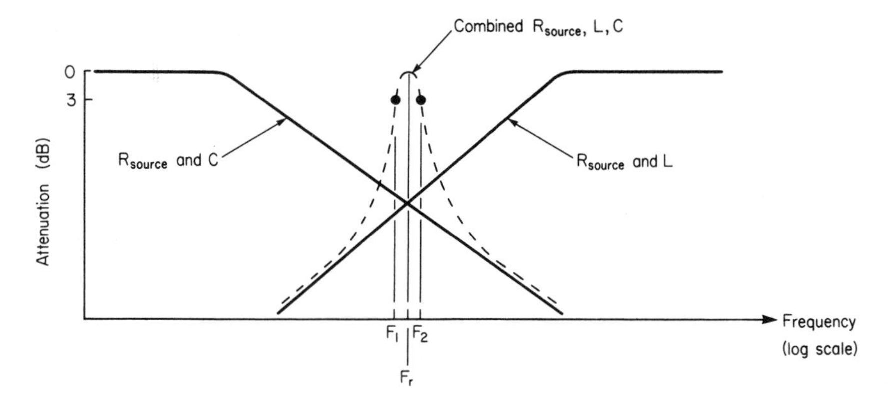

Figure 3-7 Parallel tuned-circuit response with a larger capacitor and a smaller inductor.

frequencies, the attenuation increases very rapidly, much faster than the slopes shown in Figure 3-6. This fast attenuation change is the result of the two, almost equal reactances, changing in opposite directions. At lower levels on the attenuation curve, the slope settles down to the more normal 6-dB slope. We are now far enough away from the resonant frequency that only one reactive component is significant; the other will have a very high reactance at that time.

So far we have found that the single-pole filter can provide us with a wide range of bandwidths and also a wide range of attenuation slopes, but the two are not independent. The shape of the curve depends on the relative values of the reactive components and, most important, the source resistance.

We can use one term to completely describe the shape of this simple circuit; this is the *loaded Q*. The higher the loaded Q, the narrower the passband and the steeper the initial side slopes or skirts. The 3-dB band width is given by

$$\begin{aligned} BW_{3\text{ dB}} &= \frac{\text{center frequency (Hz)}}{\text{loaded } Q} \\ &= f_2 - f_1 \end{aligned} \tag{3-1}$$

The loaded Q can be calculated from the reactance values at the resonant frequency, assuming that both components are lossless.

$$Q_{\text{loaded}} = \frac{R_{\text{source}}}{X_L} \quad \text{or} \quad \frac{R_{\text{source}}}{X_C} \tag{3-2}$$

(*Note:* This equation will be redefined latter to include all values of resistance that affect the loaded Q.)

Equation (3-2) indicates that for a high-Q narrow-band filter we must use small inductances and large capacitances or keep the source resistance high.

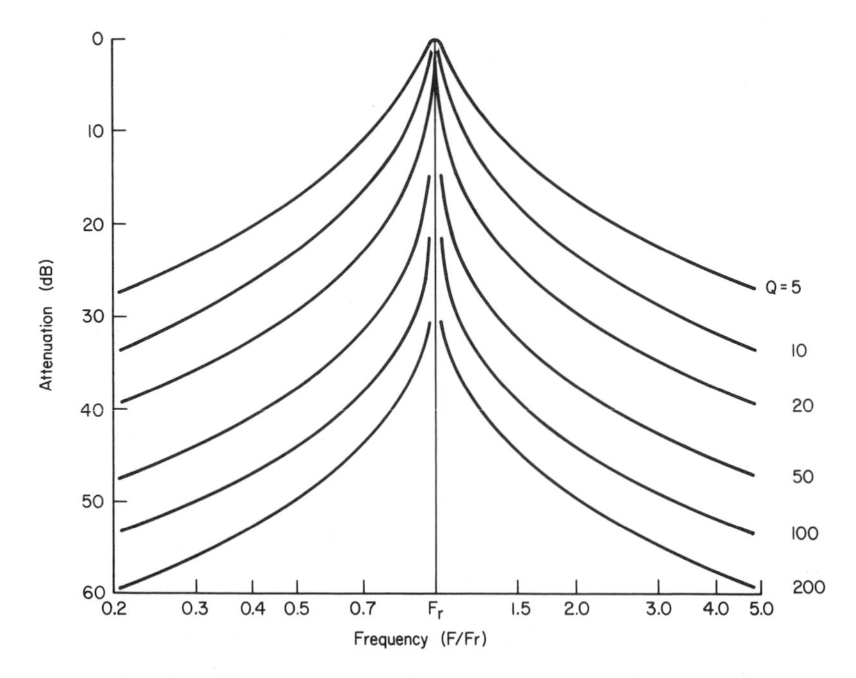

Figure 3-8 Single-pole filter response for loaded Q's from 5 to 200.

A general set of frequency-response curves for various loaded Q's is provided in Figure 3-8. If the Q of the circuit is known, this graph will indicate the attenuation that will be produced at frequencies above and below resonance. The process can also be reversed so that if the desired attenuation is known at some frequency outside the passband, the necessary value of loaded Q can be determined.

At "resonance," the output voltage will be in phase with the input since both reactances cancel out. As the frequency increases above resonance, the output will lag the input by an increasing amount and at the upper 3-dB frequency will reach $-45°$. At higher frequencies, the phase angle will slowly approach a maximum of $-90°$. Similarly, below resonance, the phase angle below will increase until the output leads the input by 45° at the lower 3-dB frequency and then will continue to increase, gradually approaching 90° at a very low frequency. A general set of phase-response curves is shown in Figure 3-9 for different values of loaded Q.

To see what we have accomplished so far, study the following example.

EXAMPLE 3-1

Design a single-pole bandpass filter for a center frequency of 1.5 MHz that will provide an attenuation of at least 35 dB at 3.0 MHz. The source resistance is 1000 Ω and the load resistance is much higher and can be ignored.

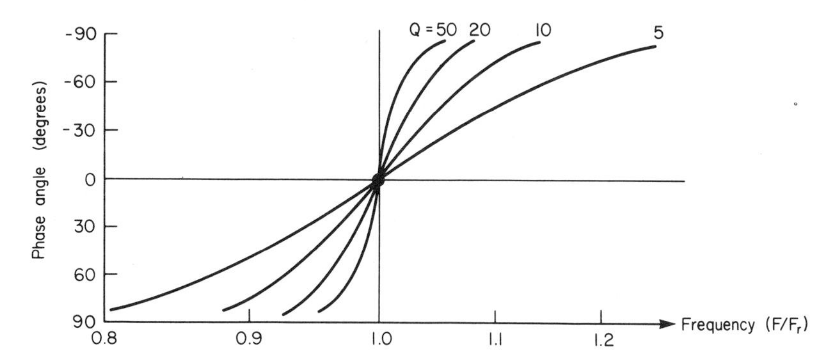

Figure 3-9 Phase shift in a single-pole bandpass filter for loaded Q's from 5 to 50.

Solution

The required value of loaded Q can be found using Figure 3-8. Since the attenuation of 35 dB must occur at twice the center frequency we can locate our point on the chart using $2F_r$ on the horizontal scale and -35 dB on the vertical scale. The line that appears to come closest to our point is a loaded Q of approximately 40.

Now we can find our reactances, using Equation (3-2):

$$X_L = X_C = \frac{R_{\text{source}}}{\text{loaded } Q} = \frac{1000}{40} = 25\ \Omega$$

This is the reactance of the capacitor and inductor at the center frequency of 1.5 MHz. The values of the components themselves are

$$X_L = 2\pi f L, \qquad L = \frac{X_L}{2\pi f} = \frac{25}{2\pi \times 1.5 \times 10^6} = 2.65\ \mu\text{H}$$

$$X_C = \frac{1}{2\pi f C}, \qquad C = \frac{1}{2\pi f X_C} = \frac{1}{2\pi \times 1.5 \times 10^6 \times 25} = 4240\ \text{pF}$$

Assuming we had ideal components, the circuit would look as shown in Figure 3-10.

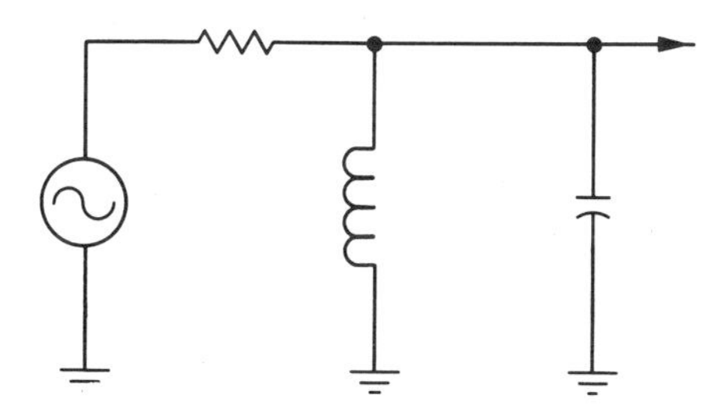

Figure 3-10

Since the design depended on an approximate reading taken from the chart, perhaps we had better check to see if we have the correct attenuation at 3.0 MHz. At this frequency, $X_L = +j50\ \Omega$ and $X_C = -j12.5\ \Omega$. Their equivalent parallel combination will be

$$\frac{(+j50)(-j12.5)}{+j50-j12.5} = -j16.67\ \Omega$$

The equivalent circuit at 3.0 MHz is shown in Figure 3-11.

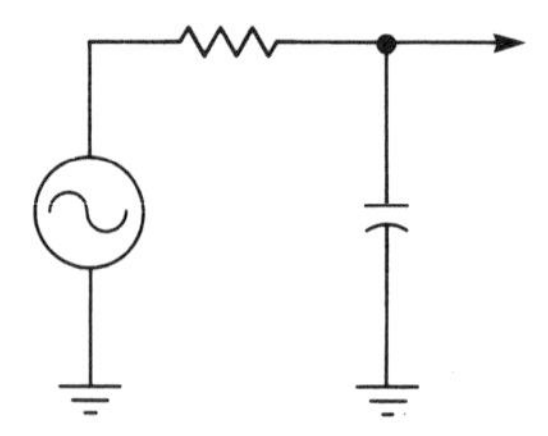

Figure 3-11

The attenuation can now be calculated:

$$\text{attenuation} = 20\log\left|\frac{-j16.67}{1000-j16.67}\right| = -35.6\text{ dB} \tag{3-3}$$

Therefore, our attenuation has met our requirements and the value of loaded Q read from the chart was accurate.

The resulting bandwidth of this design is

$$\text{bandwidth} = \frac{\text{center frequency}}{\text{loaded } Q} = \frac{1.5\text{ MHz}}{40} = 37.5\text{ kHz}$$

If a narrower bandwidth was necessary, a higher value of loaded Q could be used. However, this might not be too easy to do when the practical components are considered. If a wider bandwidth was necessary, the Q could not be lowered, as this would reduce the attenuation at 3.0 MHz. This requirement would then need something more than a single-pole design.

3-3.2 Component Losses

When designing filter circuits with practical components, the resistive losses in the capacitor and the inductor must be considered. If the losses are significant, the following effects will be noticed:

1 The loaded Q of the tuned circuit will never be higher than the Q of the poorest component.

2 The filter will have some loss at the resonant frequency.

3 The resonant frequency will be shifted slightly by the losses.

4 The phase shift of the filter will not be zero at the resonant frequency.

First, we will take a look at the resonant frequency of a parallel tuned circuit. When the Q of each component is greater than 10, we have the most common situation. The resonant frequency is given by

$$F_r = \frac{1}{2\pi\sqrt{LC}} \quad \text{(hertz)} \tag{3-4}$$

The output of the filter will be at its maximum at this frequency and will be in phase with the input.

When the Q of any component drops below about 10, slight changes start to become apparent and become more so as the component Q is reduced still further. The resistances being considered are shown in Figure 3-12. Any parallel resistance in the components themselves or due to external loading does not produce these frequency shifts.

When the resistance is in series with the inductance, as shown in Figure 3-12(a), the resonant frequencies are shifted down by an amount depending on the circuit Q. The frequency where the impedance of the parallel circuit is maximum, and therefore the peak frequency of the filter, is given by

$$F_{\text{peak}} = F_r \times \sqrt{1 - \frac{1}{4Q^2}} \tag{3-5}$$

There is still some phase shift at this frequency. The phase angle is zero at

$$F_{\text{zero phase}} = F_r \times \sqrt{1 - \frac{1}{Q^2}} \tag{3-6}$$

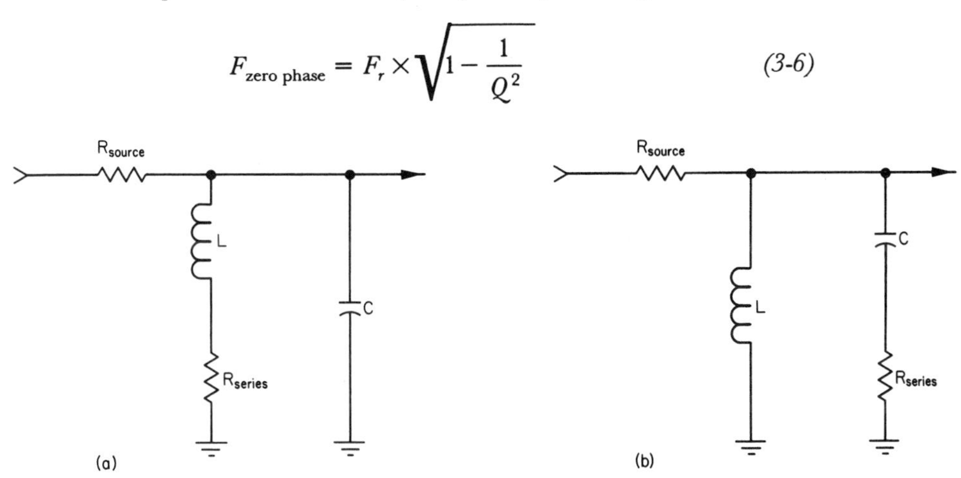

Figure 3-12 Series losses in (a) the inductor and (b) the capacitor that can alter the resonant frequency of a parallel tuned circuit.

[The Q term used in these two equations and the following two is the Q at the frequency given by Equation (3-4) and is the ratio X_L/R_s or X_C/R_s, depending on where the series resistance is.]

When the resistance is in series with the capacitor, as shown in Figure 3-12(b), the resonant frequencies are shifted up by an amount depending on the Q of the arm containing the resistance. The peak frequency is given by

$$F_{\text{peak}} = \frac{F_r}{\sqrt{1 - \dfrac{1}{4Q^2}}} \tag{3-7}$$

The frequency at which zero phase shift occurs is given by

$$F_{\text{zero phase}} = \frac{F_r}{\sqrt{1 - \dfrac{1}{Q^2}}} \tag{3-8}$$

The other effect of the component losses is an insertion loss caused by the filter even at its peak frequency. At resonance, the parallel tuned circuit will not have an infinite impedance but would have some large value of equivalent parallel resistance. The value of this equivalent resistance can be found from a series-to-parallel transform. The two equivalent circuits are shown in Figure 3-13.

Figure 3-13 Equivalent parallel circuit for a resistor and an inductor in series.

The equivalent parallel circuit is only valid at the frequency for which it is calculated, since it involves the component reactance, and this is continually changing with frequency. The exact values are given by

$$L_p = \left(1 + \frac{1}{Q^2}\right)L_s \tag{3-9}$$

$$R_p = (1 + Q^2)R_s \tag{3-10}$$

If the Q's of the circuits are higher than 10, the equations can be simplified to (with $< 1\%$ error)

$$L_p = L_s \tag{3-11}$$

$$R_p = Q^2 R_s \tag{3-12}$$

At resonance, then, the inductive and capacitive reactances cancel, and the impedance of the practical tuned circuit becomes a single parallel resistance. This resistance forms a voltage divider with the source resistance producing an insertion loss at resonance. The equivalent circuit is shown in Figure 3-14(a). The insertion loss at this single frequency can be calculated using the voltage-divider equation:

$$\text{insertion loss} = 20\log\left(\frac{R_p}{R_p + R_{\text{source}}}\right) \tag{3-13}$$

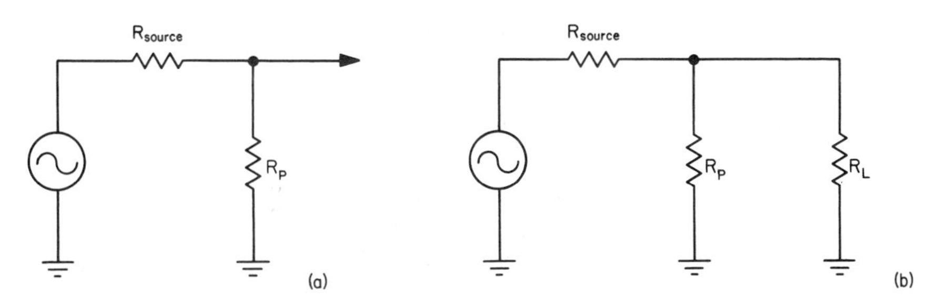

Figure 3-14 Practical tuned circuit at resonance, without (a), and with (b) a load resistance.

When used in a more practical situation, the tuned circuit would operate with both a load resistance and a source resistance. Both of these resistances must be considered for the loaded Q calculation and the insertion loss. The simplest procedure is to consider a Thévenin equivalent for the source and load resistance. This gives a single equivalent source resistance equal to the parallel combination of the original source resistance and the load resistance.

Some examples will illustrate the effects that the component losses have on the overall operation of the single tuned circuit as a filter.

EXAMPLE 3-2

Find the peak frequency, the insertion loss at this frequency, and the frequency of zero phase shift for the filter circuit shown in Figure 3-15.

Solution

First we find the resonant frequency as given by Equation (3-4). This is the frequency of both the peak and zero phase shift if the resistor was not in series with the inductor.

$$F_r = \frac{1}{2\pi\sqrt{LC}} = \frac{1}{2\pi\sqrt{1.5\times10^{-6}\times0.0033\times10^{-6}}} = 2.262\text{ MHz}$$

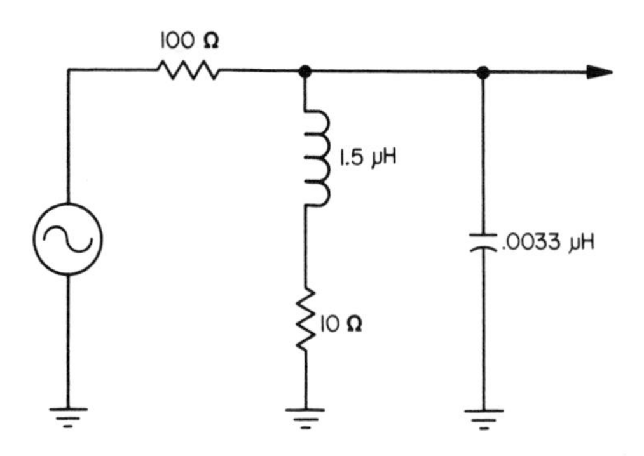

Figure 3-15

Next, we find the reactance of the inductor and the Q of the inductor in series with its resistance.

$$X_L = 2\pi F_r L = 2\pi \times 2.262 \times 10^6 \times 1.5 \times 10^{-6} = 21.32\ \Omega$$

$$Q = \frac{X_L}{R_{\text{series}}} = \frac{21.32}{10} = 2.132$$

The peak frequency of the filter is given by Equation (3-5).

$$F_{\text{peak}} = F_r \times \sqrt{1 - \frac{1}{4Q^2}} = 2.262 \times \sqrt{1 - \frac{1}{4(2.132)^2}} = 2.199\ \text{MHz}$$

The frequency of zero phase shift is given by Equation (3-6).

$$F_{\text{zero phase}} = F_r \times \sqrt{1 - \frac{1}{Q^2}} = 2.262\sqrt{1 - \frac{1}{(2.132)^2}} = 1.998\ \text{MHz}$$

(Notice that there is about a 10% difference between the peak frequency and the frequency of zero phase shift.)

Next, to find the insertion loss of this filter at the peak frequency, we will convert the inductor and resistor in series to an equivalent parallel combination. At the peak frequency,

$$X_L = 2\pi F_{\text{peak}} L = 2 \times 2.199 \times 10^6 \times 1.5 \times 10^{-6} = 20.73\ \Omega$$

$$Q = \frac{X_L}{R_{\text{series}}} = \frac{20.73}{10} = 2.073$$

(We had values for X_L and Q before but not at the frequency we are now interested in. It is necessary to recalculate them at the peak frequency if the parallel equivalent circuit is to be accurate.)

The values for the equivalent circuit are found using Equations (3-9) and (3-10).

$$L_p = \left(1 + \frac{1}{Q^2}\right) L_s = \left[1 + \frac{1}{(2.073)^2}\right] \times 1.5 = 1.849\ \mu\text{H}$$

$$R_p = (1 + Q^2) R_s = \left[1 + (2.073)^2\right] \times 10 = 52.97\ \Omega$$

The equivalent circuit at the peak frequency is shown in Figure 3-16.

The insertion loss can be found using the two values of resistance, but the two reactances do not exactly cancel, so their equivalent must also be used. The two resistances in parallel result in a value

$$R_s \| R_p = \frac{1}{(1/100) + (1/52.97)} = 34.63\ \Omega$$

The inductive and capacitive reactances in parallel result in the equivalent value

$$X_{L_p} = 2\pi F_{\text{peak}} \times L_{\text{parallel}} = 2\pi \times 2.199 \times 10^6 \times 1.849 \times 10^{-6}$$

$$= +j25.55\ \Omega$$

$$X_C = \frac{1}{2\pi F_{\text{peak}}} \times C = \frac{1}{2\pi \times 2.199 \times 10^6 \times 0.0033 \times 10^{-6}}$$

$$= -j21.93\ \Omega$$

$$X_{\text{equiv}} = X_{L_p} \| X_C = \frac{1}{\left(\frac{1}{+j25.55} + \frac{1}{-j21.93}\right)}$$

$$= -j154.8\ \Omega \quad \text{(capacitive)}$$

The insertion loss is then

$$IL = 20\log\left|\frac{52.97}{52.97 + 100}\right| + 20\log\left|\frac{-j154.8}{34.63 - j154.8}\right| = 9.0\text{ dB}$$

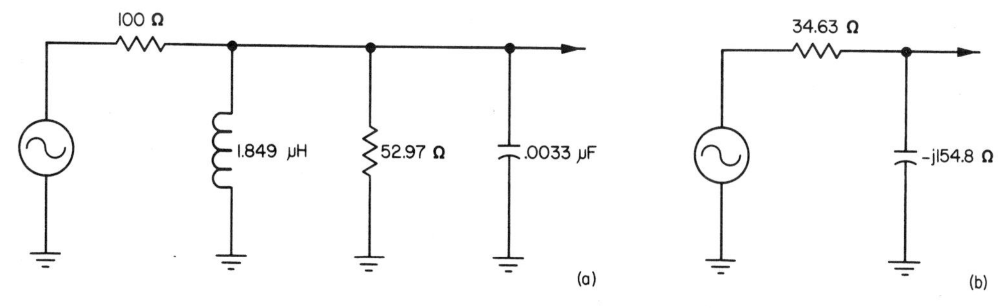

Figure 3-16

The complete amplitude and phase responses of this filter are plotted in Figures 3-17 and 3-18. The amplitude response is normal at the peak and above but shows a constant attenuation at the lower frequencies caused by the limiting value of resistance in the inductance arm. The phase response also shows a change in the same area. The phase returns to zero at the lower frequencies as the inductive reactance becomes insignificant compared to the series resistance.

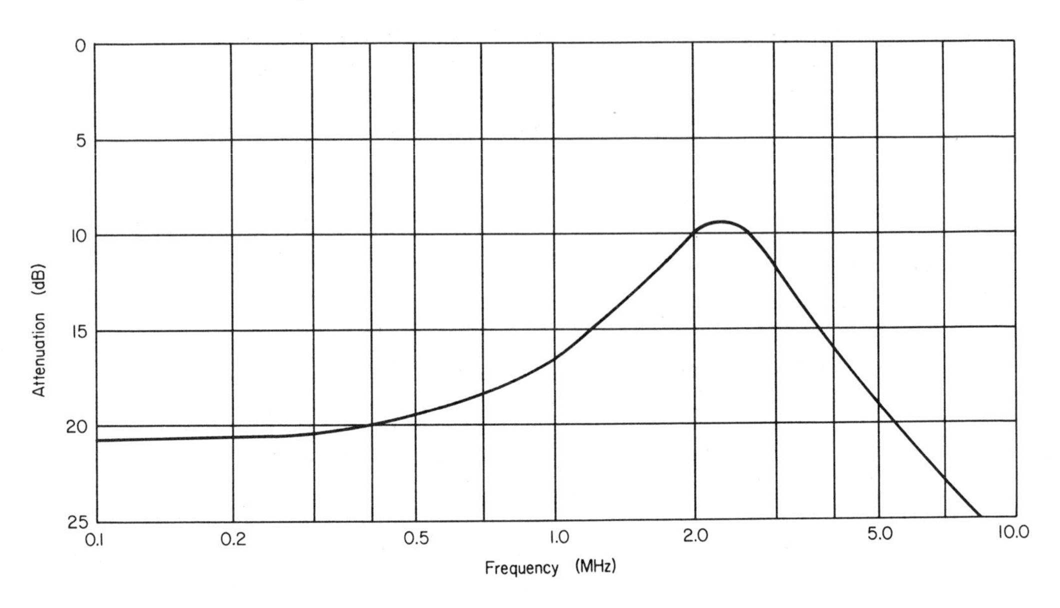

Figure 3-17 Amplitude response of sample filter.

3-4 OVERALL EFFECTS OF PRACTICAL COMPONENTS

As the frequency decreases below resonance, the reactance of the inductance should drop at a constant rate; however, with a practical inductor, the limiting factor is the series loss resistance. The attenuation curve will start to flatten out at some level of attenuation determined by the relative sizes of the loss resistance and the source resistance. Since this loss resistance is caused by the wire and core losses, its value will not be constant with frequency, and so some slope will still be seen on the curve. This portion of the curve is mainly due to resistive changes, and so the overall phase shift will be closer to 0° than to the normal 90° for a purely inductive circuit.

The practical capacitor will possess some inductance in its wire leads, and at some high frequency this will form a series-resonant circuit, producing a greater attenuation than expected. Above this frequency the capacitor will be "inductive," so the curve will start to climb again. A good designer will often use this resonant dip to provide added attenuation at perhaps an "image"

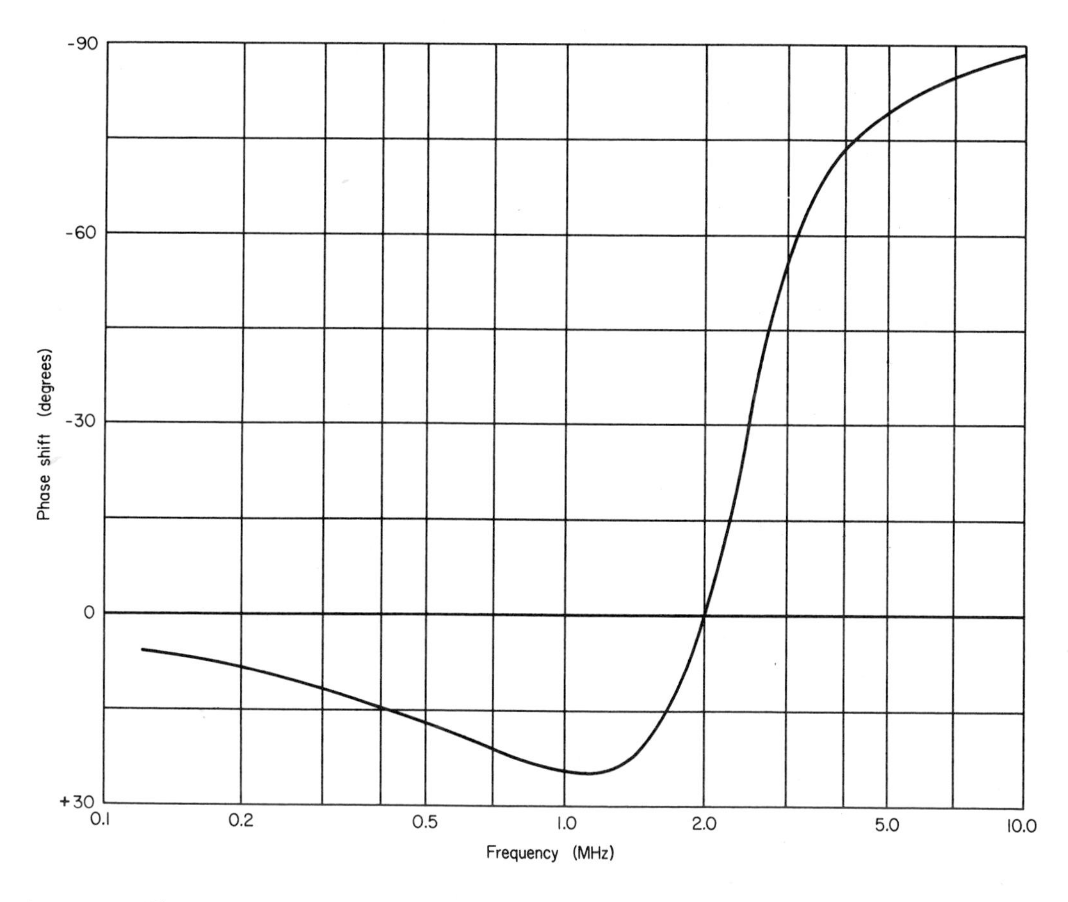

Figure 3-18 Phase response of sample filter.

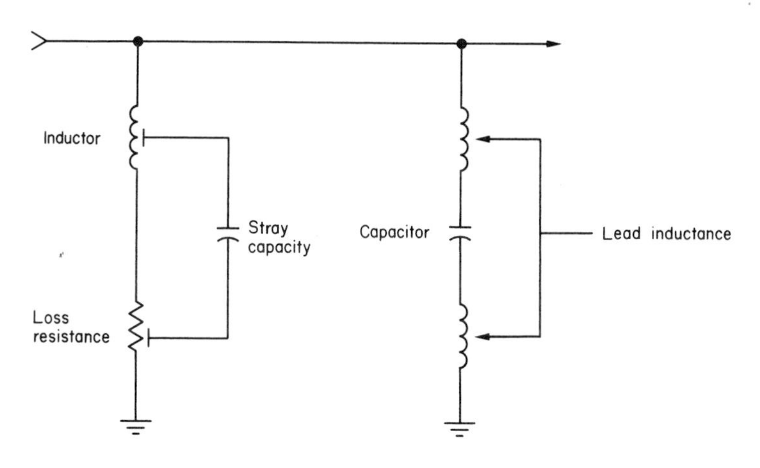

Figure 3-19 Equivalent circuit of a practical single-pole filter.

frequency. The lead length will then be carefully adjusted to provide the right resonant point.

Inductive self-resonance is not as serious, but the value of stray winding capacitance and also external stray capacitance must be considered when selecting the final value of resonating capacitance. In some cases, the stray capacitance provides all that is needed and no added components will be required.

The equivalent circuit of a practical single-pole bandpass filter is shown in Figure 3-19, and the frequency response for this circuit is shown in Figure 3-20.

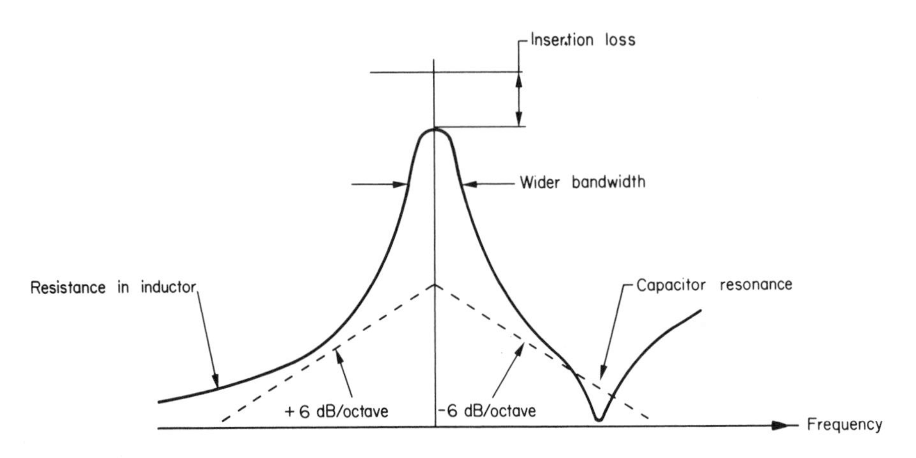

Figure 3-20 Frequency response of a practical single-pole filter showing component self-resonance and losses.

3-5 IMPEDANCE TRANSFORMATION

Often, it is necessary to operate a tuned circuit at a higher impedance level than the rest of the circuit or to use the tuned circuit as an impedance matching network between the source and load resistances. The two common methods are shown in Figure 3-21. A third method uses separate windings to obtain the same results as the tapped inductance, and the same formula applies as long as the extra winding is tightly coupled to the main winding.

For the tapped inductance as shown in Figure 3-21(a), the source resistance is stepped up to an equivalent value:

$$R'_{\text{source}} = R_{\text{source}} \times \left(\frac{n_2}{n_1}\right)^2 \quad \text{for } Q_L > 10 \qquad (3\text{-}14)$$

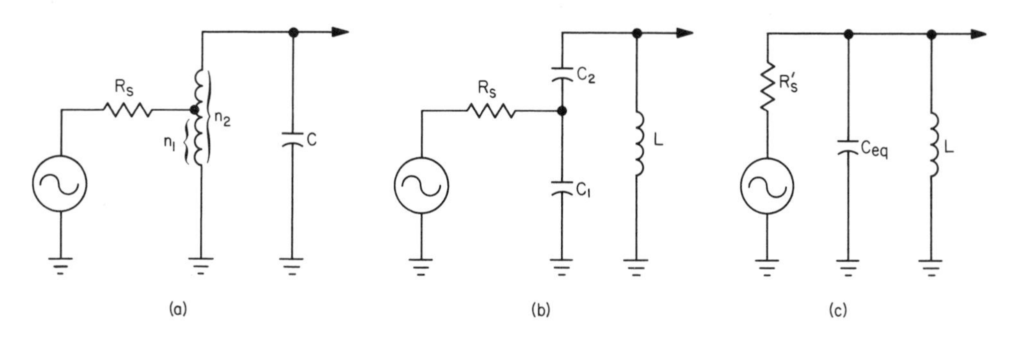

Figure 3-21 Impedance matching with tuned circuits using (a) a tapped inductor and (b) a split capacitance. The equivalent circuit for both is shown in (c).

For the split capacitor as shown in Figure 3-21(b), the source resistance is stepped up to an equivalent value:

$$R'_{\text{source}} = R_{\text{source}} \times \left(1 + \frac{C_1}{C_2}\right)^2 \qquad \text{for } Q_L > 10 \qquad (3\text{-}15)$$

For the tapped capacitor circuit, the equivalent capacitance that resonates with the inductance is given by

$$C_{\text{equivalent}} = \frac{C_1 \times C_2}{C_1 + C_2} \qquad (3\text{-}16)$$

In any application of a tuned circuit where calculations of loaded Q, bandwidth, insertion loss, and so on, are being made, all resistances must be considered. The three principal ones are:

1 The source resistance or its transformed value.
2 The load resistance or its transformed value.
3 The total parallel resistance that represents all the losses in the capacitors and inductors.

EXAMPLE 3-3

Design a single-pole bandpass filter for a center frequency of 10.7 MHz with a 3-dB bandwidth of 500 kHz. The source resistance is 150 Ω and the load resistance is 2200 Ω. The filter will be used to step up the source resistance to match the load resistance. The inductor used will have a Q of 75 and a stray winding capacitance of 8 pF (Figure 3-22).

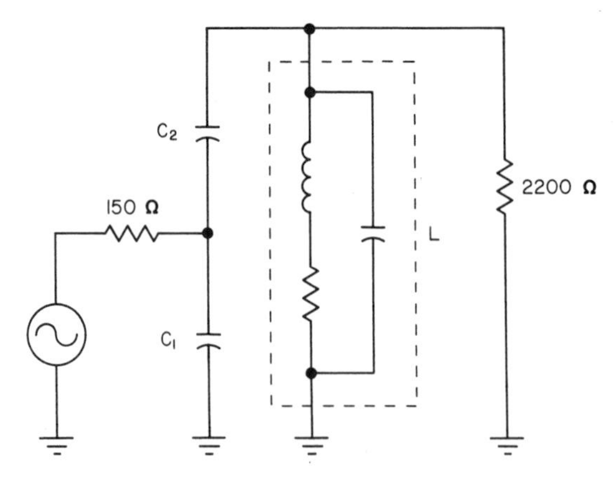

Figure 3-22

Solution

The tapped capacitors will transform the 150-Ω source resistance up to a value of 2200 Ω. The tuned circuit will then appear to be loaded by the two resistances (source and load) for a combined value of 1100 Ω.

The required loaded Q is

$$Q_{\text{loaded}} = \frac{F_r}{\text{BW}} = \frac{10.7\ \text{MHz}}{0.5\ \text{MHz}} = 21.4$$

To find the component reactances, we must consider the total parallel resistance. This, however, includes the parallel coil loss resistance, whose value is not known at this point. Therefore, we must solve two simultaneous equations:

$$R_{\text{parallel coil losses}} = Q_{\text{coil}} \times X_L$$

$$X_L = \frac{1100 \| R_{\text{parallel coil losses}}}{Q_{\text{loaded}}}$$

Solving, we obtain

$$R_p = 2755\ \Omega \quad \text{and} \quad X_C = X_L = 36.73\ \Omega$$

Our equivalent circuit at this point looks as shown in Figure 3-23.

Now we can find our component values:

$$C_{\text{equivalent}} = \frac{1}{2\pi F_r X_c} = \frac{1}{2\pi \times 10.7 \times 10^6 \times 36.73} = 405\ \text{pF}$$

$$L = \frac{X_L}{2\pi F_r} = \frac{36.73}{2\pi \times 10.7 \times 10^6} = 0.546\ \mu\text{H}$$

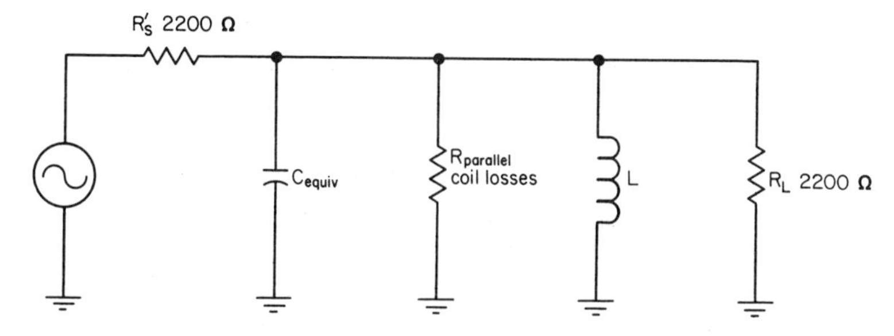

Figure 3-23

The value of capacitance must be reduced by the 8-pF stray capacitance in the coil:

$$C_{\text{equivalent}} = 405 - 8 = 397 \text{ pF}$$

This equivalent capacitance has to be split into the two final values for the circuit. This involves solving two simultaneous equations. From the step-up ratio,

$$\frac{R_s'}{R_s} = \left(1 + \frac{C_1}{C_2}\right)^2$$

From the two capacitors in series,

$$\frac{C_1 \times C_2}{C_1 + C_2} = 397 \text{ pF}$$

Solving, we obtain our component values:

$$C_1 = 1526 \text{ pF} \quad \text{and} \quad C_2 = 539 \text{ pF}$$

To see how much insertion loss this filter will produce,

$$\begin{aligned} \text{insertion loss} &= 20 \log\left(\frac{2755}{2755 + 1100}\right) \\ &= 2.92 \text{ dB} \end{aligned}$$

3-6 COUPLING THE SINGLE-POLE BANDPASS FILTER

Although simple to design and use, the single-pole bandpass filter is not suitable for all bandpass applications. A designer may have problems designing for very narrow bandwidths where high values of loaded Q and therefore even higher values of component Q are required. For some applications, the skirts may not be steep enough. This is especially true when wide bandwidths and therefore low loaded Q's are needed. Finally, the passband is not very flat but rather has a continual rolloff from the center frequency toward each of the cutoff frequencies.

To obtain narrower bandwidths and steeper skirts, several identical stages may be cascaded with one-way coupling between each filter stage. This one-way coupling can be provided by tubes, transistors, or ICs, where, at least in theory, the signal only passes through in the forward direction. (The reverse signal flow through a transistor will be examined in Chapter 6; in some cases the reverse flow is very serious.) Figure 3-24 illustrates this method of coupling tuned circuits.

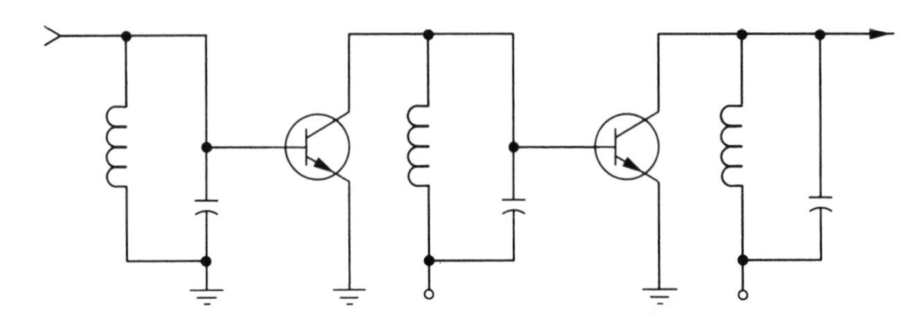

Figure 3-24 One-way coupling of single-pole filters.

If all the tuned circuits are similar, have the same loaded Q, and are set to the same frequency, the bandwidth will be reduced, depending on the number of stages. For n stages all the same,

$$BW_n = BW_1\sqrt{2^{1/n}-1} \tag{3-17}$$

This *bandwidth shrinkage* must be allowed for when selecting the loaded Q's and bandwidths of the individual resonant circuits. As an example, if three identical tuned circuits are used ($n=3$), the overall bandwidth will be 51% of the bandwidth of one circuit. To obtain a particular value of overall bandwidth, then, the individual bandwidth would have to be 1.96 times the overall bandwidth ($1/0.51=1.96$).

If all circuits are set to the same frequency, the technique is called *synchronous tuning*, and the overall bandwidth, as we have seen, becomes narrower.

A different result can be obtained if each tuned circuit is set to a different frequency; this is called *stagger tuning*. The overall bandwidth will now be wider, the bandwidth can be flatter and the skirts steeper, producing a better shape factor. As with synchronous tuning, the overall response is simply the sum of the individual responses in decibels. But now we have a choice of the resonant frequency and the Q of each circuit. This allows us to custom-make almost any shape of response desired as long as we can still obtain the circuit using practical components. The combined result of three coupled circuits is shown in Figure 3-25.

The response of each individual stage is adjusted so that adjacent responses touch at the correct points that will give the flatest bandpass. The two outer curves have a higher loaded Q to produce steeper skirts on the overall result.

Notice that the combined result shows considerable attenuation. This happens because only one peak touches the 0-dB line at a time and the other two are producing some attenuation. The combined result must be the sum of all attenuations at each frequency. In actual use, the coupling amplifiers will provide some gain, and this will be added to the overall results. The distribution of the gain between the different stages does not change the shape of the curve but might affect the overall noise figure and overload characteristics. In any case, the gain of the stagger-tuned amplifier will be lower than the same amplifier without the filters.

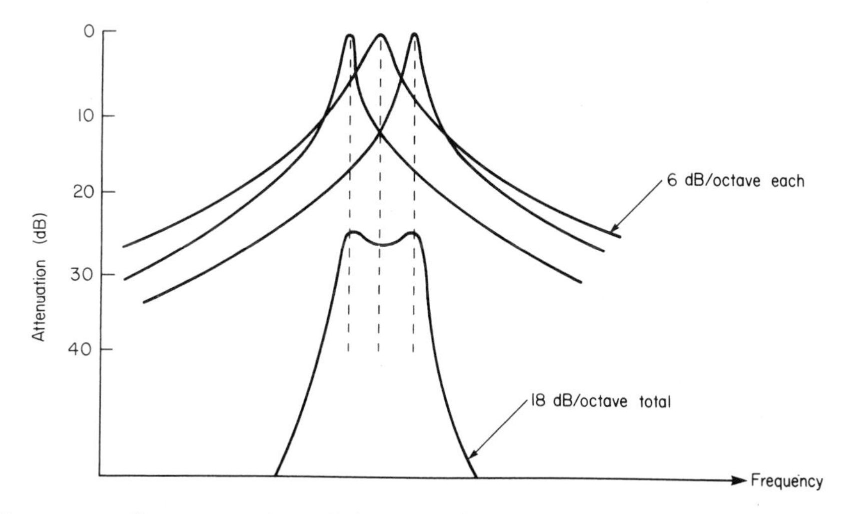

Figure 3-25 Stagger tuning of three resonant circuits to produce a wider, flatter bandpass and steeper skirts.

3-7 INDUCTIVE AND CAPACITIVE COUPLING OF RESONANT CIRCUITS

When capacitors and inductors are used to couple two or more resonant circuits together, as shown in Figure 3-26, we obtain a two-way path that produces response curves quite different from that of a single resonant circuit. The results are similar to the stagger-tuned amplifier with the wide, flat passband and the steep skirts, but in this case all tuned circuits are set to the same frequency.

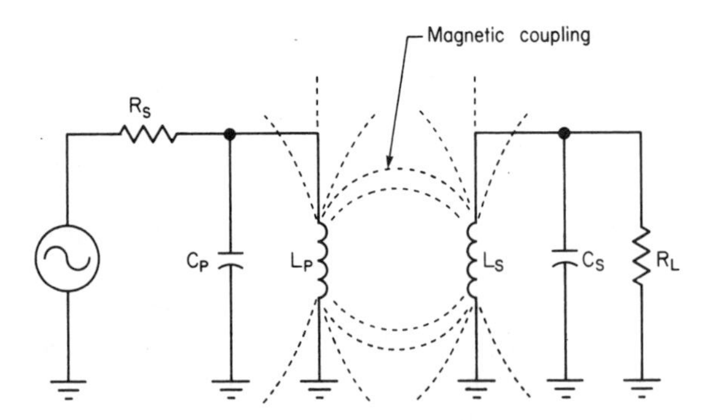

Figure 3-26 Inductive coupling of two identical resonant circuits occurs when magnetic fields overlap.

Since the coupling component is reactive, it will have an effect on the shape of the curve. The passband may show some tilt if it is very wide and the skirts will be asymmetrical.

One way to produce this two-way coupling is to position two coils physically close together so that part of their magnetic fields overlap. The overall frequency response then depends on the amount of magnetic coupling and can change from a single, sharp peak to a broad response with two peaks occurring when the coils are very close. This change is caused by the different reactances that the one parallel circuit places in parallel with the other circuit through the coupling field. The set of response curves obtained is shown in Figure 3-27.

As the coils are moved closer together, notice that the low-frequency side of the response curve remains fairly constant while the high-frequency side moves farther away as the coupling increases. The closer coupling is decreasing

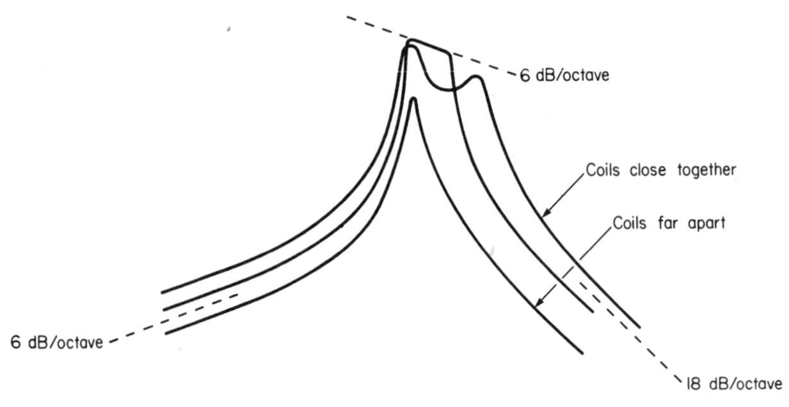

Figure 3-27 Overall response of the inductively coupled circuits depends on the separation of the coils.

the mutual inductance, thereby maintaining resonance to higher frequencies. The 6-dB slope in the passband is noticeable only on very wide bandwidths and can be corrected for by slight adjustment of loaded Q's and resonant frequencies.

The equivalent circuits for frequencies well above and below resonance are shown in Figure 3-28. At low frequencies, the equivalent circuit is the same as a single shunt, tapped inductance, and so the low-frequency slope must be 6 dB/octave, as shown in Figure 3-27. At high frequencies, the equivalent circuit resembles a three-element low-pass filter, and so the final attenuation slope must settle down to 18 dB/octave, again as shown in Figure 3-27.

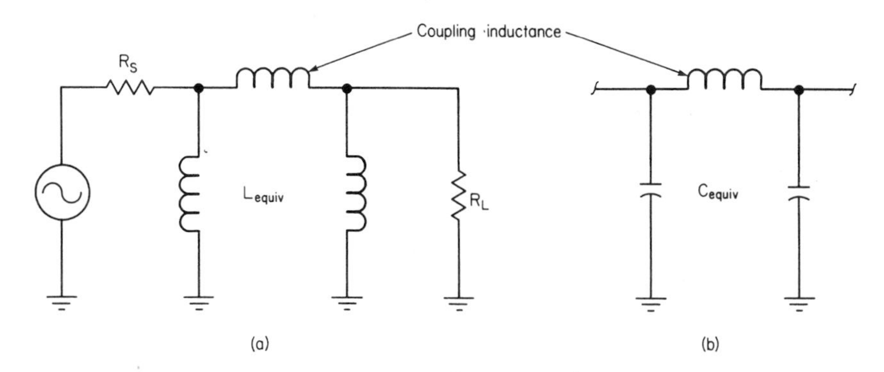

Figure 3-28 Equivalent circuits for inductively coupled resonators at frequencies (a) below the passband and (b) above the passband.

The proper design of the inductively coupled circuit is rather difficult since the amount of magnetic coupling must be calculated, and this depends on the shape of the coils, their spacing, and any core materials used. Fortunately, this is a commercial product that is readily available as IF transformers for frequencies such as 455 kHz, 4.5 MHz, and 10.7 MHz. A number of different impedances are available to match the more popular applications. These impedances are not the input and output impedance of the transformer but rather the source and load impedance that must be used to obtain the correct loaded Q and therefore the proper flat-topped shape to the response.

If it is necessary to design from scratch, select loaded Q's to give approximately one-half the required bandwidth and then experiment with the separation required to give the degree of coupling required. Care must be taken to be sure that both coils are tuned to the same frequency. This can be done when the coils are still farther apart than optimum by simply tuning the circuits for maximum output at a frequency slightly lower than the desired center frequency. Figure 3-27 shows that only a single peak will occur for this undercoupled condition, making the adjustment easy.

A commercial 455-kHz IF transformer is shown in Figure 3-29. This is one of the larger cans designed for vacuum-tube circuits, where source and load

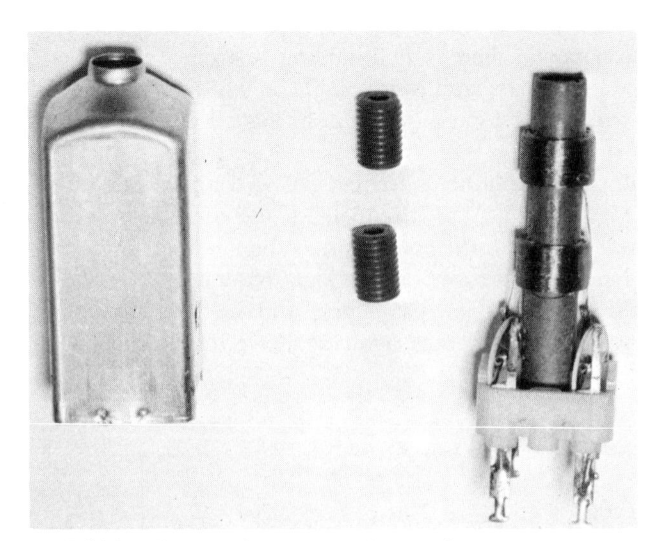

Figure 3-29 455-kHz IF transformer with its aluminum case and the two ferrite tuning slugs.

resistances are much higher than for bipolar transistor circuits. A ferrite slug for tuning is inside each coil, and the resonating capacitors are mounted near the plastic base. The equivalent circuit of the transformer is shown in Figure 3-30 with the values for each component. The Q of each coil alone is 68, and when the external source and load resistances are added, the loaded Q for each circuit drops to about 40. The values of these resistances were found experimentally to give the flattest passband. Normally, the manufacturer's literature will suggest the proper values.

Figure 3-31 shows the overall amplitude response of the transformer and compares it to the amplitude response of one of the tuned circuits from the can using the same source resistance. The difference in the width and flatness of the passband and the steepness of the skirts is readily apparent.

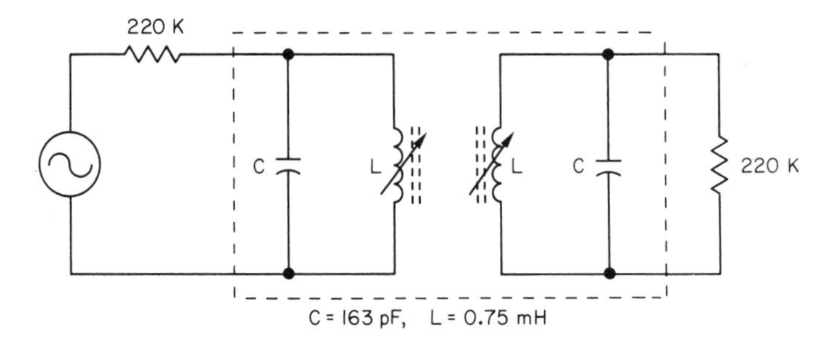

Figure 3-30 Equivalent circuit of the IF transformer with the source and load resistances used to obtain the amplitude and phase response curves shown in Figures 3-31 and 3-32.

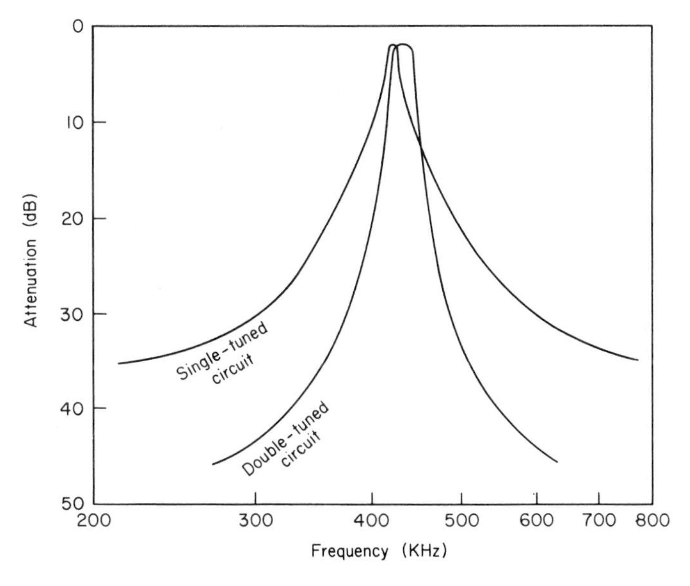

Figure 3-31 Amplitude response of the double-tuned IF transformer compared to the response obtained using only one of the resonant circuits.

The phase response of this filter is shown in Figure 3-32. The phase shift was measured with a good dual-beam oscilloscope. Notice that the phase shift is not symmetrical about the center frequency. By referring back to Figure 3-28, we see that the equivalent circuit below the passband is a single reactive component, and so the phase shift, in this frequency range, must approach 90°

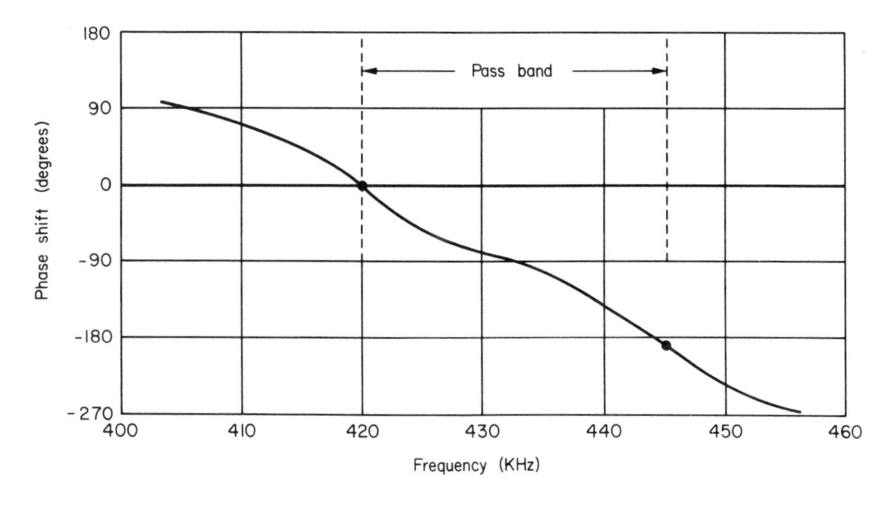

Figure 3-32 Phase response for the passband of the 455-kHz double-tuned IF transformer.

as the frequency decreases. Above the passband, three reactive components are involved, so a final phase shift of 270° in the other direction must be reached.

The phase shift across the passband is not linear, so the envelope delay would vary with sideband frequency. It is this variation that can cause distortion of modulated waveforms. To reduce the variation, the coupling between the coils must be reduced. This undercoupling will round the passband slightly but will straighten out the phase response and so flatten the delay curve. It can be accomplished by moving the coils farther apart, if they are free to move, or by using a lower value of source and load resistance on the transformer.

3-8 CAPACITOR COUPLING OF RESONANT CIRCUITS

Using a capacitor to couple two resonant circuits together, as shown in Figure 3-33, produces the same benefits as the inductive coupling; flat passband and steep skirts, but the amplitude and phase-response curves are mirror images of those obtained with the inductive coupling.

The equivalent circuits for frequencies outside the passband are shown in Figure 3-34. Below the passband (and, in fact, at any frequency below the resonant frequency of the parallel circuit) the equivalent circuit of (a) applies. This is a three-element high-pass filter, so the final attenuation slope must settle down to 18 dB/octave and the final phase angle must reach 270°. Above the parallel-resonant frequency, the filter behaves as a single tapped capacitor as in (b), so a final slope of 6 dB/octave and a final phase angle of −90° must be reached.

The general frequency responses that would be obtained for different sizes of coupling capacitor are shown in Figure 3-35. Comparing this with Figure 3-27 for the inductively coupled filter shows that the two are mirror images of

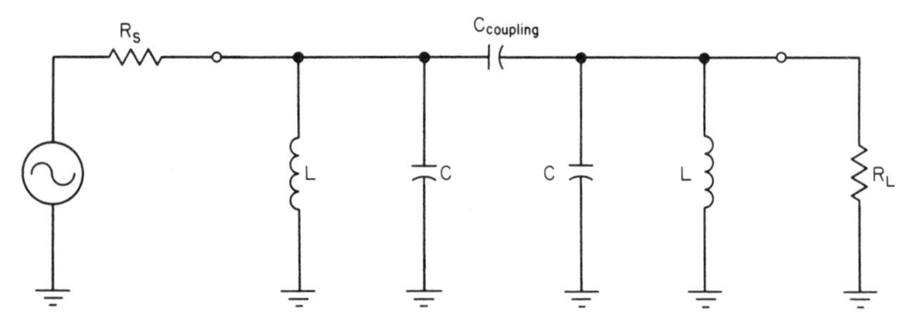

Figure 3-33 Capacitive coupling of two resonant circuits.

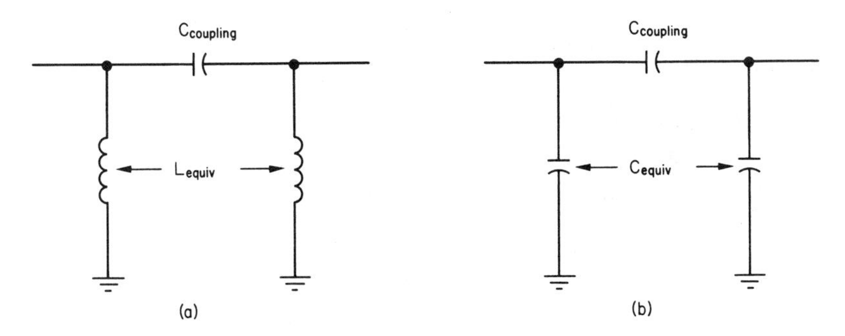

Figure 3-34 Equivalent circuits for the capacitively coupled filter at frequencies (a) below the passband and (b) above the passband.

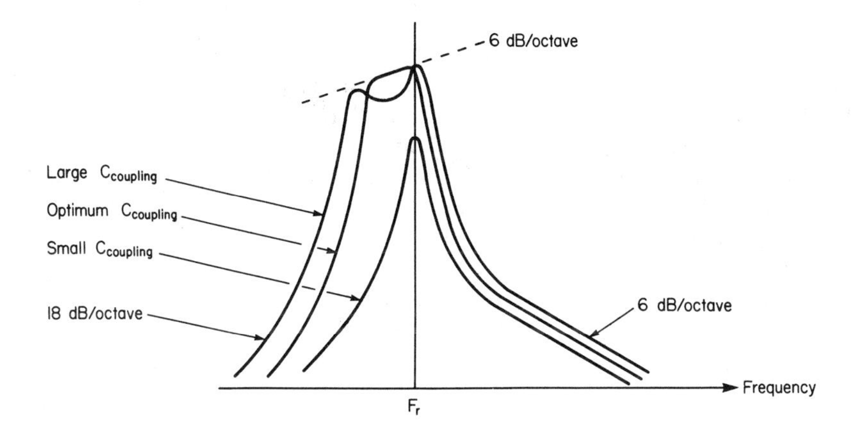

Figure 3-35 By changing the size of the coupling capacitor, the bandwidth and the flatness of the passband can be altered. A small coupling capacitor produces an undercoupled response and a large value produces an overcoupled response.

each other. Some advantage can be obtained by using the two different circuits in tandem to obtain very symmetrical results, especially for very wide passbands.

One advantage of the capacitor coupling is the ease with which it can be adjusted. Trying to alter the physical spacing of the inductively coupled coils is not an easy task, but turning a variable capacitor with an alignment tool to obtain the desired response is fast and simple. All other factors being equal, the capacitor coupling is preferred for individual circuit construction, and the inductive coupling is more suited to mass production of a circuit where its lower cost and easier alignment (two adjustments versus three) can lower total circuit costs.

To design the capacitor-coupled filter, we need to know that for the flattest bandpass:

1 The bandwidth (3 dB) of the coupled filter is approximately $\sqrt{2}$ times the bandwidth of one of its tuned circuits.

2 The value of the coupling capacitor is

$$C_{\text{coupling}} = \frac{C}{Q_{\text{loaded}}} \tag{3-18}$$

where C is the resonating capacitor of the parallel circuit.

3 It should be remembered that the resonant frequency of the parallel circuit is not at the center frequency of the filter but is roughly halfway between the center frequency and the upper 3-dB frequency.

EXAMPLE 3-4

Design a capacitively coupled bandpass filter for a center frequency of 2.5 MHz with a 3-dB bandwidth of 100 kHz. The Q of the inductors used will be 100 and the source and load resistances will be 1000 Ω each (Figure 3-36).

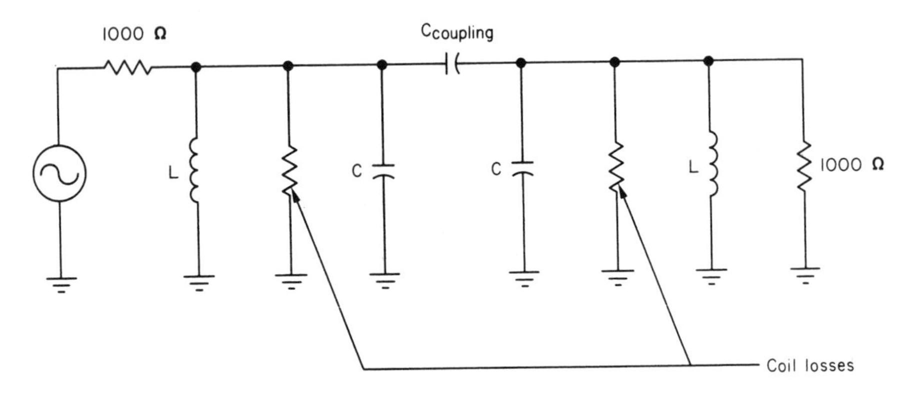

Figure 3-36

Solution

The resonant frequency of the parallel circuits is estimated to be 2.5 MHz + $\frac{1}{2} \times 50$ kHz = 2.525 MHz (see Figure 3-35). The bandwidth of each parallel circuit will be $1/\sqrt{2} \times 100$ kHz = 70.71 kHz. Therefore, the loaded Q must be

$$\frac{F_r}{\text{BW}} = \frac{2.525 \text{ MHz}}{70.71 \text{ kHz}} = 35.71$$

Now we can find the reactances of the parallel components:

$$X_L = X_C = \frac{1000\ \Omega \| R_{\text{losses}}}{35.71}$$

where: $R_{\text{losses}} = Q \times X_L = 100 \times X_L$

Solving these two simultaneous equations, we obtain

$$X_C = X_L = 18.0\ \Omega$$

Since we now know the reactance of these two components at 2.525 MHz, we can now find their actual values.

$$L = 1.13\ \mu\text{H} \quad \text{and} \quad C = 0.0035\ \mu\text{F}$$

The value of the coupling capacitor is found to be

$$C_{\text{coupling}} = \frac{C}{Q_{\text{loaded}}} = \frac{0.0035\ \mu\text{F}}{35.71}$$
$$= 98.0\ \text{pF}$$

The design is now complete. The amplitude response of the finished filter is shown in Figure 3-37.

When building a capacitively coupled filter, it must be remembered that there is a good possibility that magnetic coupling will also occur between the two coils. The amount will depend on the orientation and spacing of the inductors and whether any shielding is used. To minimize any stray coupling, the coils should be kept far apart and mounted at right angles to each other or used with metal cans for shielding. Coils wound on toroid forms are easier to use since most of the magnetic field is contained within the circular core.

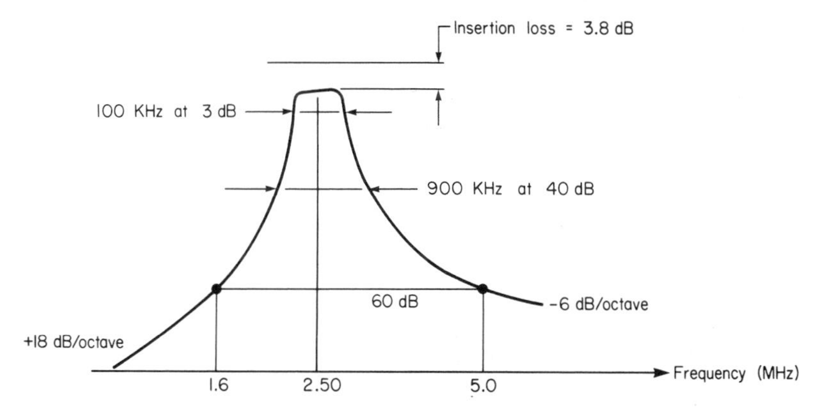

Figure 3-37

If magnetic coupling does exist, it will either add to the capacitive coupling or partially cancel it, depending on the relative directions in which the two coils are wound. The coupling capacitor in such a case would have to be increased or decreased in value to offset the stray coupling.

In some cases the two different types of coupling can be put to good use. A small amount of capacitive coupling could be added to a standard IF transformer to change its characteristics. This would be easier and quicker than attempting to reposition the coils. It must be remembered, though, that this *two-element coupling* will result in different skirt characteristics than the single-element coupling we have been discussing. The skirts will become more symmetrical, but they could be raised or lowered depending on the relative polarities of the two coupling methods.

3-9 CRYSTAL FILTERS

The limiting factor in designing any narrow bandpass filter is always the availability of high-Q components. The loaded Q of any resonant circuit will always be less than the Q of the components used to make the circuit and, if the insertion loss is to be minimized; the difference must be large. With practical limits on inductor Q's of 100 to 200, our loaded Q that we can work with must be less than 50 to 100, and this will place a limit on how narrow a filter we can design at any given frequency.

Thin slices of quartz and various ceramics can be used to make piezoelectric resonators with much higher Q's than any LC resonator. Ceramic can have a Q of 200 to 2000, with a temperature drift of 0.1%/°C. Quartz, depending on the axis of its cut from the main crystal, can have Q's of 5000 to 100,000 and a temperature drift of only 0.001%/°C. These specifications have made quartz and ceramic very popular for the design of IF filters for high-quality communication receivers and FM stereo broadcast receivers.

The simplest approach to a crystal filter is to simply place a resonator between the source and load resistances as shown in Figure 3-38. A narrow passband will occur at the series-resonant frequency of the crystal with a bandwidth and insertion loss that depend on the crystal's equivalent values and the size of the source and load resistances.

EXAMPLE 3-5

Using the component values shown in Figure 3-38, find the series- and parallel-resonant frequency, bandwidth, and insertion loss.

Solution

The series-resonant frequency is

$$\frac{1}{2\pi\sqrt{LC}} = \frac{1}{2\pi\sqrt{25\times10^{-3}\times0.008\times10^{-12}}} = 11.254 \text{ MHz}$$

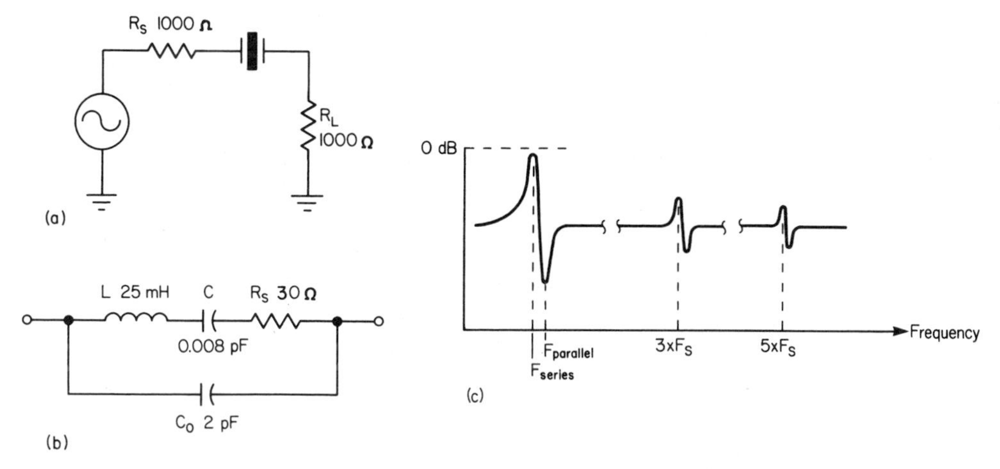

Figure 3-38 The simplest crystal filter is a resonator in series with the source and load resistances (a); the equivalent circuit of the crystal is shown in (b), and the resulting frequency response is given in (c).

The parallel-resonant frequency is

$$f_{\text{series}} \times \left(1 + \frac{C}{2C_0}\right) = 11.254 \times \left(1 + \frac{0.008 \text{ pF}}{2 \times 2 \text{ pF}}\right) = 11.276 \text{ MHz}$$

The difference between series- and parallel-resonance is only 22:51 kHz (0.2%). The Q of the crystal itself is

$$\frac{2\pi f_s L}{R_{\text{series}}} = \frac{2\pi \times 11.254 \times 10^6 \times 25 \times 10^{-3}}{30} = 58{,}925$$

and the loaded Q is

$$\frac{2\pi f_s L}{R_{\text{source}} + R_{\text{series}} + R_{\text{load}}} = \frac{1.7677 \times 10^6}{1000 + 30 + 1000} = 871$$

If the parallel-resonant frequency did not exist, the passband would be symmetrical and have a width of

$$\frac{f_{\text{series}}}{Q_{\text{loaded}}} = \frac{11.254 \text{ MHz}}{871} = 12.9 \text{ kHz}$$

But the steeper slope on the high-frequency side would reduce this bandwidth somewhat. The insertion loss would be

$$20 \log\left(\frac{1000 + 1000}{1000 + 1000 + 30}\right) = 0.13 \text{ dB}$$

At frequencies above the main passband, we find additional, lower-level passbands. These occur at the overtone resonant frequencies of the crystal and must be eliminated if the filter is to be useful. The solution is to include a simple *LC* filter, as shown in Figure 3-39.

The bandwidth of the tuned circuit will be much wider than that of the crystal, so no great change in the overall peak response of the filter will result. At the overtone frequencies, however, the *LC* filter will now provide a fair amount of attenuation and so lower the spurious peaks. The loaded *Q* would be chosen to provide whatever attenuation is required at the spurious frequencies. For the worst-case design, the parallel value of the load and source resistance must be considered when calculating loaded *Q*.

The next step in crystal filter complexity is the half-lattice circuit shown in Figure 3-40. Along with the center-tapped transformer, the two crystals form a bridge circuit, so the amount of signal that passes depends on the degree of unbalance of the bridge.

The two crystals used have slightly different series-resonant frequencies, and this would tend to suggest the double-hump response curve shown in Figure 3-40(b). However, if great care is taken with the

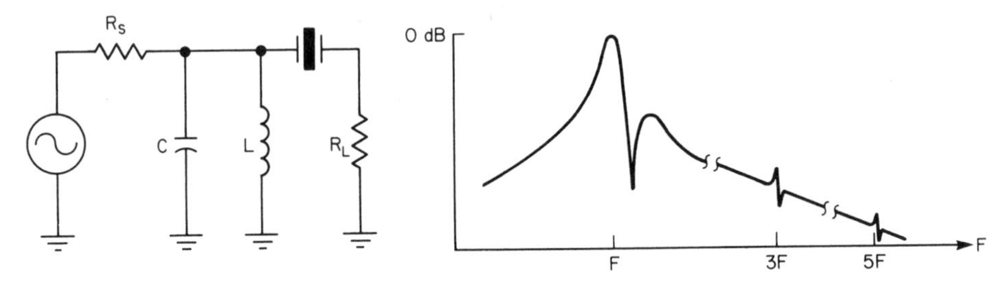

Figure 3-39 Use of an *LC* filter to reduce the effects of spurious and overtone resonance in a crystal filter.

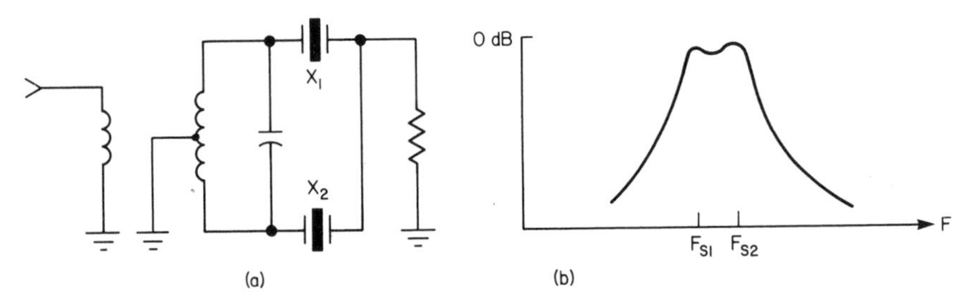

Figure 3-40 Half-lattice crystal filter (a) and an approximation of its frequency response (b).

design, some significant differences will be noted. First, let us take a look at what happens to the reactance of a crystal when a small capacitor is shunted across its terminals.

The reactance curves of Figure 3-41 show that the series-resonant frequency remains unchanged and that the parallel-resonant frequency moves lower (it can never go below the series-resonant frequency). Both below the series frequency and above the parallel frequency, the curve is capacitive and shows less reactance than the curve for the crystal without the added capacitor; this is an important point.

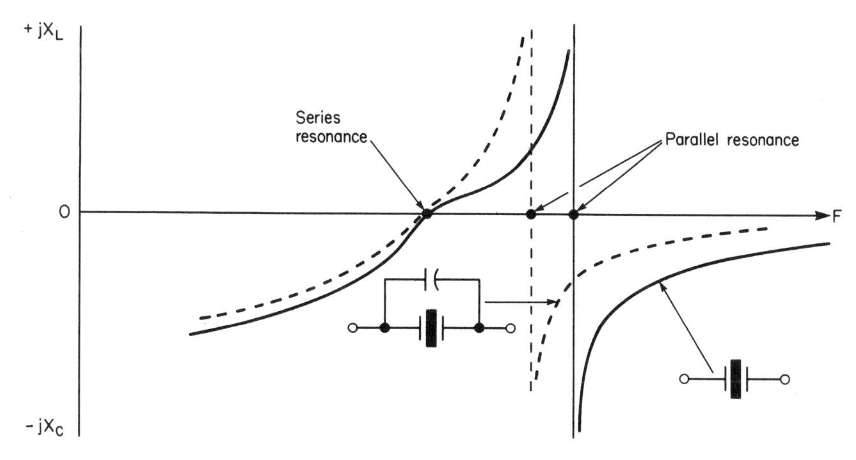

Figure 3-41 Reactance of a quartz crystal versus frequency, compared with that of the identical crystal shunted with a small capacitor.

Now we select two crystals, one series resonant at the desired center frequency and the other parallel resonant at the same frequency. A small capacitor is then placed across the terminals of the higher-frequency crystal to lower its capacitive reactance. The size of this capacitor must be obtained experimentally, but it will be about one-tenth the internal shunt capacitance of the holder. The reactances of the two crystals are shown in Figure 3-42(a). Notice how the series frequency of the higher-frequency crystal lines up with the parallel frequency of the lower-frequency crystal. This requirement places a natural limit on the bandwidths that can be obtained at any particular frequency, but some adjustment can be made by using series and shunt inductances with the crystals to spread the frequencies apart. This is best left to professional filter designers.

Meanwhile, back at Figure 3-42, we may also see that there are two frequencies where the curves of the two crystals cross. These are labeled f_∞ and represent conditions of balance for the bridge since both crystals have identical capacitive reactance thanks to the added capacitance across the one crystal. The filter should show deep nulls at both these

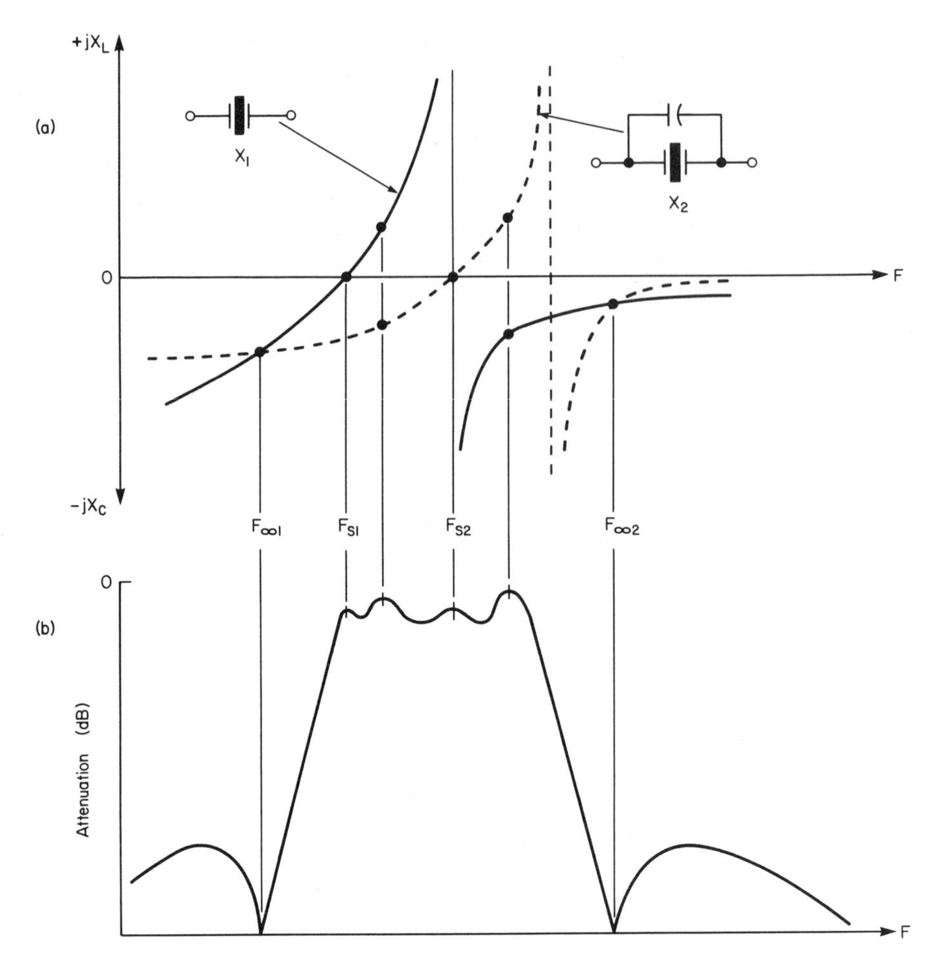

Figure 3-42 Reactance of the two crystals in the half-lattice filter (a) and the resulting bandpass response (b).

frequencies. Within the passband there is one series-resonant frequency for each crystal, and so two points are shown with very little insertion loss since the input would be connected to the output through the very low value of the crystal series resistance. Two additional points occur in the passband where the two crystals have equal but opposite reactances. This is a resonant condition that can produce high peaks, and these must be controlled to keep the peaks at the same level as the other two points in the passband. This is done by setting the source and load resistances to the "image impedance" value:

$$\text{image impedance} = \sqrt{X_1 \times X_2} \tag{3-19}$$

where X_1 and X_2 are the equal but opposite equivalent reactances of the crystals at a frequency that is the geometric mean of the series- and parallel-resonant frequencies. This all simplifies down to one equation:

$$R_S = R_L = 2\pi \Delta f L \tag{3-20}$$

where: Δf = spacing between the series-and parallel-resonant frequencies (Hz)

L = series inductance of the crystal,
typically: 5 H at 500 kHz,
2 H at 1 MHz,
0.8 H at 2 MHz,
0.2 H at 5 MHz,
30 mH at 10 MHz.

At frequencies well away from both the null frequencies and the passband, the equivalent circuit of Figure 3-43(a) also applies. The attenuation now depends on how close these two reactances can be maintained over a wide range of frequencies. The greater the difference, the poorer the attenuation. Much of the fine art of crystal-filter design involves the matching of these reactances for good values of ultimate attenuation. The center-tapped transformer and its resonant capacitor can also help improve the attenuation at frequencies well out from the null points.

This requirement for careful matching of capacitances and transformers makes the crystal-lattice filter more expensive than the newer monolithic ladder filters. To see how serious the matching problem is, consider Figure 3-44. The three curves show the normal response of a half-lattice filter and then the effects of a ±5% change in either the location of the center tap of the transformer or the ratio of the total parallel capacitance of each crystal.

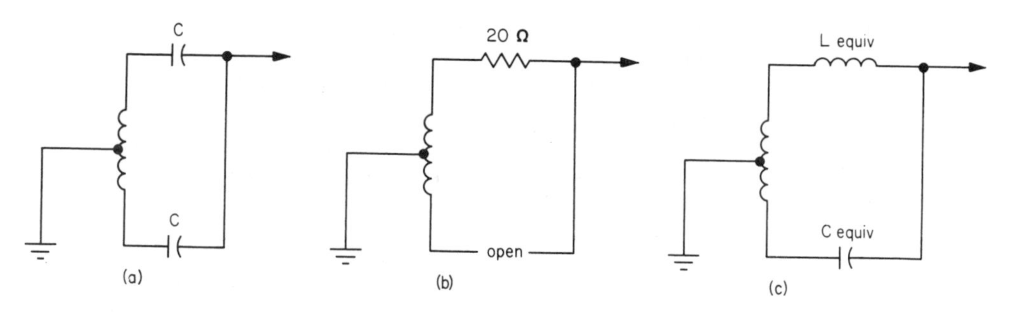

Figure 3-43 Equivalent circuits for the half-lattice crystal filter, (a) at the two null frequencies, (b) at one of the series-resonant frequencies, and (c) at a frequency within the passband where the reactance of the one crystal is equal and opposite to the reactance of the other crystal.

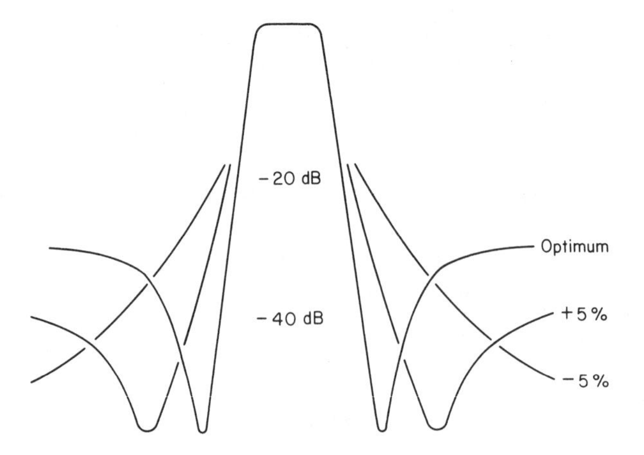

Figure 3-44 Normal amplitude response of a half-lattice filter and the changes that result if the location of the transformer center tap or the parallel capacitance ratio changes by ±5%.

EXAMPLE 3-6

Assume that we have a 1.0-MHz series-resonant quartz crystal that we would like to use for a half-lattice crystal filter.

Solution

For the average crystal, the ratio of the internal parallel to series capacitance is about 400:1, so we can estimate the parallel-resonant frequency of this crystal.

$$f_p = f_s\left(1 + \frac{C_s}{2C_p}\right) = 1.0\text{ MHz}\left(1 + \frac{1}{2\times 400}\right) = 1.00125\text{ MHz}$$

This should be verified by measuring the two resonant frequencies using a good signal generator and frequency counter, as shown in Figure 3-38. A second crystal can now be purchased that is similar to the first and series resonant at 1.00125 MHz, which is the parallel-resonant frequency of the other crystal. The equivalent circuit for each crystal will look as shown in Figure 3-45, with slight value changes for the two frequencies.

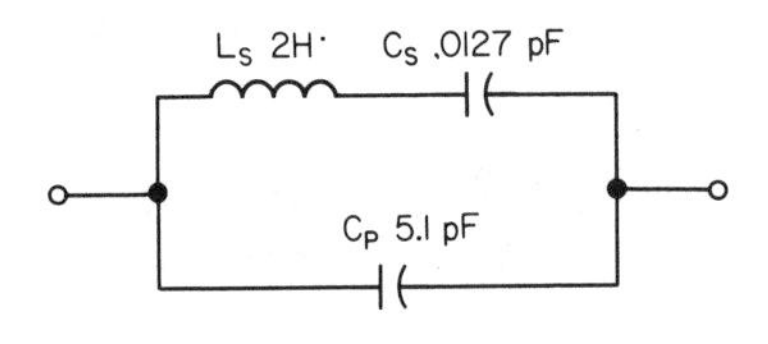

Figure 3-45

The internal parallel capacitance is 5.1 pF, so a capacitor with a lower value than this must be placed across the terminals of the higher-frequency crystal.

To find our source and load resistances, we need the equivalent reactances at one of the equal reactance points. One of these frequencies is

$$\sqrt{1.00000 \times 1.00125} = 1.00062 \text{ MHz}$$

Considering only the series components of the crystals, at this frequency we can obtain the reactances shown in Figure 3-46.

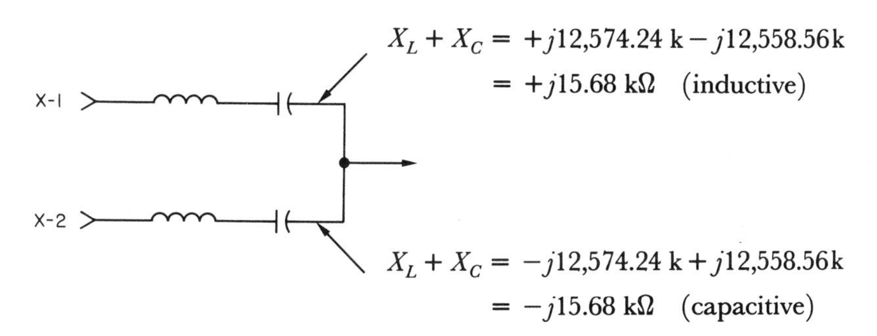

Figure 3-46

The source and load resistances will now be

$$R_S' = R_L = \sqrt{+j15.68\text{ k}\Omega \times -j15.68\text{ k}\Omega} = 15.68\text{ k}\Omega$$

The input transformer must be designed with a turns ratio that will step up the source resistance (we will assume 75 Ω) to 15.68 kΩ on each side of the center tap (Figure 3-47). The load resistance would also be chosen as 15.68 kΩ. A tuned circuit could be used with the load to provide extra attenuation at frequencies well above and below the passband, but the load must remain resistive or very slightly capacitive in the passband.

The final circuit could look as shown in Figure 3-48.

The input transformer was designed for a loaded Q of 30 to provide additional attenuation at the overtone and spurious frequencies. To minimize insertion loss, the Q of the coil itself would have to be about 150. This can be achieved by winding the coil on a toroid form. The center tap may be difficult to properly position, and so tapped capacitors across the secondary could be used instead, each having a value of 340 pF.

The final frequency response depends to a great extent on the matching of the crystals and, in particular, the relative values of the

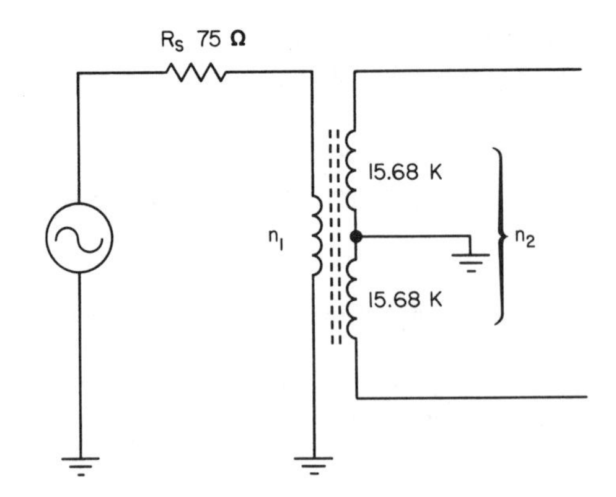

Figure 3-47

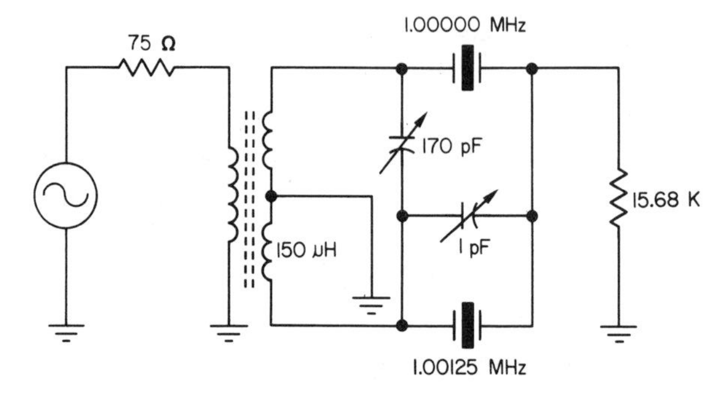

Figure 3-48

internal shunt capacitors and the extra capacitance across the higher-frequency crystal. The approximate response might be as shown in Figure 3-49.

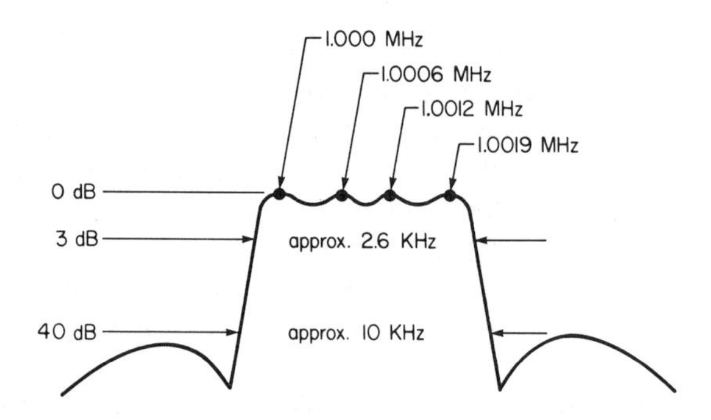

Figure 3-49

For improved shape factor and higher values of ultimate attenuation, the half-lattice design can be cascaded. Figure 3-50(a) is a photograph of an eight-pole crystal filter,[1] and its equivalent circuit and recommended terminating resistances are shown in Figure 3-50(b). The eight crystals are mounted along the edges of the small printed circuit board, and the center-tapped toroidal inductors are in the cans mounted crossways on the board. All the metal cans are bonded together for shielding, and the entire assembly is enclosed with a metal can soldered to the base.

The amplitude response of this filter is given in Figure 3-50(c). With a 6-dB bandwidth of 30 kHz and a 60-dB bandwidth of 54 kHz, the filter has a shape factor of $54/30 = 1.8$, which is quite good but not outstanding. More exotic 12-pole designs can obtain shape factors as low as 1.15 : 1, but of course there would be a drastic difference in the price of the two. The ultimate attenuation of the filter shown is about 100 dB, but the odd spurious response brings the attenuation up to only 80 dB at points.

3-10 MONOLITHIC FILTERS

Monolithic crystal filters offer a number of advantages over crystal-lattice filters. They are smaller in size, lower in cost, do not require balancing transformers, and are more reliable. Two or more coupled resonators are formed by depositing thin-film metal electrodes on a single quartz or ceramic disc, as shown in Figure 3-51. The results are very similar to the inductively coupled tuned circuits in Figure 3-26, but in this case the equivalent circuit shows series-resonant circuits, and their Q will be much higher than the LC resonators. The shunt inductance represents the acoustic coupling that takes place between the two resonators. The shunt capacitance at the input and output terminals represents the interelectrode or "holder" capacitance that occurs in all crystals.

The filter configuration obtained with this structure is a ladder network. The frequency response will, therefore, not show the two characteristic nulls at the ends of the skirts that the lattice network does. (Other forms of coupling resonators together in a ladder network do produce nulls at the series- and parallel-resonant frequencies of shunt and series crystals, respectively.) Figure 3-52 shows the attenuation curve for a 10.7-MHz commercial unit and also the improvement obtained by connecting two pole units in tandem.

Two final words on crystal filters. First, whether you are using quartz or ceramic, remember that the material mechanically vibrates with an amplitude

[1]Even though this circuit has 10 resonant circuits (eight crystals and two LC circuits), it is still referred to as an eight-pole filter since it is only the crystals that are significant within the passband. The two LC circuits are significant only at much higher and lower frequencies.

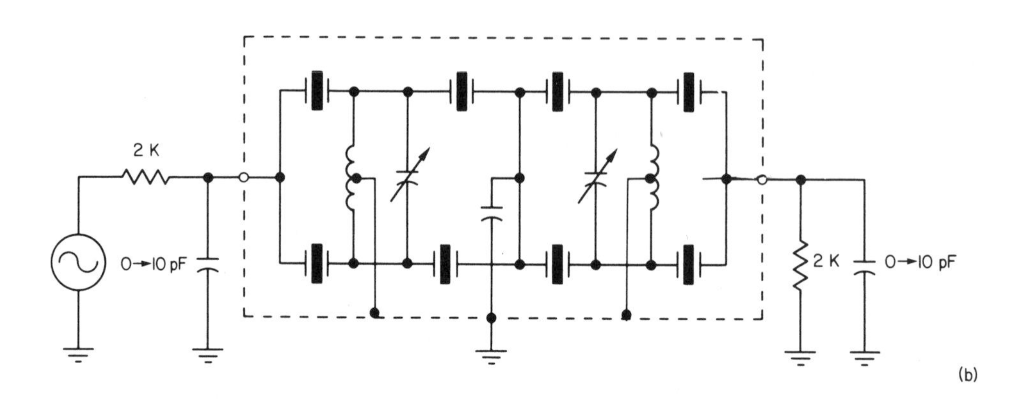

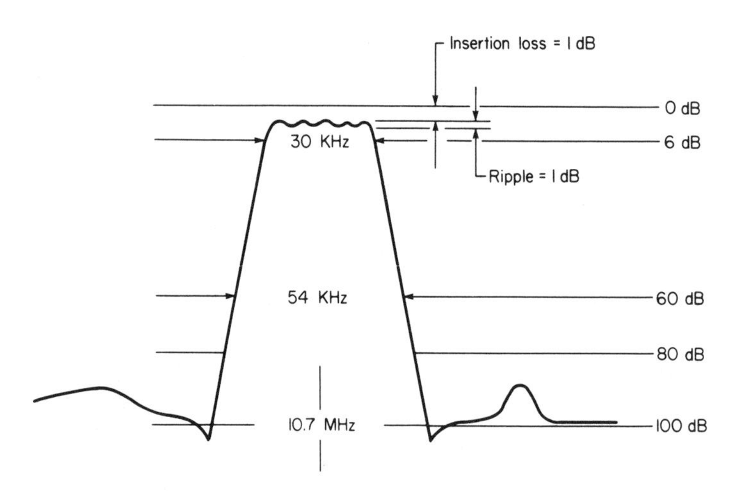

Figure 3-50 Commercial 10.7-MHz crystal-lattice filter. Photograph is shown in (a) and its circuit diagram and recommended terminating impedances are shown in (b). The amplitude response for frequencies close to the center frequency is shown in (c).

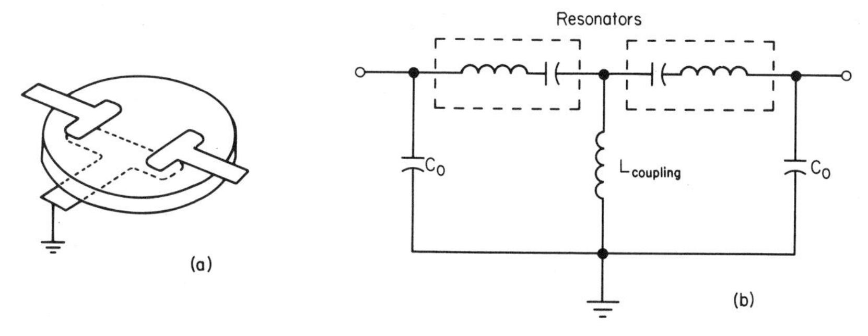

Figure 3-51 Electrode arrangement on a quartz monolithic filter (a) and the equivalent circuit of the resulting two-pole filter (b).

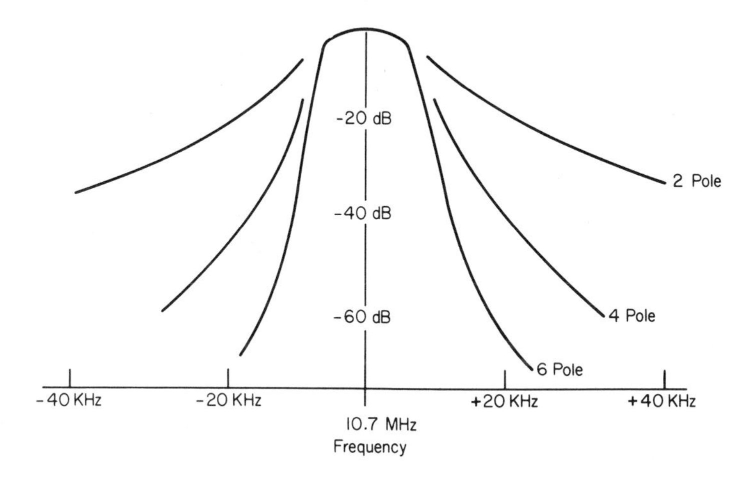

Figure 3-52 Frequency response of a two-pole monolithic crystal filter and also the improvement with units in tandem.

proportional to the applied signal level. For strong signals, the vibrations can reach nonlinear conditions and actually cause amplitude distortion of the signal passing through. At even higher levels the crystal can crack and be permanently destroyed. Manufacturers therefore specify maximum signal power levels such as 10 mW or its equivalent, +10 dBm.

Second, the terminating impedances must be accurately set, as they play an important part in the overall characteristics of the filter. Often a tuned circuit, if it is part of the termination, will have to be set so that it is slightly capacitive at the center frequency in order to obtain the desired response. Monolithic filters do not normally contain tuned *LC* circuits inside the package, so their use as part of the termination is recommended in order to further reduce spurious passbands. (The input impedance of many monolithic filters rises outside the passband because they behave as series-resonant circuits. This higher impedance could increase the gain of the driving transistor, thereby raising the level of a spurious response in the crystal.)

QUESTIONS

1 Design a single-pole bandpass filter for a center frequency of 3.1 MHz and with a 3-dB bandwidth of 180 kHz. The source resistance is 2200 Ω and the load resistance is 3300 Ω.

2 The filter designed in Question 1 was built using an inductor with an unloaded Q of 65 at the operating frequency. What will the final bandwidth be and what insertion loss will occur at the peak of the response curve?

3 Design a tuned circuit similar to that of Figure 3-21(a) that will provide proper matching between a 75-Ω source and a 560-Ω load. A 3-dB bandwidth of 75 kHz is required at a center frequency of 1.35 MHz. The tapped inductor will have an unloaded Q of 93.

4 Design a capacitively coupled two-pole bandpass filter for a center frequency of 4.5 MHz with a 3-dB bandwidth of 150 kHz. The source and load resistances are both 2200 Ω and the inductors are assumed to be lossless.

5 Explain why a half-lattice crystal filter will have deep nulls on each side of the passband and monolithic crystal filters will not.

6 Write a computer program to calculate the amplitude and phase response of the filter of Example 3-2. Cover the frequency range from 210 kHz to 10 MHz in 35 logarithmic steps. (Each new frequency is obtained by multiplying the previous one by a constant.)

REFERENCES

Henney, K. 1959. *RadioEngineering Handbook*. New York: McGraw Hill.

Kennedy, G. 1970. *Electronic Communication Systems*. New York: McGraw Hill.

Lang, J. June 1964. Narrow Band Crystal Filter Design. *Electro Technology*.

Sauerland and Blum. November 1968. *Ceramic* IF *filters for consumer products. IEEE Spectrum*.

Terman, F. E. 1943. *Radio Engineers' Handbook*. New York: McGraw Hill.

chapter 4

filters

In this chapter, we will discuss the role played by each component in a filter in order to gain a feel for the various capabilities of these circuits. The characteristics of a range of standard designs will then be examined and we will see how to adopt these to suit specific requirements. A more formal description of filter design, using the s-plane, is included as Appendix 4A to this chapter for readers interested in this approach.

There are four general types of filters: *low pass, high pass, bandpass,* and *band rejection*, named according to the range of frequencies passed and rejected. The amplitude response for each of these types is shown in Figure 4-1. The low-pass type is shown in more detail, since we will be concentrating on this filter more. The general terms used, however, apply to all four filter types.

4-1 SOME SIMPLE CIRCUITS

We will begin by looking at three circuits, each containing various numbers of reactive components (capacitors and inductors), and examining the changes in the amplitude and phase responses as the relative sizes of the components change in each circuit. From this we can then make some general comments that will apply to all filters.

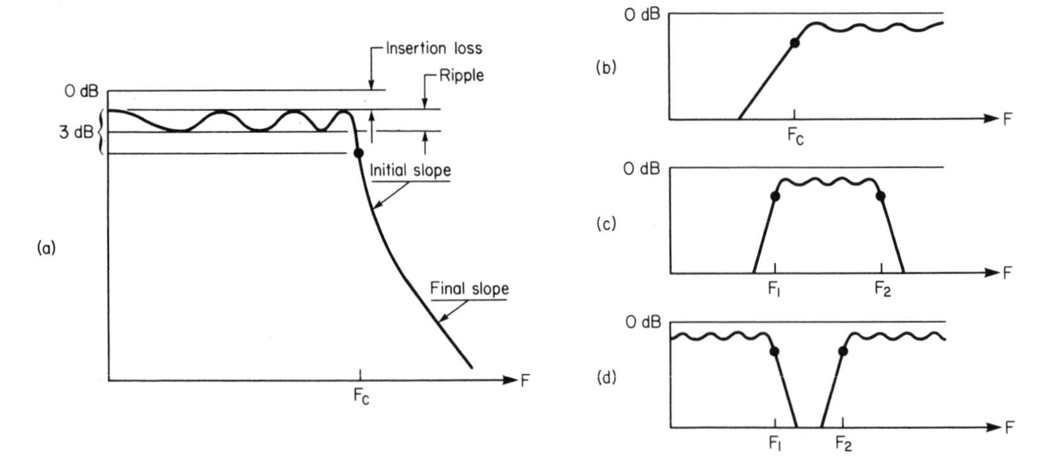

Figure 4-1 General amplitude response of the four filter types: low pass (a), shown with details of the response specifications; high pass (b); bandpass (c); band rejection (d).

4-1.1 Single Pole

First, we will consider the circuit shown in Figure 4-2. This is a one-pole (see the appendix to this chapter), low-pass filter that contains only one reactive component, the capacitor.

As in Chapter 3, here also we will consider the insertion loss caused by placing the filter in between the source and the load terminals. The notation 0 dB will indicate that the maximum *available* power from the source is being delivered to the load and a 3-dB insertion loss means that only one-half the *available* power is reaching the load. The insertion loss and phase response of the single-pole filter are shown in Figure 4-3.

The amplitude-response curve shows a flat passband that gradually rolls off to a constant 6 dB/octave slope in the stopband. The phase shift between

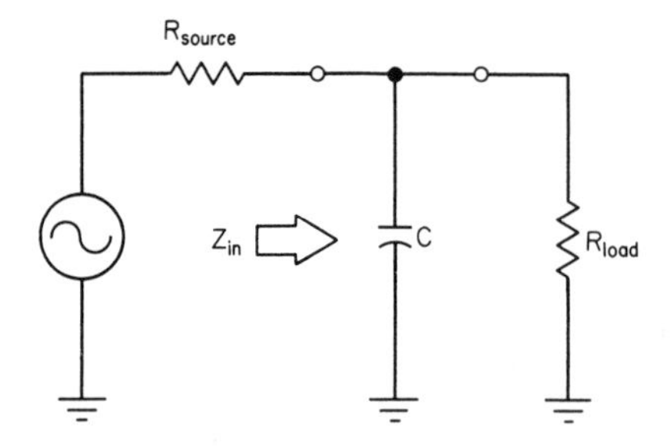

Figure 4-2 Single-pole, low-pass filter consisting of a capacitor in parallel with the source and load terminals.

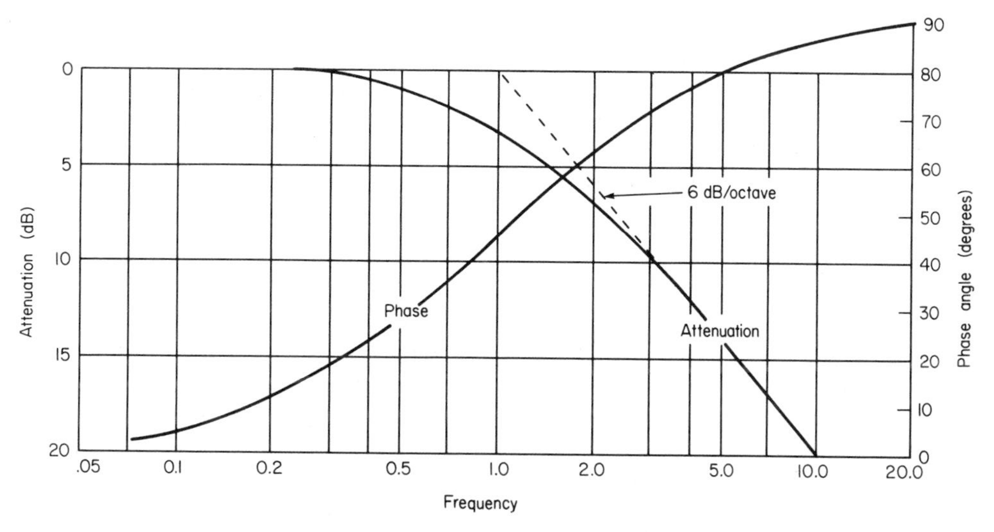

Figure 4-3 Insertion loss and phase shift between source and load for the single-capacitor low-pass filter.

the source and load voltages is zero at very low frequencies and increases slowly at first and then more rapidly as the cutoff frequency is approached. At this frequency, the load voltage lags the source voltage by 45°. As the input frequency increases further, the phase angle continues to increase at a progressively slower rate and finally approaches a maximum of 90°.

At very low frequencies the reactance of the capacitor is very high, so most of the source current goes to the load. If the source and load resistances are equal, maximum power will be transferred from the source to the load, resulting in a 0-dB insertion loss. As the frequency increases, the capacitor reactance drops and the source must then supply additional current to the capacitor. This increases the voltage drop across the source resistance, so the output voltage drops. The drop at the output has therefore been caused by a changing of the input impedance of the filter (the reactance of the capacitor in parallel with the load resistance Z_{in} of Figure 4-2). The 3-dB point occurs when the reactance of the capacitor is equal to the parallel combination of the source and load resistances.

$$F_{\text{cutoff}} = \frac{1}{2\pi RC} \tag{4-1}$$

where: $R = R_{\text{source}} \parallel R_{\text{load}}$

The final slope of 6 dB/octave is the result of the capacitor reactance being so low that it dominates the input impedance, and since the reactance halves each time the frequency is doubled (an octave), the output voltage then drops at the same rate. The maximum phase shift of 90° occurs at high frequencies where the current drawn from the source is mainly capacitive.

Notice that once the cutoff frequency has been decided upon, then for a given source and load resistance, the designer has no further choice. The attenuation and phase-shift curves can only have the one shape noted, and the unique value of the required capacitor is given by Equation (4-1).

4-1.2 Two Poles

Let us now look at a filter circuit that consists of two reactive elements, as shown in Figure 4-4. The two-pole filter is capable of a faster rate of attenuation in the stopband than the single-pole filter just described. The reactance of the series inductance is directly proportional to frequency and the reactance of the shunt capacitor is inversely proportional to frequency, so a final attenuation slope of 12 dB/octave is now possible. Resonance can also occur with these two reactive components, and so a peak in the response curve is possible. What effect this has on the final amplitude response depends on the loading produced by the source and load resistances. This filter circuit has therefore left us with a choice; the cutoff-frequency requirement will determine the approximate resonant frequency of the L and C, but we still are left with a choice of the amount of resistive loading to use. In other words, the loaded Q of the circuit is up to the designer.

The amplitude responses that result for different loaded Q's are shown in Figure 4-5 (and the corresponding phase responses in Figure 4-6). For Figure 4-5, the series-resonant frequency is constant and the horizontal axis is calibrated relative to this frequency. The changes in Q are obtained by using different values of source and load resistances. For a low Q such as 0.5, the passband is very flat and then slowly increases to a final attenuation slope of 12 dB/octave in the stopband. For higher Q's, an initial insertion loss appears at the lower frequencies and decreases near the resonant frequency to 0 dB. Above resonance, the loss increases very rapidly at first and then settles down to a final slope of 12 dB/octave in the stopband.

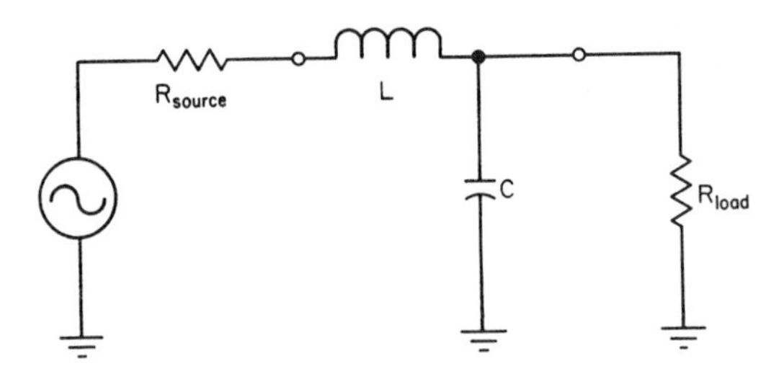

Figure 4-4 Two-pole low-pass filter consisting of an inductance in series with the source resistance and a capacitor in parallel with the load resistance.

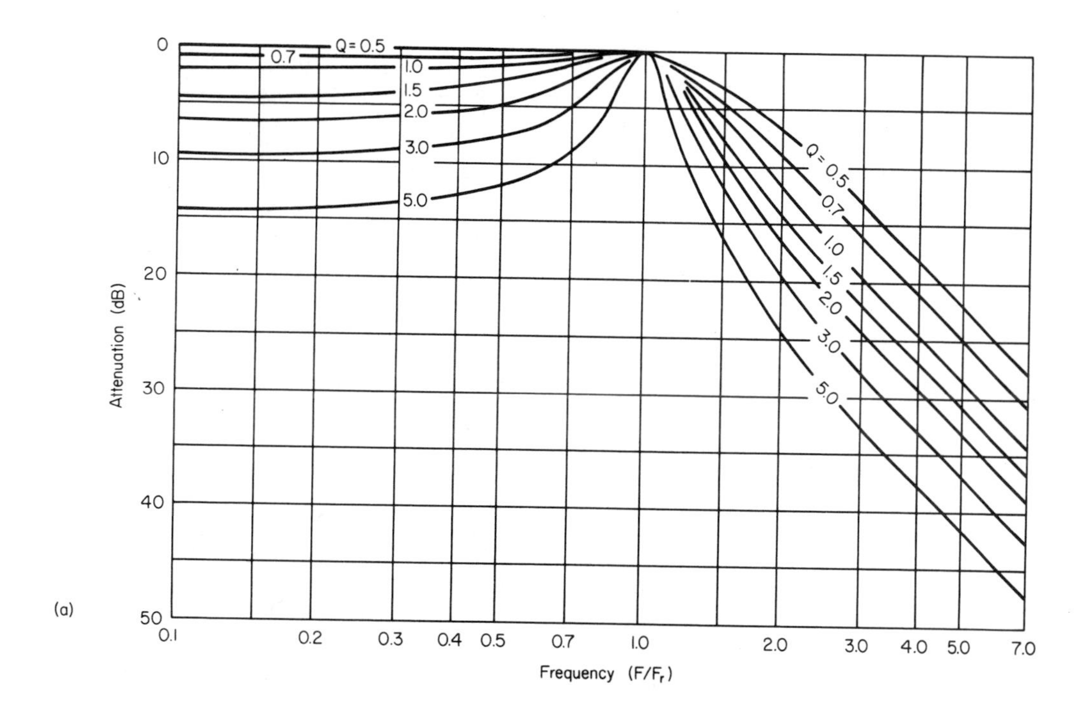

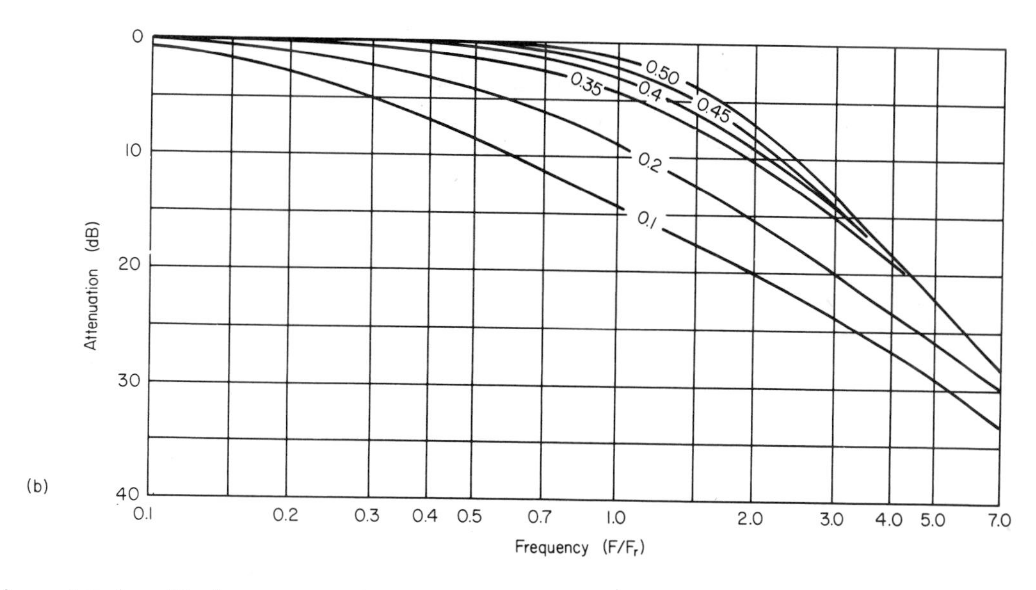

Figure 4-5 Amplitude response for a two-pole low-pass filter with (a) total $Q_L \geqslant 0.5$ and (b) $Q_L \leqslant 0.5$. The Q_L in all cases is defined at the series resonant frequency, $F/F_r = 1.0$.

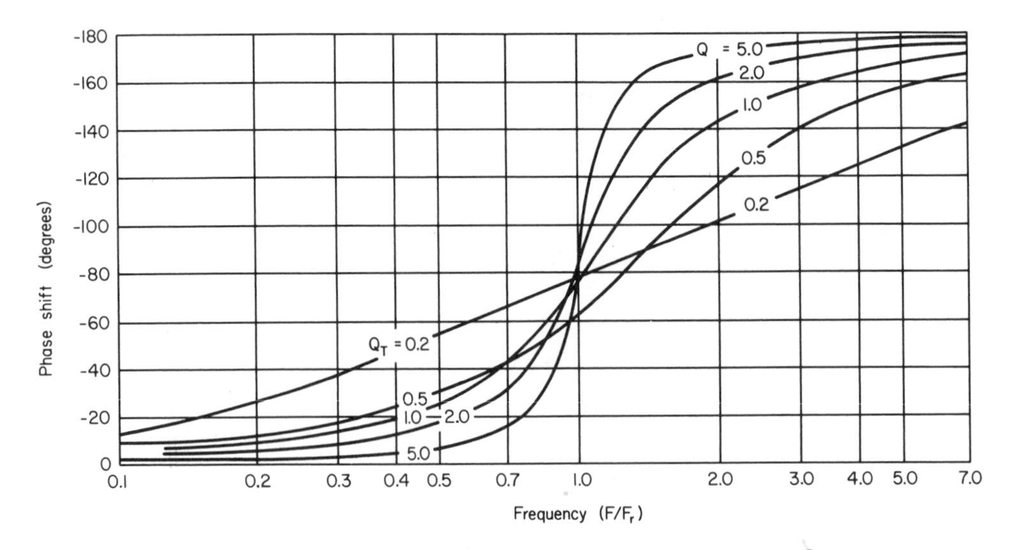

Figure 4-6 Phase response of the two-pole low-pass filters having total loaded Q's from 0.2 to 5.0.

The peak formed by the higher Q circuits can be used effectively as a bandpass filter to pass the fundamental frequency of a signal and attenuate the harmonics. For a given bandwidth, the attenuation on the high side of the peak increases at twice the rate of the parallel-resonant bandpass filters of Figure 3-5, even though the same number of components are involved. The attenuation on the low-frequency side is terrible, but then harmonics do not exist below the fundamental frequency.

The series-resonant frequency of the two-pole filter is

$$F_r = \frac{1}{2\pi\sqrt{LC}} \tag{4-2}$$

The total Q of the circuit depends on two separate circuits. For the series circuit consisting of the source resistance and the inductance,

$$Q_1 = \frac{X_L}{R_{\text{source}}} \quad (\text{at } F_r) \tag{4-3}$$

For the parallel circuit consisting of the load resistance and the capacitance,

$$Q_2 = \frac{R_{\text{load}}}{X_C} \quad (\text{at } F_r) \tag{4-4}$$

The total loaded Q of this circuit is then

$$Q_{\text{total}} = \frac{Q_1 \cdot Q_2}{Q_1 + Q_2} \tag{4-5}$$

The curves of Figure 4-5 are all drawn with the requirement that the insertion loss must be 0 dB at some point in the passband. For total loaded Q's of 0.5 and above, this is accomplished by setting $Q_1 = Q_2$. The result is that the source and load resistances will be unequal values (except for $Q_T = 0.5$). For total loaded Q's of 0.5 and less, the source and load resistances have the same value. Any conditions other than these will increase the loss at all frequencies and no part of the curve will touch the 0-dB line.

The peak appears when the total Q of the circuit is greater than 0.5 and increases in height as the Q increases. If used as a bandpass filter, the 3-dB bandwidth inside the peak is approximately

$$\text{BW} \approx \frac{F_{\text{peak}}}{Q_{\text{total}}} \tag{4-6}$$

The peak occurs at

$$F_{\text{peak}} = F_r \sqrt{1 - \frac{1}{4Q_T^{\,2}}} \tag{4-7}$$

Let us take a closer look at the source and load resistances. As explained, they determine the Q's of the two arms of the filter and therefore the total Q of the filter. For a total Q greater than 0.5, the Q of the series arm must be the same as the Q of the shunt arm. The result is that the source resistance is smaller than the load resistance, and so an insertion loss is created at the lower frequencies. However, the peak of the passband will reach back up to 0 dB. For total Q's of 0.5 and less, there is no peak, and so zero insertion loss in the passband can only be achieved if the source and load resistances are kept equal. A little experimenting will show that it is impossible to obtain a total Q greater than 0.5 if the two resistances are kept equal.

If these restrictions on the source and load resistances are not observed, the insertion-loss curve will be shifted down by some constant value, producing a loss even at the peak frequency. The overall shape of the curve will still be approximately the same as the original curve, having the same Q, although for the lower Q's the change in shape will be more noticeable. Figure 4-7 shows two sets of curves. The first has a total Q of 1.5, but the Q's of the series and shunt arms are varied. The second set has a total Q of 0.2, but the source and load resistances are changed, so they are no longer equal. The lower Q filter shows a more noticeable change in shape in addition to the increased insertion

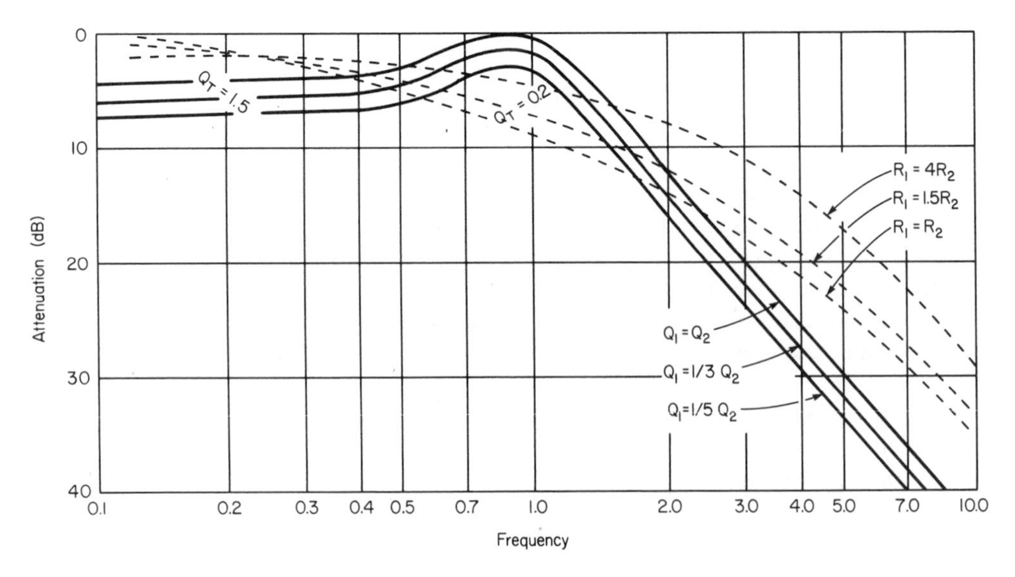

Figure 4-7 Increase in insertion loss for a two-pole filter not properly terminated. The arms of the high-Q filter do not have equal Q's, and the source and load resistances of the lower Q filter are not equal.

loss. The actual 3-dB cutoff frequency could be higher or lower than the LC resonant frequency, depending on the component values used (i.e., the loaded Q). The final attenuation slope, in all cases, is the same, 12 dB/octave. But the initial slope near the 3-dB point can be higher or lower than this, depending on the loaded Q.

The phase responses of Figure 4-6 also show considerable dependence on the loaded Q of the circuit. The "high-Q" filter shows very little phase shift over the lower-frequency portion of the passband, but then has a rapidly increasing phase angle nearer the resonant frequency. The delay time required for low-frequency signals to pass through the high-Q filter would be small, but higher frequencies would experience longer delays, and this variation in delay time is quite often intolerable, since it leads to delay distortion of complex waveforms.

For filters with lower Q's the delay becomes more nearly constant, and for $Q_T = 0.41$, the delay is the same for all frequencies within the passband. (Remember that the horizontal axis of Figure 4-6 is logarithmic, so a linear phase shift with frequency actually shows up as a curved line on the graph.) The phase angle always passes through 90° above the undamped resonant frequency rather than at it. The 90° frequency is given by

$$f_{90^\circ} \approx f_r\left(1 + \frac{1}{8Q_T^{\,2}}\right) \qquad (4\text{-}8)$$

For filters with a total loaded Q of 0.5 and less,

$$f_{90^\circ} = f_r \times 1.414 \tag{4-9}$$

The maximum phase shift will always be 180°, since two reactive elements are involved.

4-1.3 Three Poles

Finally, let us look at three reactive elements that form a three-pole low-pass filter, as shown in Figure 4-8. For simplicity, we will consider the two capacitors to be equal and the source and load resistances to have the same value. The result is a symmetrical low-pass filter. As before, the cutoff frequency dictates the approximate undamped resonant frequency, but the designer is still left with a choice of relative component values, so a wide range of amplitude and related phase responses are available. The responses that result from circuits having loaded Q's from 10 down to 0.16 are shown in Figure 4-9. For the higher Q's, the amplitude starts, at the lower frequencies, to fall at a 6-dB/octave slope, since the two large-value capacitors act almost in parallel, the small-value inductor having an insignificant reactance at the lower frequencies. Nearer resonance, the output voltage climbs back up to the 0-dB level and then rapidly drops above resonance. The final slope settles down to 18 dB/octave now that all three reactances are significant.

$$\text{undamped resonant frequency} = \frac{1}{2\pi\sqrt{L \times C/2}} \quad \text{for equal capacitors} \tag{4-10}$$

$$\text{loaded } Q \text{ at this frequency} = \frac{R_{\text{source}}\ (\text{or } R_{\text{load}})}{X_C} \tag{4-11}$$

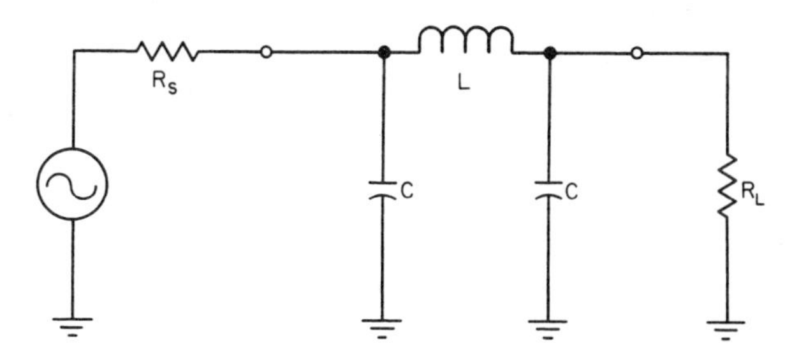

Figure 4-8 Three-element low-pass filter consisting of one series inductor and two equal-value shunt capacitors.

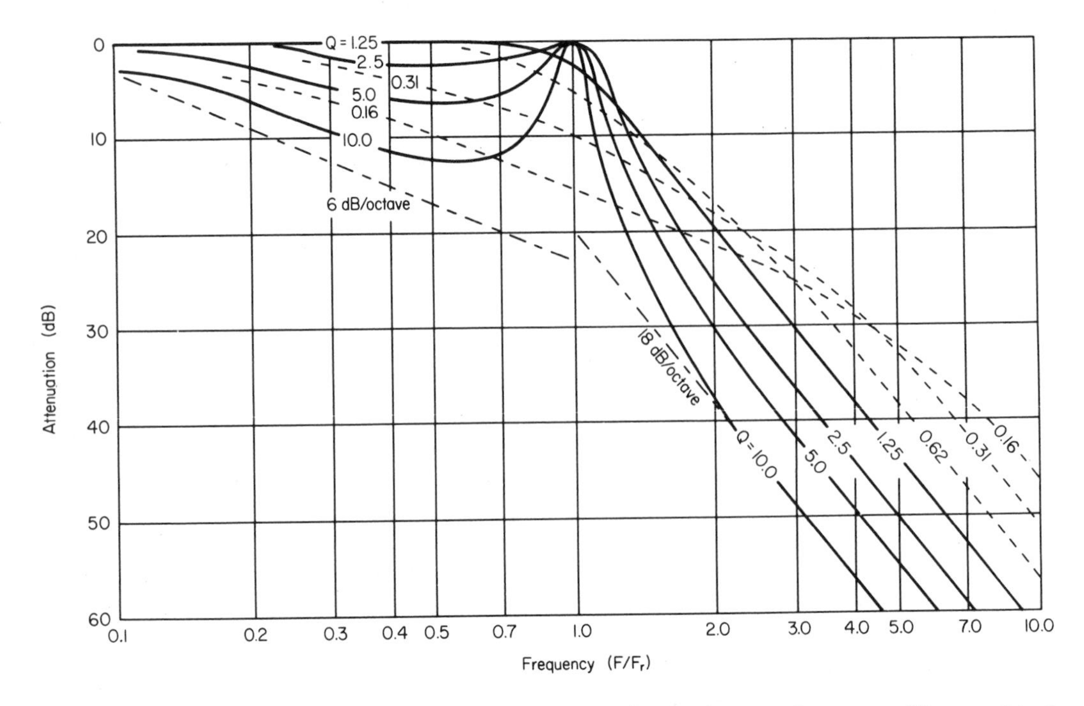

Figure 4-9 Amplitude response for three-element low-pass filters with the loaded Q changing from 10 to 0.16. The frequency scale is marked relative to the undamped resonant frequency of the three components.

The amplitude of the ripple in the passband is directly related to the circuit Q and becomes less as the Q decreases. For a Q of 1.0, the passband is as flat as possible and the 3-dB cutoff frequency is then the same as the undamped resonant frequency. For Q's lower than 1.0, no peak occurs; the amplitude drops at an initial 6-dB slope within the passband due to the single, large series inductance whose reactance is comparable to the source and load resistances. Some place above resonance, the slope increases to 18 dB/octave as the small-value capacitors finally develop a significant reactance.

The phase shift at the undamped resonant frequency is about 180°, becoming less as the Q decreases, an important characteristic in the design of oscillator circuits. Figure 4-10 also shows that the phase shift at 0.7 times the undamped frequency is always close to 90°. The phase shift changes linearly with frequency for $Q = 0.4$.

It's time to stop for awhile to look at three examples that should clarify the work we have just presented.

EXAMPLE 4-1

Find the shape of the amplitude response curve and its overall insertion loss for the circuit shown in Figure 4-11.

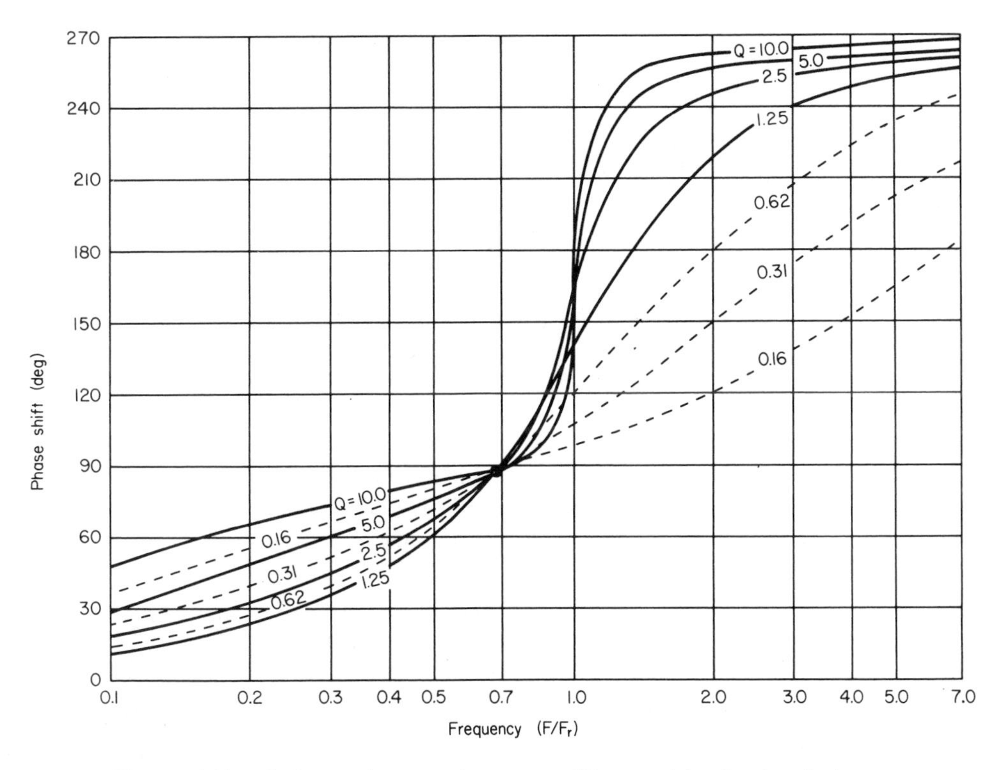

Figure 4-10 Phase shift of three-element low-pass filters with the loaded Q changing from 10 to 0.16.

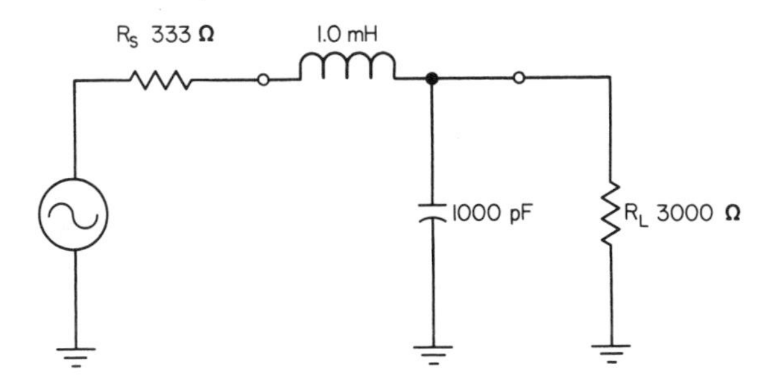

Figure 4-11

Solution

Find the undamped resonant frequency:

$$F = \frac{1}{2\pi\sqrt{LC}} = \frac{1}{2\pi\sqrt{1.0\times10^{-3}\times1000\times10^{-12}}} = 159.2 \text{ kHz}$$

The component reactances at this frequency are

$$X_C = X_L = 2\pi FL = 2\pi \times 159.2 \times 10^3 \times 1.0 \times 10^{-3} = 1000\ \Omega$$

The loaded Q of the inductance and source resistance is:

$$Q_1 = \frac{X_L}{R_{\text{source}}} = \frac{1000}{333} = 3.0$$

The loaded Q of the capacitor and load resistance is:

$$Q_2 = \frac{R_{\text{load}}}{X_C} = \frac{3000}{1000} = 3.0$$

Since the two portions of the circuit have the same Q, the insertion loss will be reduced to zero at the frequency of the peak—about 150 kHz.

The shape of the amplitude response is determined by the total Q; this is

$$Q_{\text{total}} = \frac{Q_1 Q_2}{Q_1 + Q_2} = \frac{3.0 \times 3.0}{3.0 + 3.0} = 1.5$$

The cutoff frequency for this filter depends on our definition. If we want the point at which the loss passes through a point with 3 dB more attenuation than the low frequency, we look at the $Q = 1.5$ curve in Figure 4-5. The low-frequency insertion loss is 5.0 dB. A point 3 dB below this would be 8.0 dB. The $Q = 1.5$ curve passes through 8.0 dB at a frequency 1.53 times the series-resonant frequency. For our example, this would be

$$F_{3\text{ dB}} = 1.53 \times 159.2\text{ kHz} = 243.6\text{ kHz}$$

Whether this frequency should be called the cutoff frequency or not would depend on the designer and the application. The amplitude response of this filter is shown in Figure 4-12.

EXAMPLE 4-2

The circuit shown in Figure 4-13 was found scribbled in an ancient manuscript. What are its secrets?

Solution

The series-resonant frequency can be found using Equation (4-10):

$$F_r = \frac{1}{2\pi\sqrt{LC/2}} = \frac{1}{2\pi\sqrt{1.76 \times .0022 \times 10^{-12}/2}} = 3.617\text{ MHz}$$

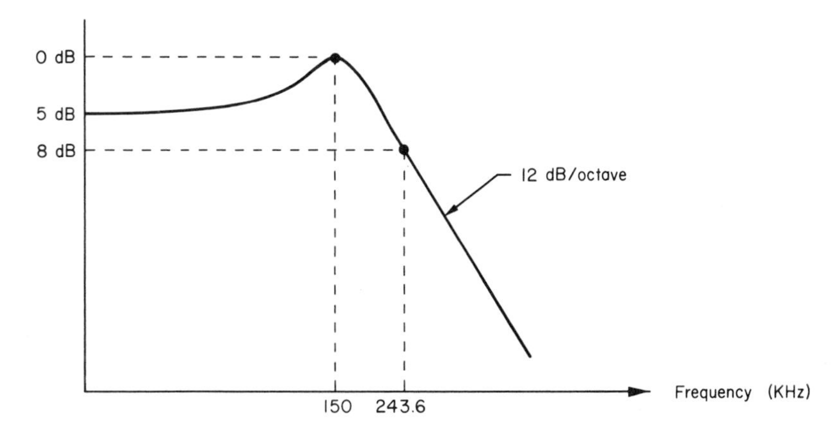

Figure 4-12

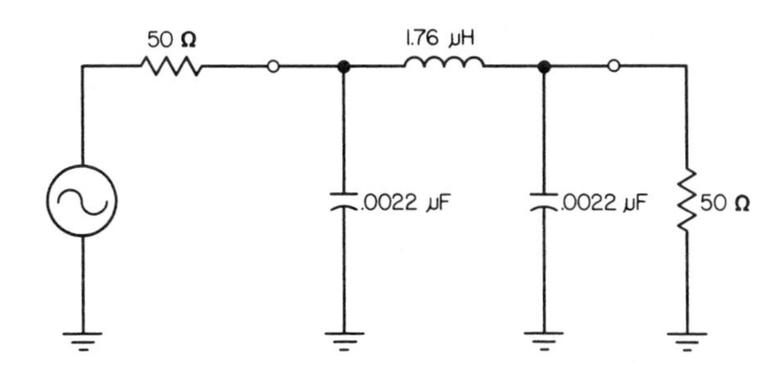

Figure 4-13

Each capacitor would have a reactance at this frequency of

$$X_C = \frac{1}{2\pi F_r C} = \frac{1}{2\pi \times 3.617 \times 10^6 \times .0022 \times 10^{-6}} = 20\ \Omega$$

Because the circuit is symmetrical, the loaded Q at each end will be the same. The Q is then

$$Q = \frac{R_{\text{load}}}{X_C} = \frac{50}{20} = 2.5$$

This identifies the curve from Figure 4-9 that describes this circuit. The passband has a 1.8-dB dip in it at one-half the resonant frequency. The curve finally passes through the 3-dB level at 1.1 times the resonant frequency. The amplitude response of this example is shown in Figure 4-14.

$$F_{3\text{ dB}} = 1.1 \times F_r = 1.1 \times 3.617 = 3.98 \text{ MHz}$$

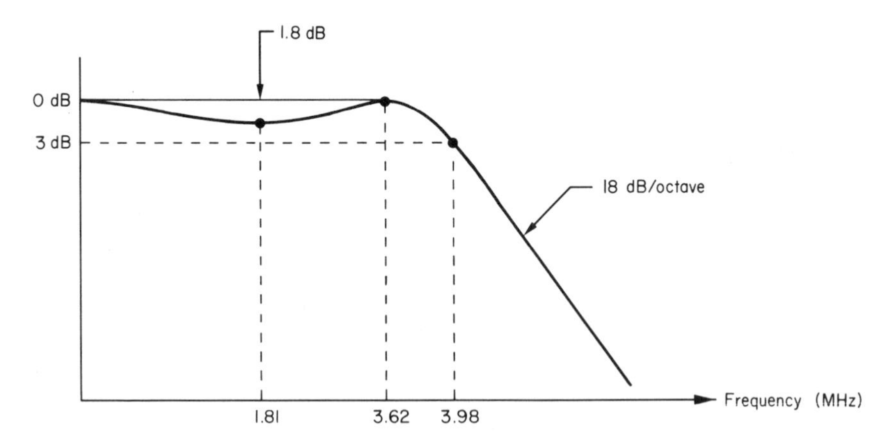

Figure 4-14

EXAMPLE 4-3

Design a two-element low-pass filter with a passband ripple of 1.0 dB, a cutoff frequency of 1.0 rad/s, and a load resistance of 1.0 Ω.

Solution

For a two-element low-pass filter, the 1.0-dB ripple will produce an insertion loss of 1.0 dB at very low frequencies, including dc, where the inductive reactance will be zero and the capacitive reactance infinity. We can then find the value of source resistance that will produce this insertion loss at dc.

$$\text{IL} = 10\log\left(\frac{\text{load power}}{\text{available power}}\right) \tag{4-12}$$

$$1.0\text{ dB} = 10\log\left[\frac{4R_S R_L}{(R_S + R_L)^2}\right] \tag{4-13}$$

This equation will produce two values for R_{source}, one higher and one lower than 1.0 Ω. The value we will use is the lower one, to obtain the same value of Q in the series arm and the shunt arm of the filter (the 1.0-dB ripple means a total Q higher than 0.5).

The value for R_{source} is found to be 0.376 Ω. The total Q of the filter can now be found:

$$Q_{\text{total}} = \frac{1}{2}\sqrt{\frac{R_L}{R_S}} = \frac{1}{2}\sqrt{\frac{1.0}{0.376}} = 0.815$$

The Q of each arm must be the same and will be twice the total Q. Therefore, the Q of each arm will be

$$Q_1 = Q_2 = 2Q_T = 2 \times 0.815 = 1.63$$

At the series resonant frequency of the two components, the reactances of the inductance and capacitance will be

$$X_L = Q_1 \times R_{\text{source}} = 1.63 \times 0.376 = 0.613\ \Omega$$

$$X_C = \frac{R_{\text{load}}}{Q_2} = \frac{1.00}{1.63} = 0.613\ \Omega$$

All that remains is to find where the 3-dB cutoff frequency is located relative to the series resonant frequency. From the curves of Figure 4-5,

$$F_{3\text{ dB}} \approx 1.36 \times F_{\text{resonant}}$$

Then for a 3-dB cutoff frequency of 1.0 rad, the resonant frequency must be

$$F_{\text{resonant}} = \frac{1.0}{1.36} = 0.735\ \text{rad/s}$$

The value of L and C can now be found:

$$C = \frac{1}{\omega X_C} = \frac{1}{0.735 \times 0.613} = 2.219\ \text{F}$$

$$L = \frac{X_L}{\omega} = \frac{0.613}{0.735} = 0.834\ \text{H}$$

These values are very similar to the values given in Table 4-2 for the 1.0-dB ripple Chebychev filter. The final design is shown in Figure 4-15.

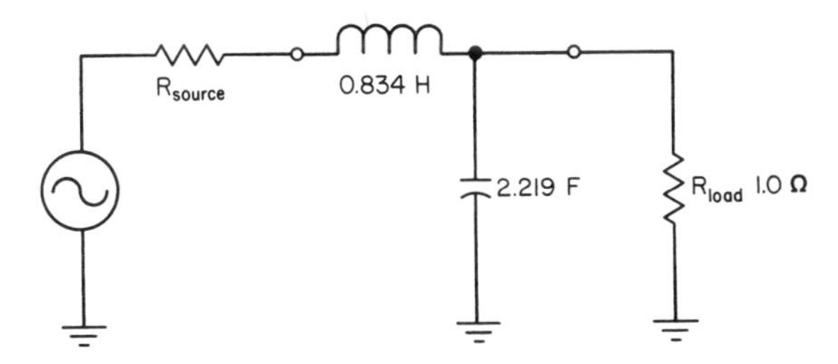

Figure 4-15

From our look at these three simple filter circuits, we can make some general comments that pertain not only to low-pass but also to bandpass, band-rejection, and high-pass filters:

1 Low-Q filters generally have a flat, ripple-free passband, but their initial attenuation slope at the beginning of the stopband is low. Their phase response is more linear with frequency, and so the delay is more nearly constant across the passband.

2 High-Q filters produce a faster initial slope in the stopband, and so have a greater attenuation at any stopband frequency than does a lower-Q design with the same number of components. The penalty for this improvement is a passband attenuation ripple and poorer phase linearity.

3 The final attenuation slope and the final phase shift of any filter depends on the effective number of reactive components in the circuit. The final slope will be 6 dB/octave (20 dB/decade) and 90° for each component. Not all components contribute to these figures. If two capacitors, for example, are placed in parallel, they can only be counted as one component. Parallel and series tuned circuits are very often used in bandpass designs and occasionally for low-pass designs. Since only one component is dominant on each side of the resonant frequency, each tuned circuit is counted as a single element (single pole).

4 The source and load resistances determine the overall Q of the filter and therefore the passband ripple, the phase linearity, and the exact cutoff frequency. If a filter is used with terminating resistances other than the values for which it was designed, the amplitude, phase, and delay characteristics will not be correct.

5 Filters can also be used as impedance matching networks. For an even number of components in the filter, the designer can select the ratio of source to load resistance (within certain limits). For an even number of components, the ratio is fixed as soon as the circuit Q is decided.

4-2 MODERN FILTER DESIGN

There are two approaches that can be used to formally design filters to meet given specifications. The older method is based on *image parameter theory*. The resulting designs are known as constant-k filters and are often used with m-derived terminating half-sections. Although the design technique is relatively straightforward, the results do not exactly match the theory. The problem is that the theory assumes complex terminating impedances that keep changing with frequency. When the filters are terminated in constant-value resistances, the characteristics change slightly.

The more modern technique assumes a constant-value terminating resistance (can be obtained most of the time, but not always) and then involves the writing of a polynomial equation that describes the shape of the required amplitude or phase response. The equation can then be manipulated to obtain the component values for the circuit (see the appendix to this chapter). The full design process is too laborious for the average designer but, fortunately, a number of the more useful designs have already been broken down into component values, and the remaining work is quite simple.

These modern filter designs bear the names of the people who either developed the polynomial equations or reduced the equations down to the component values for the circuit. In some cases this has led to some confusion, as in the case where Thompson developed the filters using the equations of Bessel. The reader can become famous if he/she takes the time to write his/her own exotic equation and then generates the corresponding component values. Of course, it would help if the filter so designed had some practical applications —but most of these areas are already well covered.

Three of the more popular families of filters will be discussed in the next section. The relative advantages and disadvantages of each will be discussed; standard curves for attenuation, delay, and input impedance will be shown; and component values will be given. For reasons that will be explained shortly, all the values given use a source and load resistance of 1 Ω and produce a 3-dB cutoff frequency of 0.159 Hz (1 rad/s). These three families cover a wide range of circuit Q's and each can consist of any number of components depending on the attenuation rates required in the stopband.

1 High-Q[1] Designs—*Chebychev*: an entire family of filters, ranging from medium Q to high Q.
2 Medium-Q Design—*Butterworth*: for a given number of components, only one design is available.
3 Low-Q Design—*Butterworth–Thompson*: an entire family of filters, ranging from medium Q to low Q. One of the lower-Q members of this family is usually referred to as the *Bessel filter*.

---

[1]For most filter applications, the passband should be fairly flat, with amplitude variations less than several dB. Therefore, references to high-Q filters would involve total loaded Q's no higher than perhaps, 1.5. Low-Q designs may use a loaded Q as low as 0.1. Some sections within more elaborate filters may use Q's higher than these.

4-3 BUTTERWORTH FILTER

This is a middle-of-the-road design having characteristics that lie between the extremes of the other two families. The circuit can have any number of components, but will always produce a passband that is flat and free from ripple. The initial attenuation slope in the stopband is steeper than any of the Butterworth–Thompson family but not as steep as any of the Chebychev family when the same number of components are used. The phase response is not very linear and produces more delay variation than the Butterworth–Thompson family but less than the Chebychev family.

The amplitude responses for Butterworth filters, consisting of from one to seven components, are shown in Figure 4-16. The cutoff frequency is set to the 3-dB level in all cases, and the frequency axis is marked relative to this frequency. The passband is very flat, and the attenuation curve quickly picks up a final slope of 6 dB/octave for each reactive component used. If a

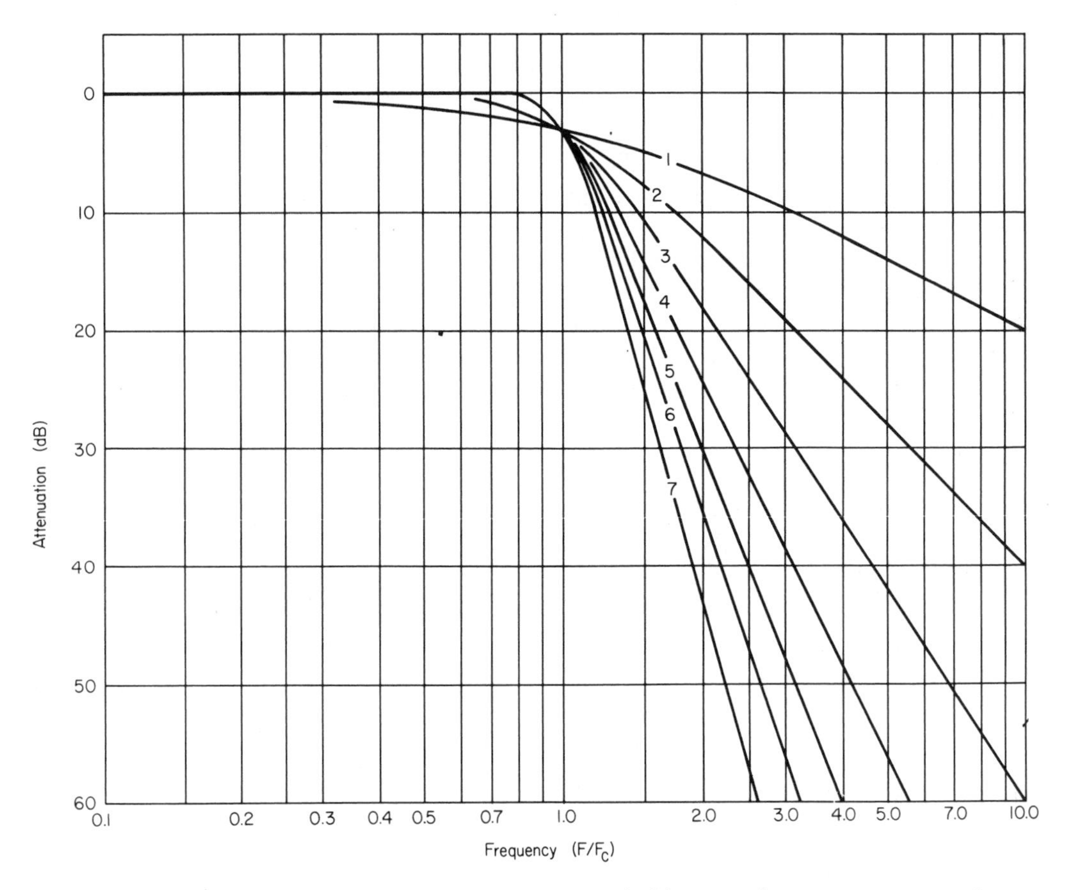

Figure 4-16 Attenuation of Butterworth filters with one to seven poles.

particular application required an attenuation of, say, 28 dB for a signal at twice the cutoff frequency, Figure 4-16 shows that a five-element filter would be required.

The corresponding delays for the Butterworth filters are shown in Figure 4-17. This is the delay experienced by a single sine wave at each frequency, as discussed in Chapter 1. The delay times that are shown are correct only for the cutoff frequency of 1 rad/s; for higher cutoff frequencies, the delay times are shortened proportionally. The variation in delay for the two- and three-element filters is fairly small, but for the higher orders, the delay variation becomes excessive for many critical applications, especially television.

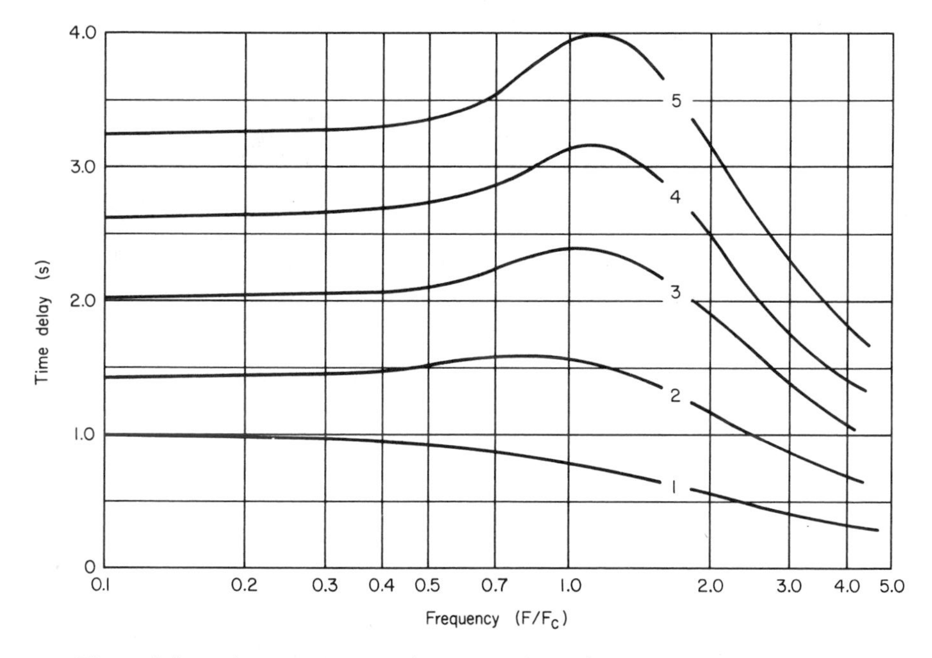

Figure 4-17 Time-delay characteristics of one- to five-element Butterworth low-pass filters with a 3-dB cutoff frequency of 1 rad/s.

The input impedance for the three-element Butterworth is plotted on a *Smith chart*[2] in Figure 4-18 for 1-Ω terminating resistances. It is interesting to compare the frequencies and attenuation circles on the Smith chart to the

[2]Developed by P. H. Smith in 1944, the chart is one of many ways to present the variations of a complex impedance with frequency. The advantages of this chart are that all impedances from zero to infinity, resistive or reactive, can be shown, and many of the operations performed on the chart require the drawing of simple circles. The chart can also be used to convert a complex impedance into the corresponding admittance, a technique we will use in the work on impedance matching. The attenuation circles drawn in Figure 4-18 are circles of constant VSWR (voltage standing-wave ratio, for those not familiar with transmission lines).

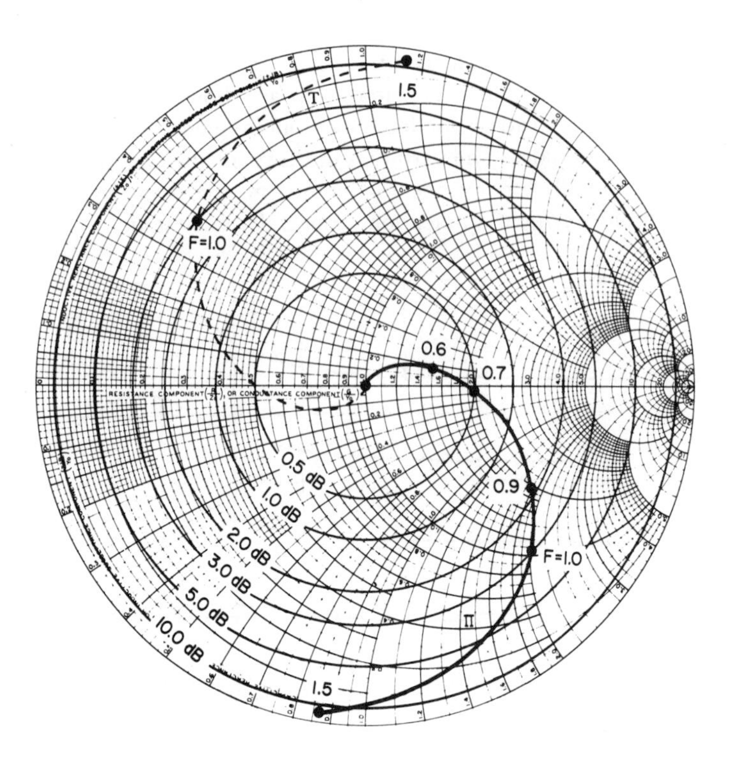

Figure 4-18 Variation of the input impedance of a three-element Butterworth low-pass filter with frequency. The filter has a load resistance of 1 Ω and a cutoff frequency of 1 rad/s (0.159 Hz). The solid curve is the Π form and the dashed curve is the T form (see Figure 4-29).

attenuation curve for the three-pole filter in Figure 4-16. The input impedance variations are therefore responsible for the change in attenuation; this fact is used in the synthesis of these filters as explained in Appendix 4A.

The component values for one- to seven-element Butterworth low-pass filters are shown in Table 4-1, and Figure 4-19 shows a circuit illustrating the placement of the components in the form of a ladder network. Notice that each component is numbered individually (i.e., component 1 is a capacitor, and component 2 is an inductor). This arrangement will help later when we change to high-pass and bandpass designs.

The 3-dB cutoff frequency of all filters is 0.159 Hz (1 rad/s) and the source and load resistances are both 1 Ω. All inductors are given in henrys and capacitors in farads.

TABLE 4-1
COMPONENT VALUES FOR BUTTERWORTH LOW-PASS FILTERS.

n	C_1	L_2	C_3	L_4	C_5	L_6	C_7
1	2.000						
2	1.414	1.414					
3	1.000	2.000	1.000				
4	0.765	1.848	1.848	0.765			
5	0.618	1.618	2.000	1.618	0.618		
6	0.518	1.414	1.932	1.932	1.414	0.518	
7	0.445	1.247	1.802	2.000	1.802	1.247	0.445

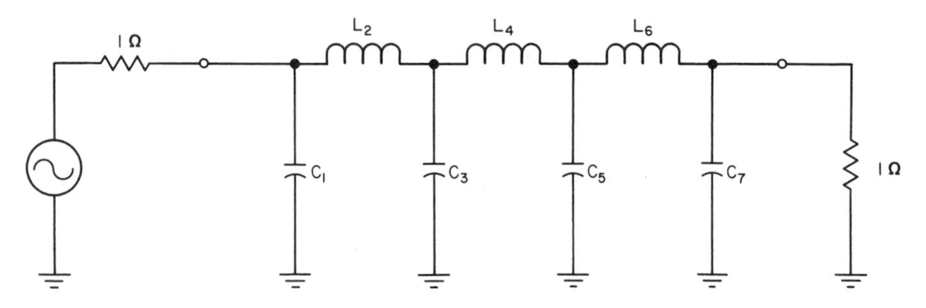

Figure 4-19 Ladder-network arrangement of components for low-pass filters.

4-5 CHEBYCHEV FAMILY OF FILTERS

The Chebychev filters are designed to produce faster attenuation slopes at the beginning of the stopband than do the Butterworth filters. This is accomplished by using a higher circuit Q, and so some attenuation ripple appears in the passband. It is the amplitude of this ripple that distinguishes the different members of the Chebychev family. For a given number of components, the designer has his/her choice of the amount of ripple to allow—the more allowed, the greater the initial slope at the beginning of the stopband and the greater the attenuation that will be produced with a given number of components. As little as 0.1 dB of ripple in a three-element filter will result in a 5-dB improvement in the stopband compared to the same-size Butterworth filter. This comparison is made in Figure 4-20. However, the higher Q also results in poorer phase linearity, so the Chebychev family is not used for constant delay applications such as television.

Chebychev filters consisting of an even number of components have another feature (or defect, depending on the application): the required source and load resistances are not equal, and so the filters also perform an impedance-matching function. The amount by which the source and load resistances

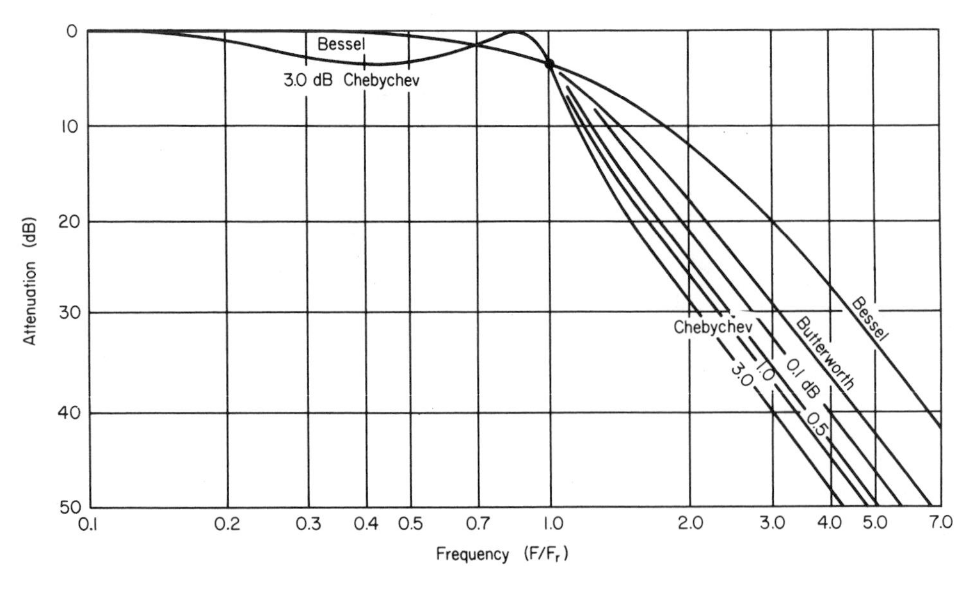

Figure 4-20 Comparison of the attenuation of six low-pass filters, all with three reactive components and the same 3-dB cutoff frequency.

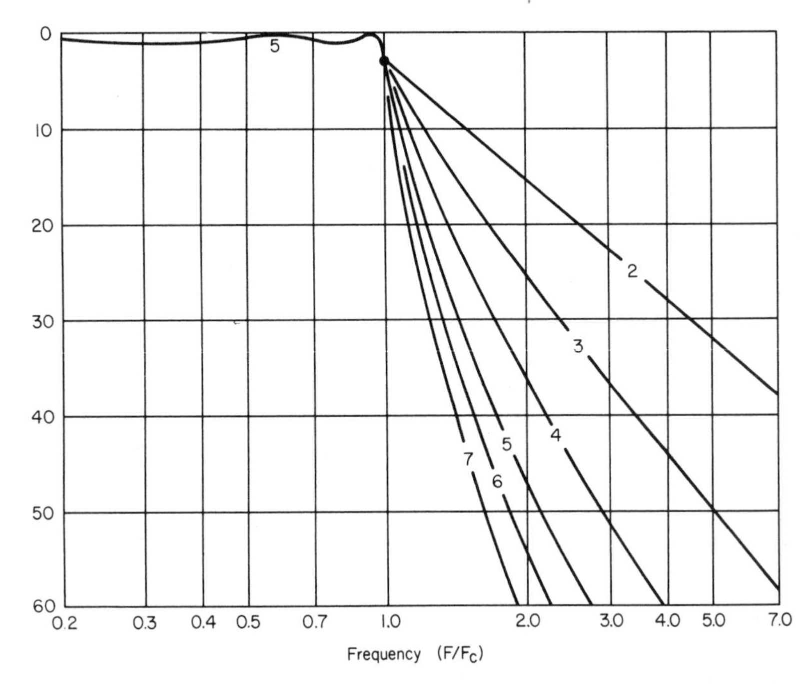

Figure 4-21 Attenuation curves for the 1.0-dB Chebychev filter using from two to seven components. Lower ripple will result in slightly less stop-band attenuation, and higher ripple slightly more.

differ is determined by the ripple amplitude in the passband unless a constant insertion loss is tolerated.

Because of the large number of combinations that could result from filters with different numbers of components and also different ripple amplitudes, only a selection of the more useful types are provided. Figure 4-20 is a comparison of three-element Chebychev filters with different ripple amplitudes and three-element Butterworth and Bessel filters. Circuits with different numbers of components would show similar attenuation improvements for the high-ripple Chebychev filters.

Figure 4-21 is a set of attenuation curves for the 1.0-dB ripple Chebychev filter having from two to seven components. (One-element filters are all the same, since there can be no choice of Q. This filter was included in the Butterworth designs.) The amplitude of the passband ripple is 1.0 dB in all cases, but the number of peaks and valleys will change with the number of components. The ripple shown is for the five-element filter. If the designer needs an attenuation of 35 dB at twice the cutoff frequency, the four-element filter could be used.

Figure 4-22 shows the delay characteristics for the 1.0-dB ripple designs. Notice that the variation is rather severe, even poorer than the Butterworth design. As mentioned before, the actual delay time depends on the cutoff

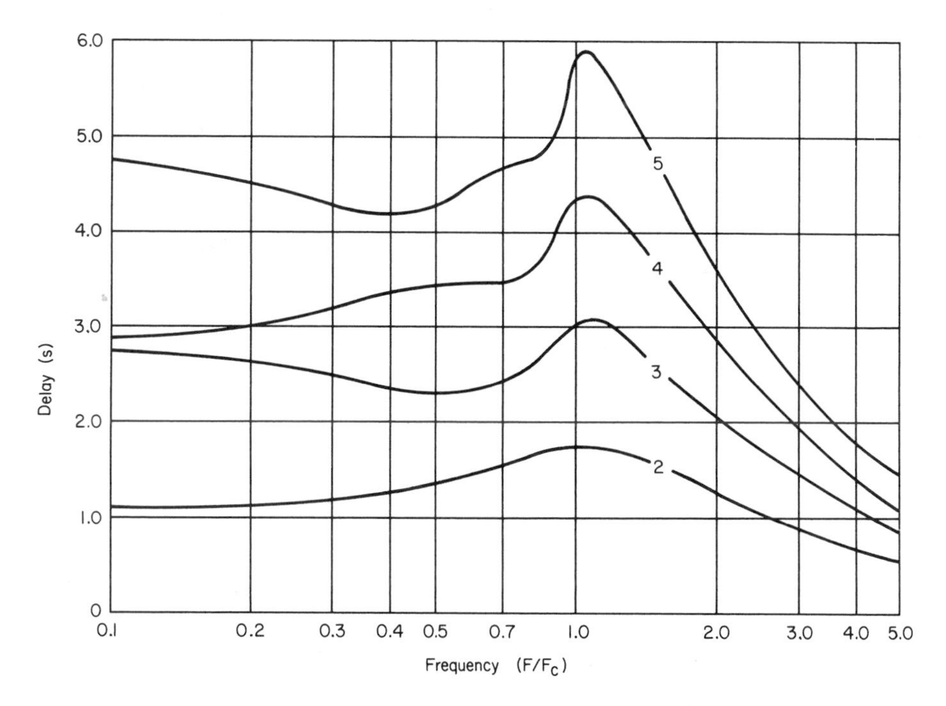

Figure 4-22 Delay characteristics for 1.0-dB Chebychev filters using from two to five components.

frequency of the final design, but the relative variation in this time will remain the same. The cutoff frequency for this graph is 0.159 Hz (1 rad/s).

The next parameter of the Chebychev filter is its variation of input impedance with frequency; this is shown in Figure 4-23 for the three-element filter with a 1.0-dB passband ripple. Notice that the impedance starts at the center of the chart ($1+j0$), moves away until it touches the 1.0-dB circle and then crosses back through the center again. This second point will correspond to the peak in the passband (see Figure 4-20). This frequency may be of interest if a low-pass filter is required for harmonics of, say, a C.B. transmitter. The middle of the C.B. band (26.960–27.410 MHz) could be set to the peak frequency, and the impedance of the transmission line would then remain unchanged when the filter was inserted.

The component values for Chebychev filters with two to seven components are shown in Table 4-2. Four values of passband ripple are included; higher ripples than 3.0 dB are possible but would have limited use as low-pass

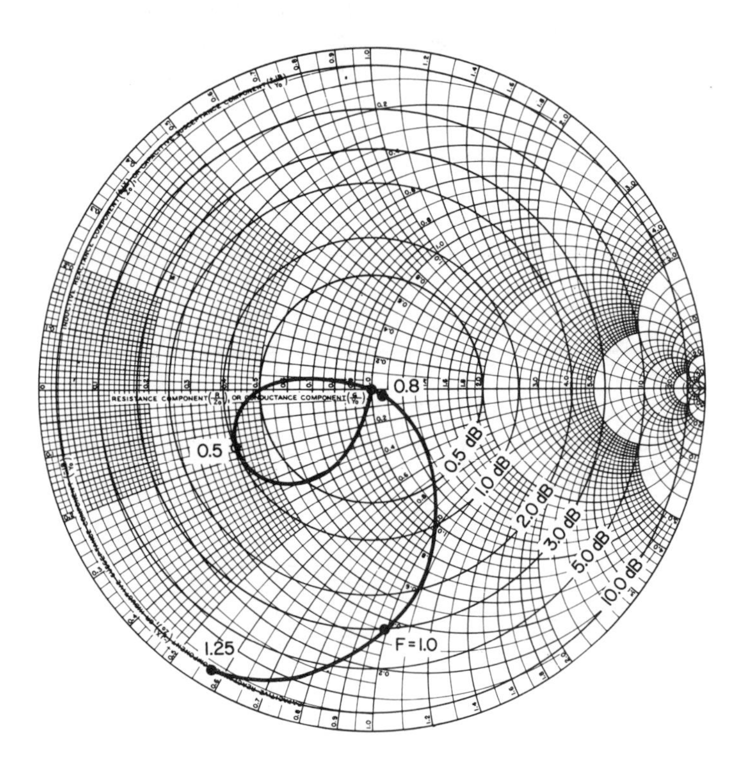

Figure 4-23 Input impedance variation across the passband of the three-element Chebychev filter with 1.0 dB of ripple. The filter is terminated with a load resistance of 1 Ω.

TABLE 4-2

COMPONENT VALUES FOR CHEBYCHEV FILTERS WITH TWO TO SEVEN COMPONENTS. THE BANDPASS RIPPLE CAN BE 0.1, 0.5, 1.0, OR 3.0 dB. THE LOAD RESISTANCE IS 1.0 IN ALL CASES AND THE SOURCE RESISTANCE IS SHOWN IN THE RIGHT-HAND COLUMN. THE CUTOFF FREQUENCY IS 1.0 rad/s IN ALL CASES.

ORDER	*RIPPLE (dB)*	C_1	L_2	C_3	L_4	C_5	L_6	C_7	R_{source}
2	0.1		1.211	1.643					0.738
	0.5		0.982	1.948					0.504
	1.0		0.835	2.219					0.376
	3.0		0.535	3.109					0.172
3	0.1	1.438	1.589	1.438					1.000
	0.5	1.862	1.280	1.862					1.000
	1.0	2.215	1.088	2.215					1.000
	3.0	3.349	0.712	3.349					1.000
4	0.1		0.981	2.225	1.642	1.330			0.738
	0.5		0.920	2.585	1.303	1.824			0.504
	1.0		0.831	2.981	1.121	2.212			0.376
	3.0		0.592	4.349	0.748	3.441			0.172
5	0.1	1.300	1.553	2.238	1.553	1.300			1.000
	0.5	1.800	1.298	2.682	1.298	1.800			1.000
	1.0	2.216	1.132	3.115	1.132	2.216			1.000
	3.0	3.481	0.762	4.538	0.762	3.481			1.000
6	0.1		0.980	2.220	1.550	2.350	1.450	1.350	0.738
	0.5		0.905	2.575	1.367	2.710	1.298	1.794	0.504
	1.0		0.836	3.031	1.190	3.161	1.140	2.224	0.376
	3.0		0.603	4.464	0.793	4.606	0.769	3.505	0.172
7	0.1	1.198	1.444	2.128	1.596	2.128	1.444	1.198	1.000
	0.5	1.737	1.271	2.664	1.357	2.664	1.271	1.737	1.000
	1.0	2.184	1.121	3.118	1.183	3.118	1.121	2.184	1.000
	3.0	3.519	0.772	4.639	0.804	4.639	0.772	3.519	1.000

filters. All the odd-order filters are symmetrical and use equal-value 1-Ω source and load resistances. The even-order filters are asymmetrical and use a load resistance of 1 Ω and a source resistance as shown in the table. C_1 is omitted for the even-order filters, so the first component is the series inductor L_2. The cutoff frequency for all designs is 1 rad/s (0.159 Hz).[3]

[3]Filter handbooks often quote the cutoff frequency for the Chebychev filters as being that frequency which produces a final attenuation in the passband equal to the design ripple, that is, the frequency where the attenuation passes through the 1.0-dB level for the last time for a 1.0-dB filter. The final results will be the same in all cases.

4-5 BUTTERWORTH–THOMPSON FAMILY

This, again, is an entire family of filters, so a large number of designs are possible for a given number of components. All designs produce passbands that are free of ripple, although some designs produce a gradual slope within the passband. The main attraction of the family is the improved delay response over the other filter designs. At one end of the family is the Butterworth filter, with its mediocre delay characteristics. At the lower-Q end of the family is the Bessel design, which produces a completely flat delay response for filters of all orders. The initial stopband attenuation for the Bessel is, however, very poor. The other members of the family, then, are compromises that give various degrees of improvement in this initial stopband slope but at the expense of increased delay variation across the passband.

Figure 4-9 showed in part that low-Q designs can have a passband slope that becomes steeper as the Q is reduced. This undesirable effect can be partially offset by making the odd-order filters asymmetrical; (i.e., selecting unequal values of shunt input and output capacitors while retaining the equal value source and load resistances). This "steps up" the source impedance so that more of the available power can be delivered to the load at the high end of the passband. The total Q of the filter can still be maintained at whatever value is desired to produce the delay variation required.

The members of this family all have pole positions on the s-plane (see Appendix 4A) that lie on a line between the poles of the Bessel and the poles of the Butterworth designs. To show how far along this line another pole lies, the letter m is used (not to be confused with m-derived filters). Then $m=0$ describes the Butterworth filter and $m=1.0$ describes the Bessel filter. An $m=0.4$ Butterworth–Thompson filter would have poles located 40% of the way between the Bessel and Butterworth designs. Figure 4-24 is the amplitude response for the $m=0.4$ and $m=1.0$ members of the family. Notice how, in all cases, the attenuation increases slowly right from the lowest frequency until the final slope is reached in the stopband. The delay characteristics for the same filters are shown in Figure 4-25. Notice the improvement in the delay flatness as the value of m increases.

The component values for the $m=0$ Butterworth–Thompson filters were given in Table 4-1 as the Butterworth design. Some values for the $m=0.4$ and 1.0 are given in Table 4-3.

4-6 FREQUENCY AND IMPEDANCE SCALING

The previous sections presented the characteristics of the three basic families of filters so that the designer could select the best one for his/her application. Component values were presented for a wide range of designs, but they were all for filters with a cutoff frequency of 0.159 Hz and a load resistance of 1.0 Ω.

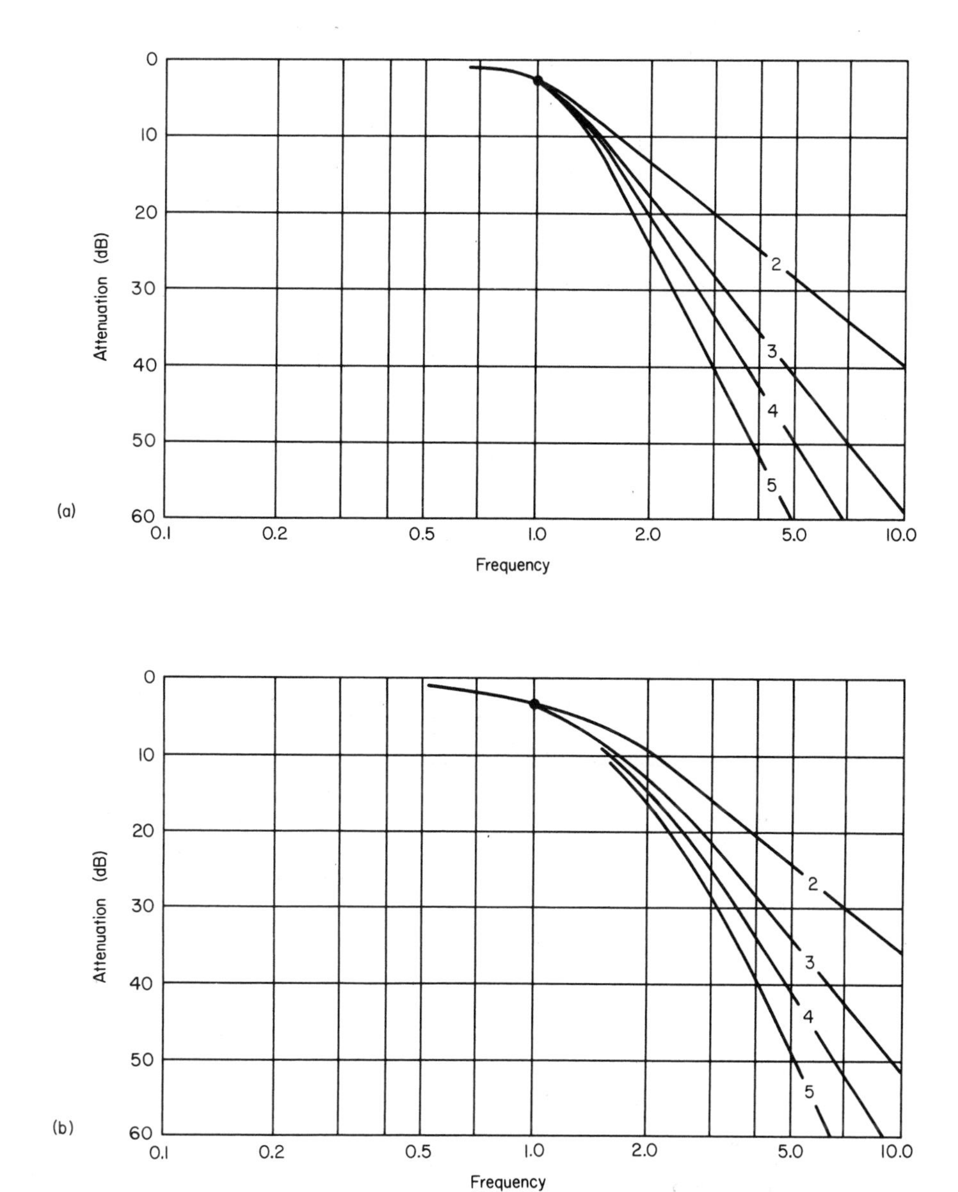

Figure 4-24 Amplitude response of Butterworth–Thompson filters with two to five elements. The $m=0.4$ response is shown in (a) and the $m=1.0$ (Bessel) is shown in (b). The $m=0$ (Butterworth) response was shown in Figure 4-16.

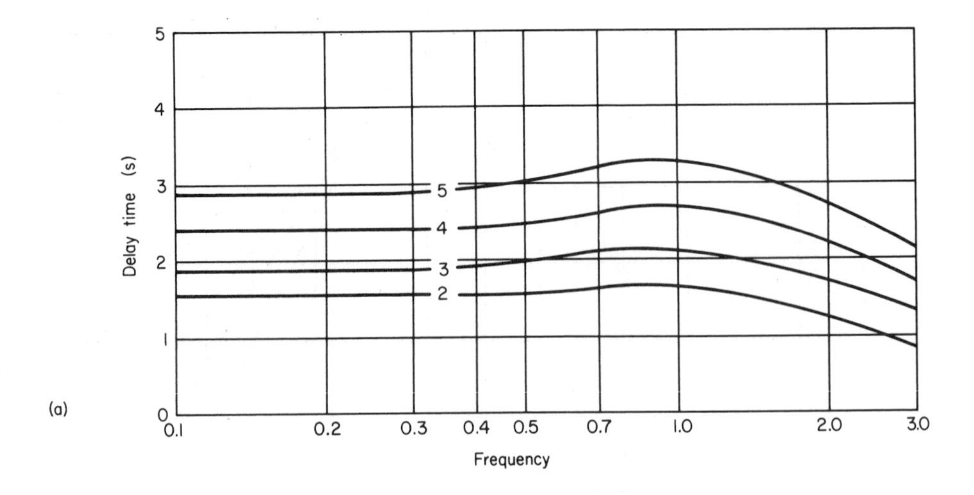

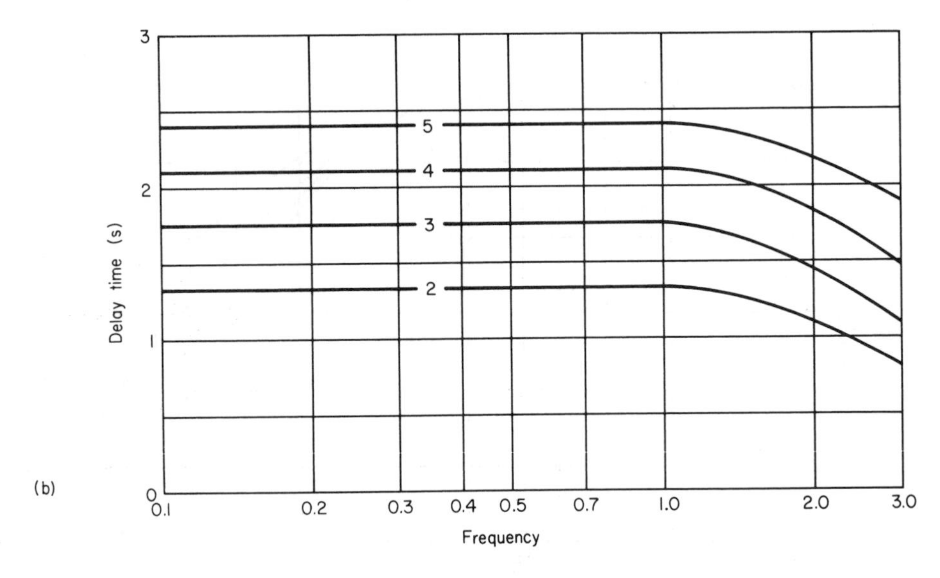

Figure 4-25 Delay characteristics of two- to five-element Butterworth–Thompson filters. The $m=0.4$ is shown in (a) and the $m=1.0$ (Bessel) is shown in (b). The $m=0$ (Butterworth) characteristic was shown in Figure 4-17.

TABLE 4-3

COMPONENT VALUES FOR $m=0.4$ AND 1.0 BUTTERWORTH–THOMPSON FILTERS. BOTH THE SOURCE AND LOAD RESISTANCES ARE 1.0 Ω AND THE CUTOFF FREQUENCY IN ALL CASES IS 1.0 rad/s. ($m=1.0$ IS ALSO KNOWN AS A BESSEL FILTER.)

ORDER	*M*	C_1	L_2	C_3	L_4	C_5
2	0.4		1.983	0.820		
	1.0		2.148	0.576		
3	0.4	1.720	1.615	0.674		
	1.0	2.203	0.971	0.337		
4	0.4		1.634	1.765	1.391	0.526
	1.0		2.226	1.075	0.668	0.231
5	0.4	1.600	1.814	1.715	1.226	0.446
	1.0	2.267	1.117	0.807	0.510	0.176

The final step in the design is then to scale the values from the charts up to the terminating resistances and cutoff frequencies necessary for the application.

The scaling is accomplished with these two formulas:

$$C = \frac{C_n}{2\pi F_c R} \qquad (4\text{-}14)$$

$$L = \frac{RL_n}{2\pi F_c} \qquad (4\text{-}15)$$

where: C_n and L_n = "normalized values" taken from the tables of component values

R = load resistance of the final design

F_c = 3-dB cutoff point of the final design

EXAMPLE 4-4

A low-pass filter is required with a 3-dB cutoff frequency of 5000 Hz. The passband must be as flat as possible, and the source and load resistances will be 150 Ω. The attenuation in the stopband must be at least 17 dB at twice the cutoff frequency.

Solution

Because of the requirement for the flat passband, the Butterworth filter must be used. Figure 4-16 shows that the stopband attenuation for the three-pole filter is 18 dB at twice the cutoff frequency, so a filter of this size will meet the design requirements.

From Table 4-1, the component values are

$$C_1 = 1.000 \text{ F}$$
$$L_2 = 2.000 \text{ H}$$
$$C_3 = 1.000 \text{ F}$$

Using the scaling formulas for the cutoff frequency of 5000 Hz and the terminating resistances of 150 Ω, the final values are

$$C_1 = C_3 = \frac{1.0}{2\pi 5000 \times 150} = 0.212\ \mu\text{F}$$

$$L_2 = \frac{150 \times 2.0}{2\pi 5000} = 9.55 \text{ mH}$$

The final filter is shown in Figure 4-26.

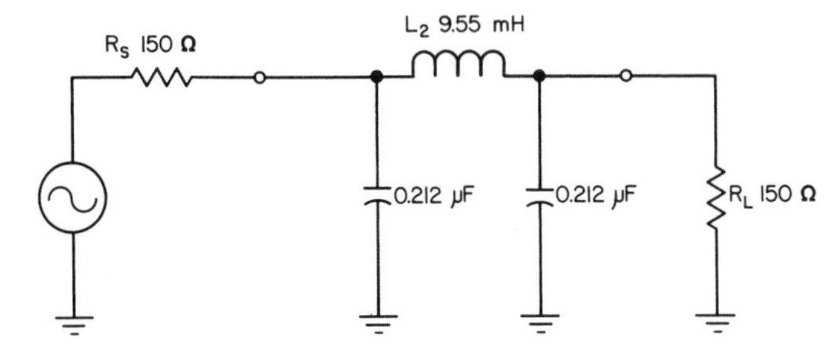

Figure 4-26

EXAMPLE 4-5

Design a two-element Chebychev low-pass filter with 1.0 dB of passband ripple, a 3-dB cutoff frequency of 1.5 MHz, and a load resistance of 75 Ω.

Solution

From Table 4-2, the normalized component values are as shown in Figure 4-27. The final values will be

$$R_{\text{source}} = 0.376 \times 75 = 28.2\ \Omega$$
$$R_{\text{load}} = 1.00 \times 75 = 75\ \Omega$$

$$C = \frac{C_n}{2\pi F_c R} = \frac{2.15}{2\pi \times 1.5 \times 10^6 \times 75} = 3040 \text{ pF}$$

$$L = \frac{RL_n}{2\pi F_c} = \frac{75 \times 0.809}{2\pi \times 1.5 \times 10^6} = 6.44\ \mu\text{H}$$

The final design is shown in Figure 4-28.

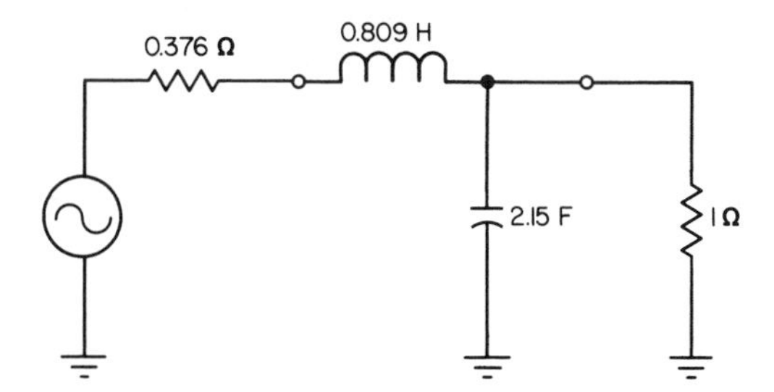

Figure 4-27

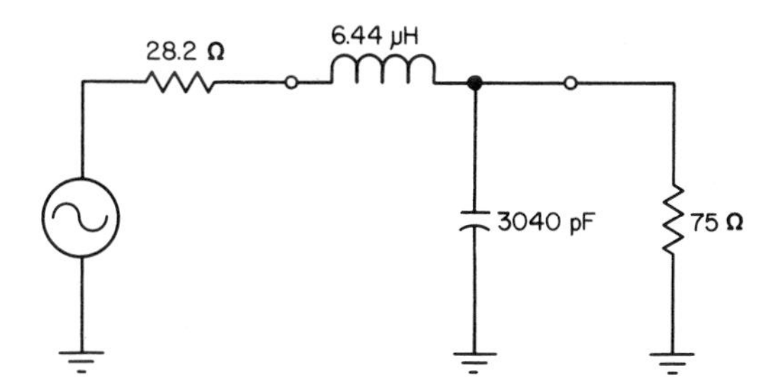

Figure 4-28

4-7 DUAL NETWORKS

For some applications, it may be desirable to construct a filter in a different form from that shown by the tables of values. For three elements, the ladder network values given by the tables result in the Π network of Figure 4-29(a). If the T-network form of (b) is more suitable for some reason, its normalized values are easily obtained by changing component 1 from a shunt capacitor to a series inductor with the same numerical value. The same simple conversion is

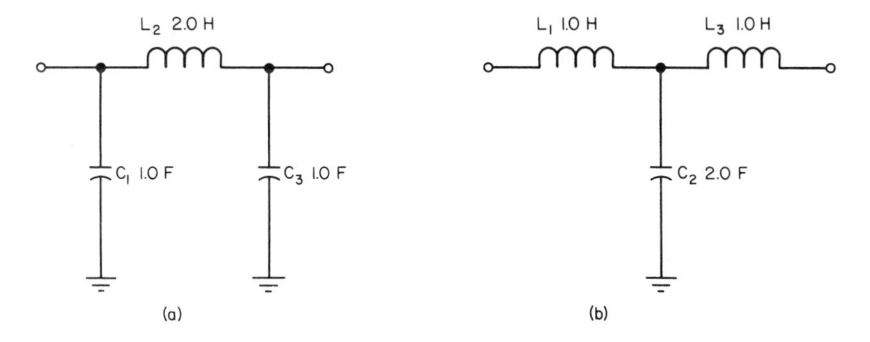

Figure 4-29 T-filter equivalent (b) of the three-element Butterworth filter that is normally used in its Π form (a).

made for the remaining components. The values used in Figure 4-29 are for the three-element Butterworth filter.

The attenuation, phase shift, and delay characteristics of both networks in Figure 4-29 will be identical; the one difference will be the input impedance. In the stopband, the Π network will have an input impedance that is lower than the load resistance and capacitive, as shown in Figure 4-18. The T network will have an input impedance that does exactly the opposite; it will be inductive in the stopband and have a higher value than the load resistance. The other advantages revolve around the ease and cost of building the two circuits and of minimizing the resistive losses in the components themselves. This usually means reducing the number of inductors. While the T network has not accomplished this for the low-pass filter, keep the technique in mind when we discuss high-pass filters.

4-8 HIGH-PASS FILTER DESIGN

High-pass filters can easily be designed using the component values given in the charts for low-pass filters. Once the type and size of the filter are determined, the high-pass values are obtained by reading each value from the chart and then using a component of the opposite type with a reciprocal value. For example, if the chart shows a 2.25-F capacitor for the first shunt position, an inductor with a value of $1/2.25 = 0.44$ H would be used instead (see Figure 4-30). (A 2.25-F capacitor and a 0.44-H inductor will have the same amount of reactance at 1 rad/s.)

To determine the attenuation that a high-pass filter will provide at any frequency in the stopband, the standard low-pass graphs are used, and the horizontal frequency axis is read as a reciprocal. For example, to find the insertion loss at 1/3 and 1/2 the cutoff frequency of a high-pass filter, use the corresponding low-pass graph and read the loss at three times and two times the cutoff frequency, respectively.

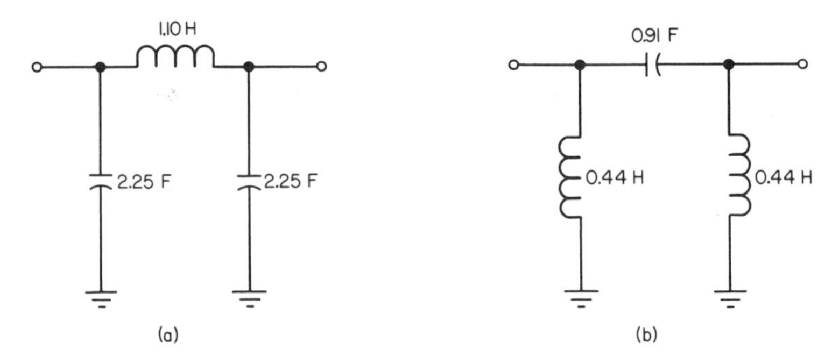

Figure 4-30 Normalized high-pass filter (b) is obtained by replacing each component of the low-pass filter (a) with the opposite component having a reciprocal value.

Thus, for a three-element Butterworth high-pass filter with a 3-dB cutoff frequency of 8.0 MHz, the attenuation at 3.0 and 5.0 MHz can be determined from Figure 4-16 as follows:

1 Three MHz is 3/8 of the cutoff frequency. On the low-pass chart, this would correspond to 8/3 times the cutoff frequency or $2.67F_c$. The attenuation is then 25.0 dB.

2 Five MHz is 5/8 of the cutoff frequency, so on the low-pass chart, the attenuation at $8/5 = 1.6$ times the cutoff frequency is found to be 12.0 dB.

EXAMPLE 4-6

Design a high-pass filter with as few inductors as possible, for a cutoff frequency of 5.5 MHz. A ripple of 1.0 dB can be tolerated in the passband and the attenuation at 3.0 MHz must be 40 dB. The source and load resistances will be 50 Ω.

Solution

Three MHz is 3.0/5.5 of the cutoff frequency for the high-pass filter. To use the low-pass charts, we use the reciprocal of this $= 5.5/3.0 = 1.83$ on the normalized frequency axis.

Using the chart for the 1.0-dB ripple Chebychev filters (Figure 4-21), we see that a filter with five reactive components will provide 42-dB attenuation at 1.83 times the cutoff frequency.

From Table 4-2, the component values are

$$C_1 = 2.22\ \text{F}$$
$$L_2 = 1.13\ \text{H}$$
$$C_3 = 3.12\ \text{F}$$
$$L_4 = 1.13\ \text{H}$$
$$C_5 = 2.22\ \text{F}$$

The position of these five components are shown in Figure 4-31(a).

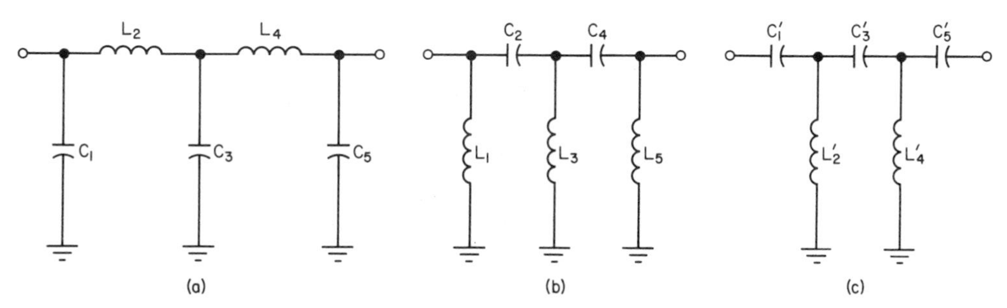

Figure 4-31

The values for the high-pass filter are found by taking the reciprocals of the low-pass values. The new values are

$$L_1 = \frac{1}{2.22} = 0.450 \text{ H}$$

$$C_2 = \frac{1}{1.13} = 0.885 \text{ F}$$

$$L_3 = \frac{1}{3.12} = 0.321 \text{ H}$$

$$C_4 = \frac{1}{1.13} = 0.885 \text{ F}$$

$$L_4 = \frac{1}{2.22} = 0.450 \text{ H}$$

But we have ended up with three inductors. This can be reduced to two if we use the *dual form* of this filter. This simply requires that the shunt inductors be replaced with series capacitors of the same numerical value, and vice versa for the series capacitors. The values now become [the component positions are shown in Figure 4-31(c)]

$$C_1' = 0.450 \text{ F}$$

$$L_2' = 0.885 \text{ H}$$

$$C_3' = 0.321 \text{ F}$$

$$L_4' = 0.885 \text{ H}$$

$$C_5' = 0.450 \text{ F}$$

We have finally obtained the normalized values for our design. All that remains is to scale up to the impedance and frequency levels specified. These values are

$$C_1' = C_5' = \frac{0.450}{2\pi \times 5.5 \times 10 \times 50} = 260 \text{ pF}$$

$$L_2' = L_4' = \frac{50 \times 0.885}{2\pi \times 5.5 \times 10} = 1.28\ \mu\text{H}$$

$$C_3' = \frac{0.321}{2\pi \times 5.5 \times 10 \times 50} = 186 \text{ pF}$$

These values will be used in the positions of circuit (c).

4-9 BANDPASS FILTERS

In a manner similar to the design of high-pass filters, it is possible to design bandpass filters. Again, the normalized low-pass values form the basis for the designs, so flat passbands, constant delays, and so on, can be realized.

The amplitude response of a particular bandpass design is approximated by adding a mirror image to the low-pass response. The zero-frequency point of the low-pass response becomes the center frequency of the equivalent passband, and the cutoff frequency now has two values, f_1 and f_2. All scaling is relative to the *half-bandwidth* of the bandpass design. This would be equal to either $(f_2 - f_0)$ or $(f_0 - f_1)$, the upper and lower half-bandwidths, respectively. An example is shown in Figure 4-32 for a three-element Chebychev filter with 1 dB of passband ripple. The passband of the normalized low-pass response extends from zero frequency to 1.0 (rad/s); at twice the cutoff frequency this filter will have an attenuation of 25 dB. For the bandpass design, the center frequency and the upper cutoff frequency are arbitrarily set to 10.0 and 11.0 MHz, respectively. This results in an upper half-bandwidth of 1.0 MHz, and so at 2.0 MHz above the center frequency (i.e., 12.0 MHz) the response should also be down 25 dB. For reasons to be explained shortly, the actual attenuation in the stopband will be somewhat less than this, as shown by the dashed lines on the bandpass response.

Note that the curve is symmetrical only as long as a logarithmic frequency scale is used. The result is that the lower half-bandwidth is slightly narrower (0.9 MHz) than the upper half. The 25-dB point on the lower side should then be at 8.33 MHz (the ratio 10 MHz/8.33 MHz is the same as the ratio 12 MHz/10 MHz). Again, the true attenuation will be slightly less. The full bandwidth of the filter will be

$$\mathrm{BW} = f_2 - f_1 \tag{4-16}$$

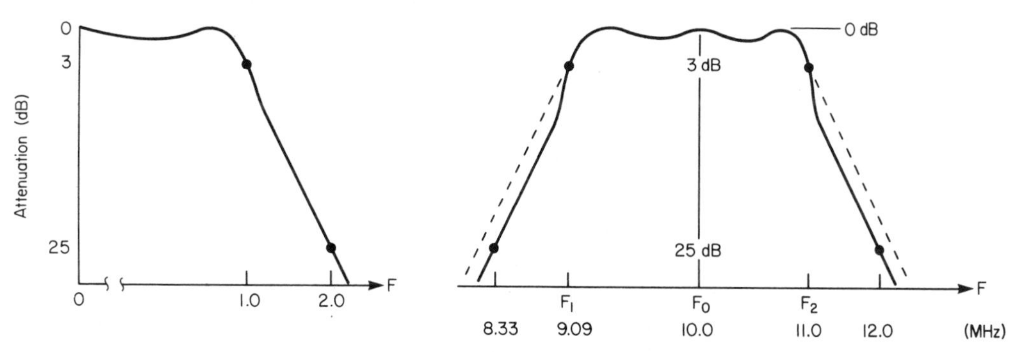

Figure 4-32 Bandpass equivalent of a 1.0-dB, three-element Chebychev filter response.

The center frequency will be related to the cutoff frequencies by

$$f_0 = \sqrt{f_1 \times f_2} \tag{4-17}$$

The transformation from low pass to bandpass involves replacing each low-pass component with a resonant circuit. The series components are replaced with series-resonant circuits and the shunt components are replaced with parallel-resonant circuits. All circuits will be resonant at the center frequency and will have the same amount of reactance at each cutoff frequency that a high- and a low-pass filter would at the same frequencies. This transformation from low-pass to bandpass is illustrated in Figure 4-33.

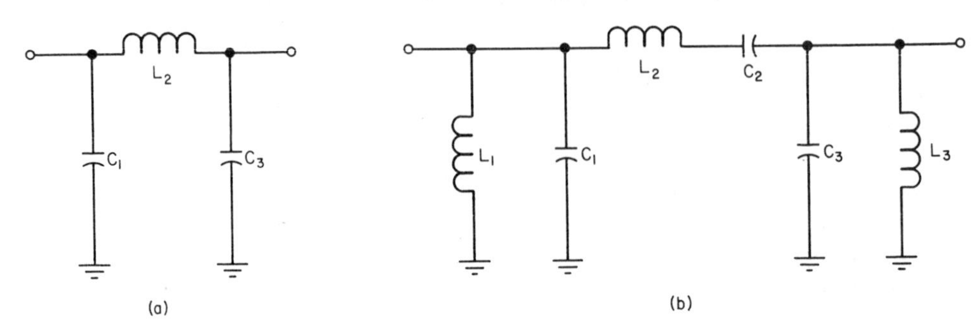

Figure 4-33 Low-pass (a) to bandpass (b) transformation for a three-element design.

The reason for the numbering system used on the original low-pass filters should now be apparent. The two components in the first shunt position are C_1 and L_1, the subscript (1) is the same and so refers to the location. In a similar manner, C_2 and L_2 are the series components of the second location, and so on.

Now, why is there a loss of attenuation in the stopband compared to what would be suggested by the low-pass response? The reason is that, because each location of the filter involves a resonant circuit, two opposite reactances are involved. The total reactance of each location will then change rapidly for frequency changes close to the resonant frequency and more slowly at frequencies farther away. Because the bandpass filter is designed to meet specified 3-dB points, the stopband suffers as the rate of reactance change slows down.

The passband will also change. The 1.0-dB ripple, or whatever ripple is chosen, will remain, but the locations of the maximums and minimums will move closer to the center frequency (f_0).

The only solution to this change in attenuation characteristic is to use a higher number of elements in the design than would be suggested by the attenuation curves of the low-pass curves. A closer look at these differences will be made when we try our next example. To find the component values for our filters, we use the formulas that follow. They include the transformations for all three factors—the load and source resistances (R), the full bandwidth between 3-dB points (BW), and the center frequency (f_0). For any parallel-resonant circuits,

$$C_p = \frac{C_n}{2\pi R \times \text{BW}} \tag{4-18}$$

$$L_p = \frac{R \times \text{BW}}{2\pi f_0^2 \times C_n} \tag{4-19}$$

(C_n is the normalized value from the low-pass charts.) For any series-resonant circuits,

$$C_s = \frac{\text{BW}}{2\pi f_0^2 L_n \times R} \tag{4-20}$$

$$L_s = \frac{R \times L_n}{2\pi \times \text{BW}} \tag{4-21}$$

(L_n is the normalized value from the low-pass charts.)

EXAMPLE 4-7

Design a bandpass filter for a 50-Ω system with 3-dB cutoff frequencies at 9.09 MHz and 11 MHz. The attenuation at 7.6 and 13.0 MHz must be at least 32 dB. One dB of passband ripple can be tolerated.

Solution

The full bandwidth

$$\text{BW} = f_2 - f_1 = 11.0 - 9.09 = 1.91 \text{ MHz}$$

The center frequency will be

$$f_0 = \sqrt{f_1 \times f_2} = \sqrt{9.09 \times 11.0} = 10.0 \text{ MHz}$$

The upper half-bandwidth

$$f_2 - f_0 = 11.0 - 10.0 = 1.0 \text{ MHz}$$

The lower half-bandwidth

$$f_0 - f_1 = 10.0 - 9.09 = 0.91 \text{ MHz}$$

On the upper attenuation skirt, the response of the filter must be down 32 dB at 13 MHz. This frequency is $(13-10)/1=3.0$ half-bandwidths away from the center frequency. From the 1.0-dB Chebychev curves (Figure 4-21) we can see that the three-element design will give us our required attenuation at three times the cutoff frequency, as long as the lower skirt is adequate. The point on the lower skirt is 2.64 half-bandwidths away from the center frequency, and again a look at the three-element chart shows that it will be adequate.

The normalized values from the low-pass chart (Table 4-2) are

$$C_n = C_1 = 2.215 \text{ F}$$
$$L_n = L_2 = 1.088 \text{ H}$$
$$C_n = C_3 = 2.215 \text{ F}$$

For the parallel-resonant circuits,

$$C_1 = C_3 = \frac{C_n}{2\pi R \times \text{BW}} = \frac{2.215}{2\pi \times 50 \times 1.91 \times 10^6} = 3691 \text{ pF}$$

$$L_1 = L_3 = \frac{R \times \text{BW}}{2\pi f_0^2 \times C_n} = \frac{50 \times 1.91 \times 10^6}{2\pi (10 \times 10^6)^2 \times 2.215} = 0.0686 \ \mu\text{H}$$

For the series-resonant circuits:

$$C_2 = \frac{\text{BW}}{2\pi f_0^2 L_n \times R} = \frac{1.91 \times 10^6}{2\pi (10 \times 10^6)^2 \times 1.088 \times 50} = 55.9 \text{ pF}$$

$$L_2 = \frac{R \times L_n}{2\pi \times \text{BW}} = \frac{50 \times 1.088}{2\pi \times 1.91 \times 10^6} = 4.53 \ \mu\text{H}$$

The circuit for this filter will be the same as that of Figure 4-33(b). The actual frequency response obtained for this circuit is shown in Figure 4-34. The dashed curve is what was suggested by the low-pass response.

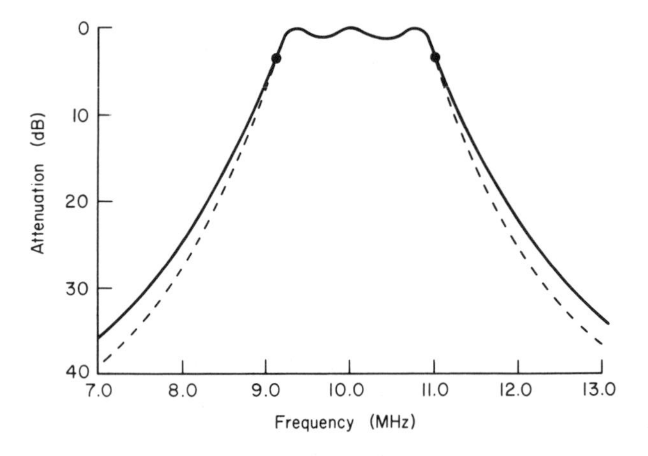

Figure 4-34

4-10 REALISTIC VALUES

With a bandpass design, it is very easy to end up with component values that are just not realistic when it comes to the practical work of building the filter. The capacitors and inductors may end up so big that they will be beyond their self-resonance frequencies at the operating frequency of the filter. Losses, particularly in inductors, may be serious enough to greatly distort the shape of the desired response curve. At the same time, some other values in the circuit may actually be too small to be practical.

These problems are usually the result of trying to design for too narrow a bandwidth relative to the center frequency. As the bandwidth is made narrower, the series inductors and the shunt capacitors become larger and the series capacitors and shunt inductors become smaller. Some variation in component size can be had by changing the source and load resistances. Smaller resistances will increase all capacitor values and lower all inductor values.

As a practical limit on narrow bandwidths,

$$\text{BW should be} > \frac{f_0}{10}$$

This of course is a very approximate limitation and, if in doubt, it is usually best to quickly calculate the component values and see if they are realistic or not. In some cases, it may be necessary to build the filter to see if losses or stray resonances are serious or not. Of course, the complete results can also be calculated if all losses and stray reactances are known, but this is best handled by a computer.

For narrower bandwidths, the coupled resonators of Chapter 3 should be considered, or, for still narrower requirements, ceramic and crystal filters are available.

4-11 BAND-REJECTION FILTERS

Filters can also be designed to reject a narrow range of frequencies and these, too, can be based on the low-pass values. A good example of an application would be the rejection of 27-MHz citizens' band frequencies at the input of a television tuner before cross-modulation could occur.

A typical response curve is shown in Figure 4-35. The curve has two points where the response is down 3 dB, and the difference $(f_4 - f_1)$ is taken as the bandwidth (BW) of the rejected range of frequencies. Two additional points (f_2 and f_3) at some particular attenuation level must also be known in order to set the steepness of the skirts. All four of these points will be symmetrically

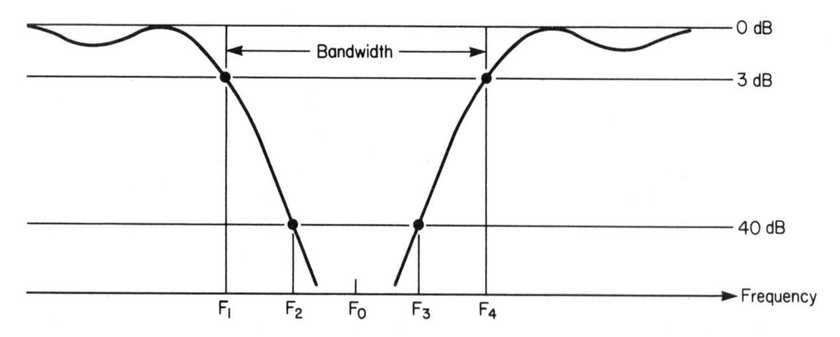

Figure 4-35 Band-rejection response curve.

located around the center frequency if the horizontal scale is logarithmic. Therefore,

$$f_0 = \sqrt{f_1 \times f_4} \tag{4-22}$$

and

$$f_0 = \sqrt{f_2 \times f_3} \tag{4-23}$$

The circuit diagram for a three-element band-rejection filter is shown in Figure 4-36. All series-and parallel-resonant circuits will be resonant at the center frequency (f_0). The number of resonant circuits is found by using the ratio $(f_4 - f_1)/(f_3 - f_2)$ for the horizontal frequency scale on the standard low-pass attenuation curves, along with the amount of attenuation required. The formulas needed for scaling are:

For the series-resonant circuits:

$$C_s = \frac{C_n}{2\pi R \times \mathrm{BW}} \tag{4-24}$$

$$L_s = \frac{R \times \mathrm{BW}}{2\pi f_0^2 \times C_n} \tag{4-25}$$

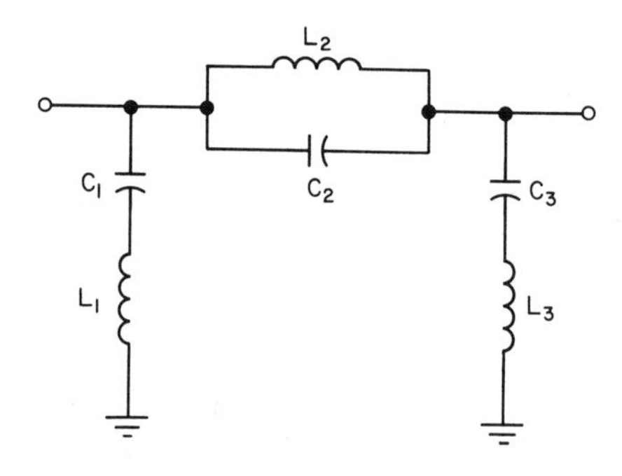

Figure 4-36 Component location for a three-pole (three resonant circuits) band-rejection filter.

For the parallel-resonant circuits:

$$C_p = \frac{\text{BW}}{2\pi f_0^2 \times R \times L_n} \tag{4-26}$$

$$L_p = \frac{R \times L_n}{2\pi \times \text{BW}} \tag{4-27}$$

4-12 DELAY TIME

For the normalized low-pass filter, the variation of delay time across the passband was shown in Figures 4-17, 4-22, and 4-25. These times range from one to several seconds, depending on filter type and number of components, and the frequency. But what happens to these values when the bandwidths are altered by scaling?

We originally saw that delay time depended on how fast the phase angle was changing with frequency:

$$\text{delay (seconds)} = \frac{1}{360} \times \frac{\text{phase shift (degrees)}}{\text{frequency (Hz)}} \tag{4-28}$$

If we increase the cutoff frequency of any low-pass filter by scaling, then the phase angle that was originally reached at 0.159 Hz will now not be obtained until the new cutoff frequency is reached. The delay time for the final filter with a new 3-dB cutoff frequency of F_c will be

$$\text{delay (seconds)} = \text{normalized delay} \times \frac{0.159}{F_c(\text{Hz})}$$

$$= \frac{\text{normalized delay}}{2\pi F_c(\text{Hz})} \tag{4-29}$$

where delay is the normalized value of time delay from the graphs. What about high-pass filters? The same formula applies here, but the normalized delay curves must first be reversed to match the high-pass form. This is accomplished by using the reciprocals of the frequencies shown on the horizontal axis of the graph.

Finally, we come to the time delay of a bandpass filter; again, we run into the same problem as with the amplitude-response scaling. In other words, accurate scaling of the envelope delay from the low-pass delay curves is not easily made.

If we were to follow along with the procedure described so far for low-pass filters, we could obtain some idea of the envelope delay by scaling with the half-bandwidths of the bandpass filter. But because the upper and lower half-bandwidths are not the same size, the delay in the lower sideband will be higher (this, of course, depends on where the center or carrier frequency is located within the passband). The apparent offset at the center frequency does not actually occur since there can be no instantaneous change in phase angle. As already explained for the amplitude response of the bandpass filters, the reactances within the passband are changing faster than they do for the equivalent low-pass filter. The phase-shift characteristics are likewise altered. Figure 4-37 shows the delay characteristics for the bandpass filter of Example 4-7. The dashed lines represent the delay obtained by scaling up the low-pass delay values. The solid line represents the true delay for this filter. Note carefully that this is the delay of the envelope that results from a carrier at the center frequency (f_0) and one sideband at the indicated frequency.

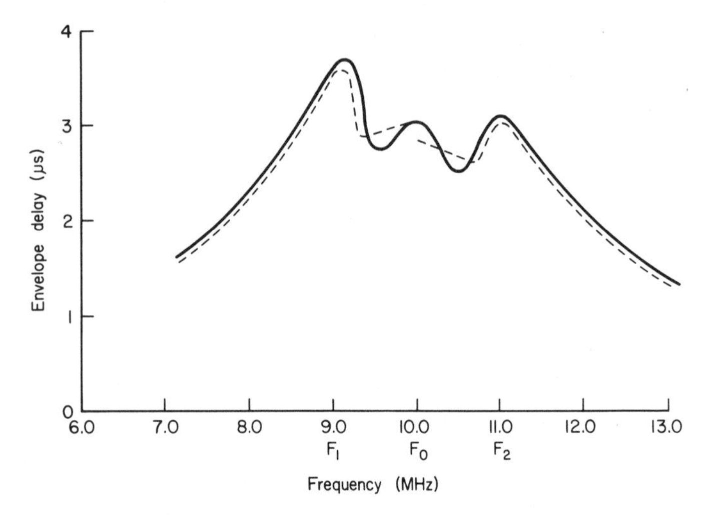

Figure 4-37 Envelope delay of the bandpass filter designed in Example 4-7. The actual delay is shown as a solid line. The predicted delay using the low-pass filter data is shown as a dashed line. The carrier frequency is located at 10.0 MHz.

QUESTIONS

1 If a low-pass filter uses 5 reactive components, what is the maximum phase shift and what is the final attenuation slope in the stopband? Under what conditions could part of the attenuation slope in the stopband exceed the final value?

2 Design a two-element low-pass filter for a source resistance of 50 Ω and a load resistance of 450 Ω. The filter must produce an attenuation of 20 dB at 175 kHz. (Hint: use Figure 4-5.)

3 Design a low-pass filter with a 3-dB point at 14.0 MHz and an

attenuation of at least 20 dB at 25.5 MHz. The source and load resistances are both 600 Ω and the passband should be as flat as possible.

4 Design a three-element high-pass filter with 1.0 dB of passband ripple and a cutoff frequency of 3.5 MHz. The source and load resistances will both be 125 Ω.

5 Design a three-element bandpass filter with 0.5 dB of passband ripple, a center frequency of 5.5 MHz, and a total bandwidth of 2.0 MHz. The source and load resistances will be 75 Ω.

6 Design a three element low-pass "T" filter with flat passband characteristics. The cutoff frequency will be 7.5 MHz and the terminating resistors each 45 Ω. Calculate the input impedance of this filter at approximately 6.0 MHz with the help of Figure 4-18.

7 Draw a graph of the delay characteristics for the filter of Question 4.

REFERENCES

Haykin, S.S. 1968. *Class notes*, McMaster University, 1968

Kuo, F.F. 1962. *Network Analysis and Synthesis*, New York: Wiley and Sons:

White, D.R.J., 1963.*Electrical Filter,* Rockville, Md. White Electromagnetics, Inc.

APPENDIX 4A FILTERS AND THE s-PLANE

This appendix is a short description of the *s*-plane, of poles and zeros, and how they relate to filter design.

4A-1 s-PLANE

Complex filters contain many inductors and capacitors, so a large number of resonant frequencies exist. Every resonance will be loaded by the terminating resistances to various degrees, resulting in various values of loaded Q's for each. The s-plane is a graphical presentation of these natural resonant frequencies and their associated loaded Q's.

Figure 4A-1 is part of the full s-plane and shows one point marked with an ×. The radial distance of this point from the origin represents the undamped resonant frequency (ω_0), and the angle (θ) up from the horizontal

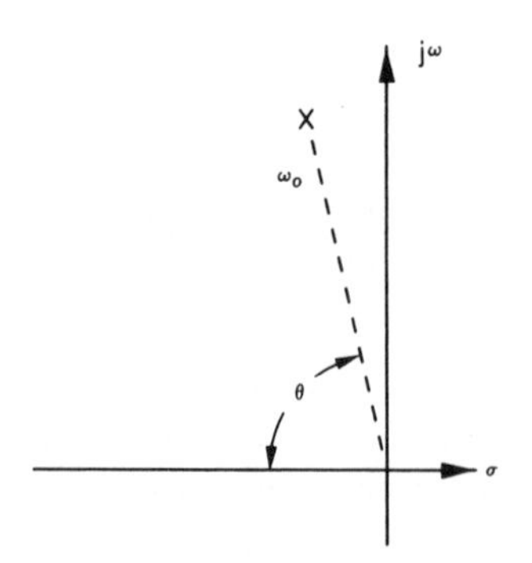

Figure 4A-1

axis represents the loaded Q:

$$Q_L = \frac{1}{2\cos\theta}$$

The × is the symbol for a *pole* and, if this s-plane represents the amplitude response of a filter, a pole indicates a frequency where signals can pass through the filter easily. This could be the result of either a series path containing a series-resonant circuit, or a parallel-resonant circuit across the path as shown in Figure 4A-2. (*Caution:* Figure 4A-1 is not yet a complete s-plane representation of the circuits of Figure 4A-2.)

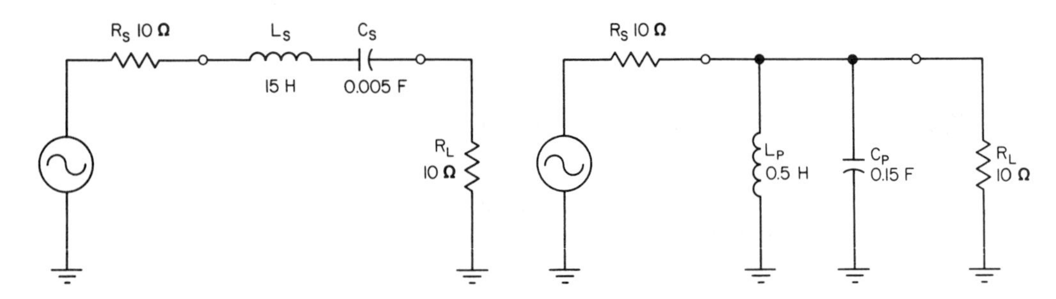

Figure 4A-2 Equivalent single-pole bandpass filters.

Before performing any calculations with the s-plane, it must be completed by adding a mirror image below the horizontal axis. One more addition is still necessary before we have a complete diagram that will represent the circuits of Figure 4A-2. Neither circuit will let any of the input signal pass through to the load when the input frequency is zero (dc). Therefore, we will place a 0 at the origin to indicate that zero signal passes through at this frequency. The complete s-plane for either of the circuits above is shown in Figure 4A-3. The vertical axis is calibrated in radians per second for convenience but could be converted to hertz by dividing by 2π.

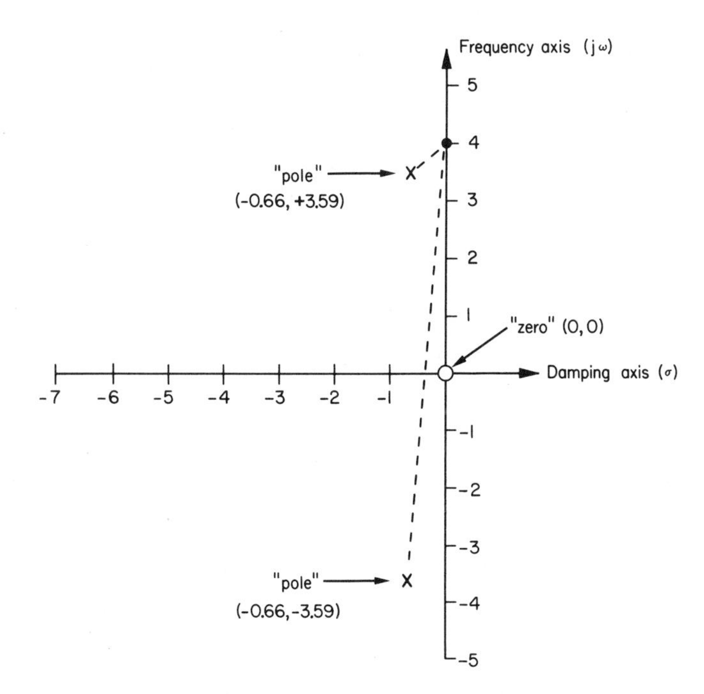

Figure 4A-3 s-Plane diagram for the single-pole band-pass filters of Figure 4A-2.

The horizontal damping axis should be calibrated in radians per second using the same scaling factor as for the vertical axis but with negative numbers. The significance of these numbers on the damping axis is only important when transient responses of the filters are of interest.

With the diagram now completed, we can calculate the gain and phase changes with frequency of the filter represented. This can be done in either of two ways, one graphical and the other mathematical. We will look at the graphical one first.

To find the *gain* (it will end up being a loss) at 4.0 rad per second, place a dot on the +4.0 point on the vertical axis and then draw lines from the dot to each of the two poles and to the zero. The magnitude of the gain can be determined from the lengths of these lines:

$$\text{gain} = \frac{\text{product of all distances to zeros}}{\text{product of all distances to poles}} = \frac{4.0}{(0.777)(7.62)} = 0.676$$

If a large number of gains are calculated at different frequencies, the curve of Figure 4A-4 will result. This is the standard tuned-circuit response plotted on a linear frequency axis. The peak would be sharper and the bandwidth narrower if the poles were located closer to the vertical axis. This, of course, would be a higher Q, since the angle up from the horizontal axis would be higher.

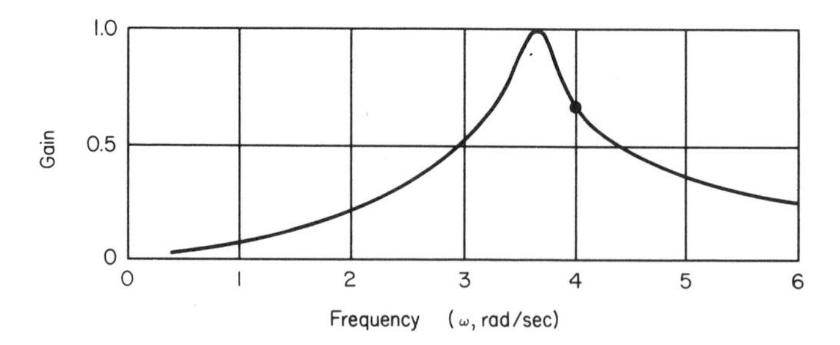

Figure 4A-4 Amplitude response calculated from the s-plane of Figure 4A-3.

The phase response is calculated just as easily by using the angles counterclockwise from a horizontal line to the poles and zeros (the pole at the origin is taken at an angle of 90°).

$$\begin{aligned}\text{phase angle} &= (\text{sum of all angles to zeros}) - (\text{sum of all angles to poles}) \\ &= 90° - (31.8° + 85°) \\ &= -26.8°\end{aligned}$$

If a large number of angles are calculated for different frequencies, the phase response that is shown in Figure 4A-5 is obtained. Again, this is shown on a linear scale although, as with Figure 4A-4, the scale could have been logarithmic.

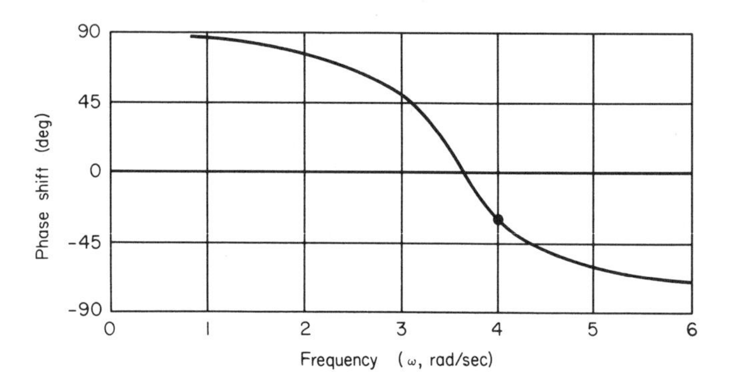

Figure 4A-5 Phase response of the filter described by the s-plane of Figure 4A-3.

The mathematical approach is just as easy. The location of each pole and zero is first described in either rectangular ($\sigma + j\omega$) or polar ($\omega_0 \angle\theta$) coordinates. The following transfer function can then be evaluated and will give the gain and phase angle simultaneously for each frequency.

$$\frac{V_{\text{load}}}{V_{\text{source}}} = \frac{\text{product of zero locations}}{\text{product of pole locations}}$$

$$= \frac{s}{(s+0.66-j3.59)(s+0.66+j3.59)}$$

If we want the magnitude and phase at any frequency, say $\omega = 4.0$ rad/s, we just substitute in $s = +j4.0$, so that

$$\frac{V_{\text{load}}}{V_{\text{source}}} = \frac{+j4.0}{(0.66-j3.59+j4.0)(0.66+j3.59+j4.0)}$$

$$= \frac{+j4.0}{(0.66+j0.41)(0.66+j7.59)}$$

$$= 0.676\angle -26.8^\circ$$

4A-2 EXAMPLES

Some examples of the amplitude responses that result from different poles and zero combinations and their locations are summarized here.

4A-2.1 Single Pole on the Horizontal Axis

This can be realized with a simple RC low-pass filter. The general form of the transfer function is

$$\frac{V_{\text{out}}}{V_{\text{in}}} = \frac{1}{RC_s+1}$$

The frequency response has a 3-dB cutoff frequency at

$$f_c = \frac{1}{2\pi RC} \qquad \text{(Hz)}$$

and will ultimately reach a final attenuation slope of 6 dB/octave (Figure 4A-6).

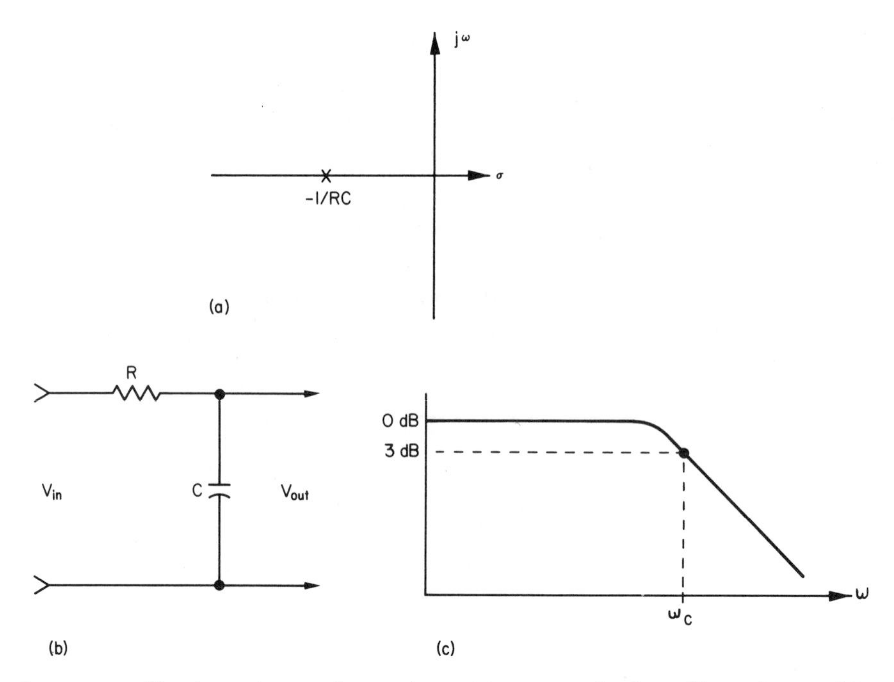

Figure 4A-6 Single pole on the *s*-plane (a), an equivalent filter (b), and its phase and amplitude response (c).

4A-2.2 Single Zero at the Origin

This is physically impossible to build by itself and will always occur with some other poles. The frequency response shows a constant positive slope of +6 dB/octave passing through 0 dB at a frequency of 1.0 rad/s (Figure 4A-7).

$$\textit{transfer function} = \frac{V_{out}}{V_{in}} = S$$

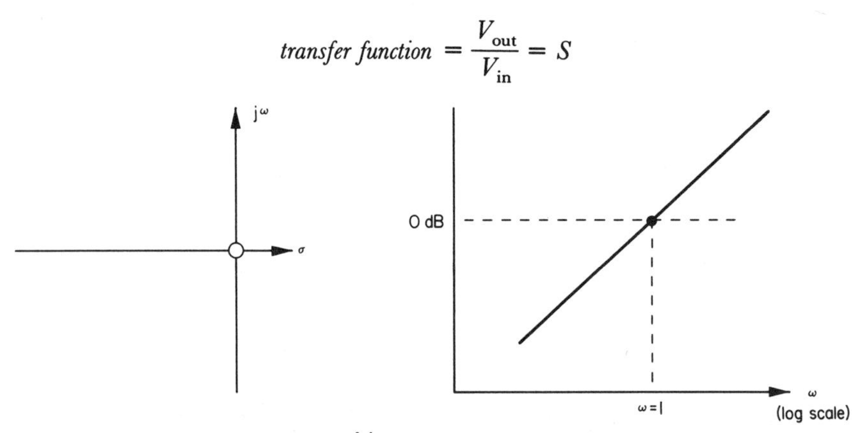

Figure 4A-7 Transfer function $= \frac{V_{out}}{V_{in}} = s$.

4A-2.3 A Pair of Complex Poles

Poles that are not located on the horizontal axis must always occur in pairs. If the location is close to the vertical axis, the Q will be high and the response will have a peak in it (Figure 4A-8). If the poles are located at a lower angle than 45° up from the horizontal, there will be no peak. The final attenuation slope will be 12 dB/octave in all cases. The frequency of the peak, if it occurs, will be given by the *peaking circle.*

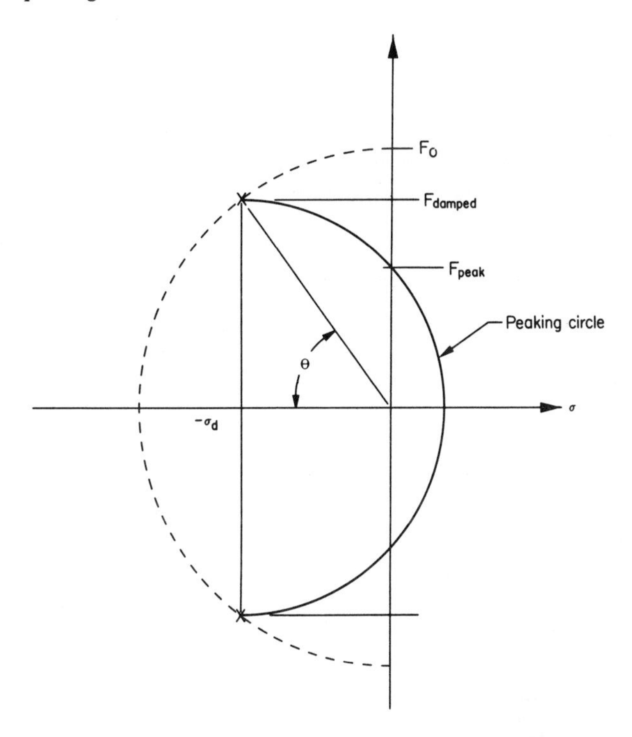

Figure 4A-8 Relative locations of the undamped frequency, the damped frequency, and the peak frequency on a two-pole *s*-plane.

The location of the complex pair of poles can be described using two different sets of coordinates:

1 Polar coordinates:
   (a) Radius equal to f_0 or ω_0.
   (b) Angle up from the horizontal axis. (For control systems work, the angular position is given by ζ (zeta), where $\zeta = 1/2Q_L$.)
2 Rectangular Coordinates:
   (a) Vertical distance equal to f_d, the damped resonant frequency, or ω_d

where: $$f_d = f_0 \sin\theta = f_0\sqrt{1 - \frac{1}{4Q^2}} \qquad \text{(Hz)}$$

(b) Horizontal distance equal to $-\sigma_d$,

where: $$\sigma_d = 2\pi f_0 \cos\theta = \frac{\pi f_0}{Q} \qquad \text{(rad/s)}$$

The *peaking circle* is drawn with the two poles defining the diameter of the circle and $-\sigma_d$ defining the center point. The point where the circle crosses the vertical frequency axis is the frequency of the peak in the response curve.

The magnitude of the response is given in Figure 4A-9 for $\omega_o = 1.0$ and a range of loaded Q's.

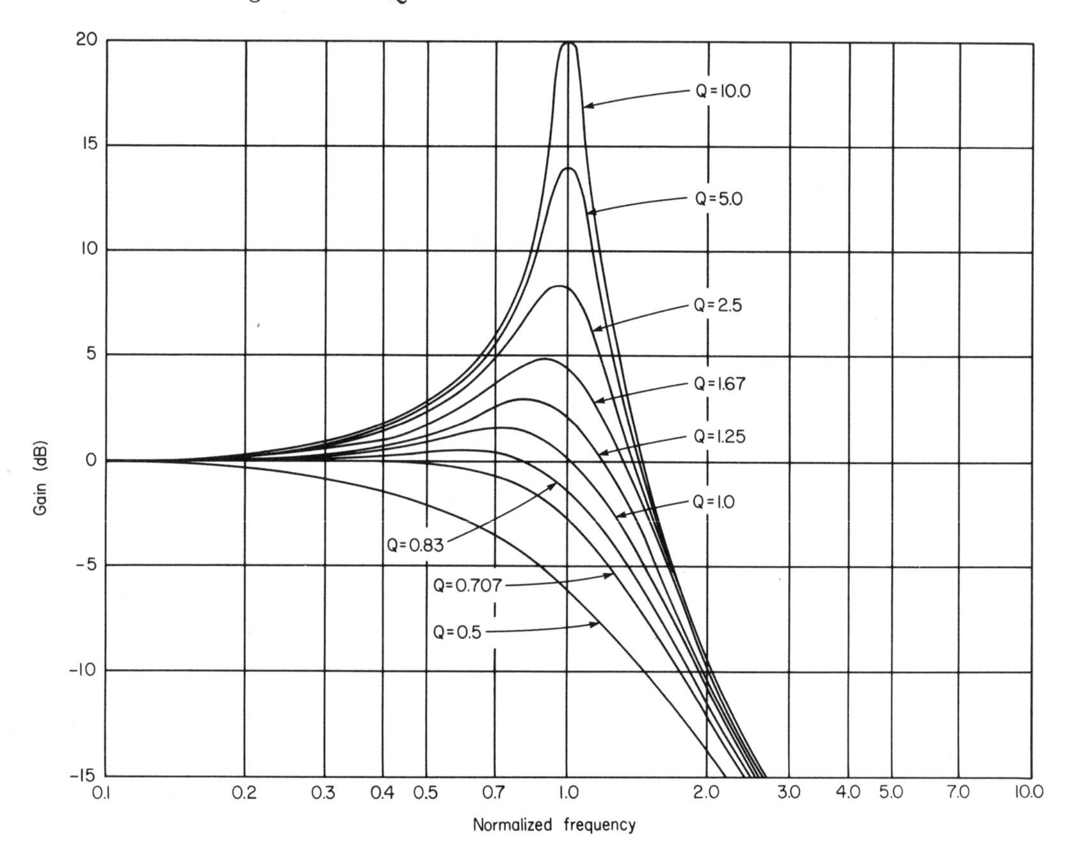

Figure 4A-9 Magnitude versus frequency response for two poles on the s-plane with various values of loaded Q.

4A-2.4 Combining Poles and Zeros

Pole and zero responses can be combined to obtain the overall response of a more complex circuit. As an example, consider a zero at the origin and a single pole on the horizontal axis as shown in Figure 4A-10. The individual responses of the pole and zero are shown in Figure 4A-11(a) and the combined response is shown in Figure 4A-11(b). The result of course is a high-pass *RC* filter and can be realized with the components of Figure 4A-12, which will have the

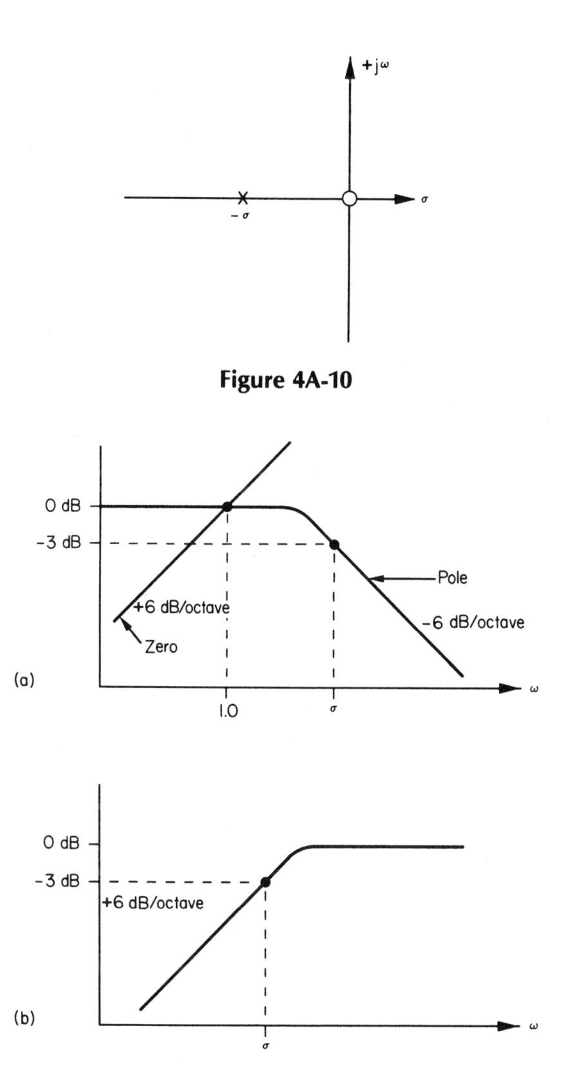

Figure 4A-10

Figure 4A-11 Individual response of the pole and zero (a) and combined response (b).

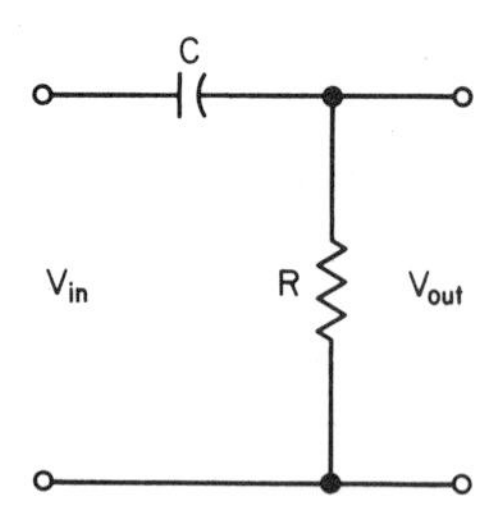

Figure 4A-12 High-pass *RC* filter with the s-plane of Figure 4A-10 and the response of Figure 4A-11(b).

transfer function

$$\frac{V_{\text{out}}}{V_{\text{in}}} = \frac{\sigma_s}{\sigma_{s+1}}$$

$$\sigma = \frac{1}{\text{RC}}(\text{rad/s})$$

$$\frac{V_{\text{out}}}{V_{\text{in}}} = \frac{RC_s}{RC_{s+1}}$$

For a second example we will consider the original circuits of Figure 4A-2 and its s-plane of Figure 4A-3. Since the *Q* of the pole pair is 2.76, we can use the corresponding curve from Figure 4A-9. To this we have to add the 6-dB slope of the zero at the origin. The total curve, as shown in Figure 4A-13(b), will have a slope of 6 dB on either side of the peak instead of the 0-dB and 12-dB slopes of Figure 4A-9. Note also that the frequency of the peak in the combined curve is at a higher frequency and corresponds to the point on the original curve with a 6-dB slope.

The combined response of (b) has been set back down to zero to represent the actual frequency response obtained with the circuits. The s-plane does not give absolute values of gain, so constants must be frequently inserted to describe specific applications.

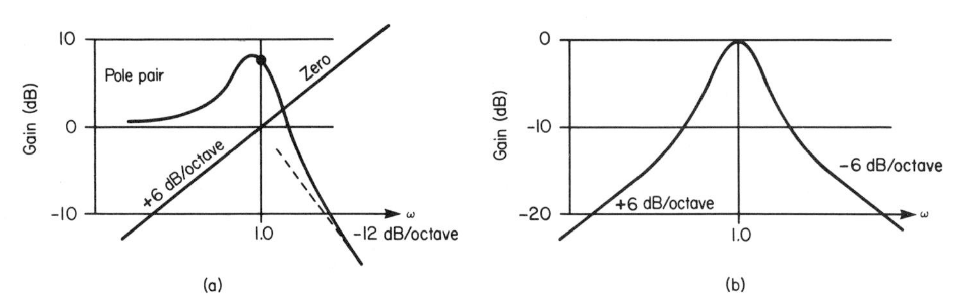

Figure 4A-13 Individual response of the pole pair and the zero (a) and the combined response (b).

4A-3 THREE-ELEMENT LOW-PASS FILTERS

The three-element low-pass filters previously discussed in Chapter 4 have an *s*-plane that generally looks like Figure 4A-14. The poles of the Butterworth filter are spaced around a perfect circle; the Chebychev poles lie on a vertical ellipse that becomes narrower and taller as the passband ripple increases. The Bessel poles lie on a horizontal ellipse.

The location of the poles for each three-element filter is given in Table 4A-1 both in rectangular and in polar coordinates. The cutoff frequency in all cases is 1.0 rad/s.

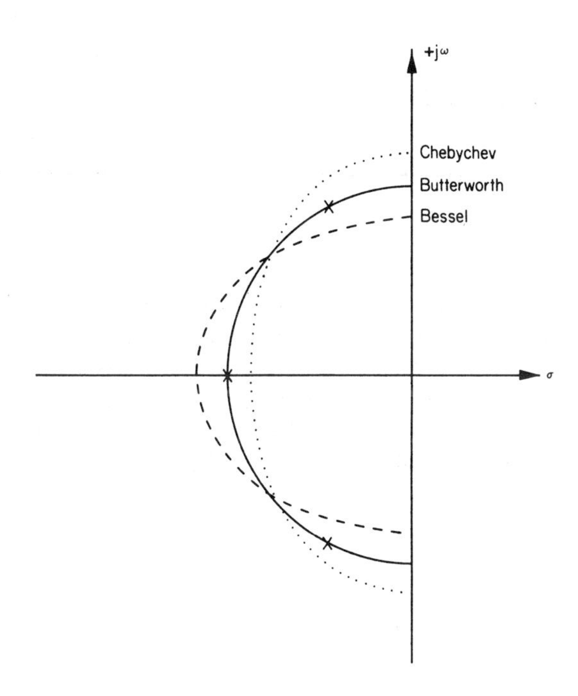

Figure 4A-14 General *s*-plane diagram for three-element low-pass filters.

4A-4 FILTER SYNTHESIS

The full description of the synthesis of filter networks is beyond the intent of this book; however, a few general comments can be made. The modern approach to synthesis requires that the designer describe exactly what he wants in mathematical form; if a particular shape of amplitude response is required, this will be described by an equation. If the delay and phase response are more

TABLE 4A-1

RECTANGULAR AND POLAR COORDINATES FOR EACH POLE OF THREE-ELEMENT LOW-PASS FILTERS.

	RECTANGULAR COORDINATES	*POLAR COORDINATES*
Butterworth	−1.0, 0.0	$1.00\angle 0^\circ$
	−0.5, +0.866	$1.00\angle 60^\circ$
	−0.5, −0.866	$1.00\angle -60^\circ$
1.0-dB Chebychev	−0.455, 0.0	$0.455\angle 0^\circ$
	−0.227, +0.888	$0.917\angle 75.6^\circ$
	−0.227, −0.888	$0.917\angle -75.6^\circ$
Bessel	−1.346, 0.0	$1.346\angle 0^\circ$
	−1.066, +1.017	$1.473\angle 43.6^\circ$
	−1.066, −1.017	$1.473\angle -43.6^\circ$

of interest, that will be described in an equation. The next step is the hard one. From the equation, the network must be designed. If the designer is lucky, he/she may recognize that the equation written is of the same general form as a ladder filter (or some other recognizable type). If so, he/she can equate his/her terms to the general terms of the ladder filter and come up with component values. In the following case, a designer wants a filter with the frequency response given by

$$\frac{V_{\text{load}}}{V_{\text{source}}} = \frac{1.0}{2.0s^2 + 2.9s + 2.0}$$

This happens to be in the same form as that of a two-element low-pass filter (Figure 4A-15) with the general equation

$$\frac{V_{\text{load}}}{V_{\text{source}}} = \frac{1.0}{LCs^2 + \left(\frac{L}{R_L} + R_s C\right)s + \frac{R_s}{R_L} + 1.0}$$

By equating the terms of the specific and general equations, the designer can

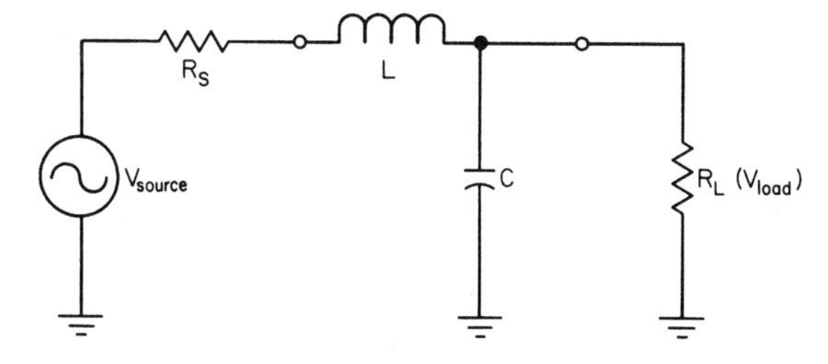

Figure 4A-15 Two-element low-pass filter.

come up with the values. First, let $R_s = 1.0\Omega$.Then

$$LC = 2.0$$

$$\frac{L}{R_L} + R_s C = 2.9$$

$$\frac{R_s}{R_L} + 1.0 = 2.0$$

Solving we get

$$R_s = 1.0\ \Omega$$
$$R_L = 1.0\ \Omega$$
$$C = 1.13\ \text{F}$$
$$L = 1.77\ \text{H}$$

If the standard form of the response cannot be recognized, both the form or order of the components and their actual values must be synthesized. For large-order filters, this type of work is best done by a computer.

The secret of network synthesis is that the full characteristics of a filter are known if the input impedance is known. We have already seen this in Chapter 4; for example, in Figure 4-18 we were able to draw circles on the Smith chart plot of the input impedance of the Butterworth filter and mark them with the attenuation of the filter. The same rings can also be used to describe the attenuation of any of the other filter types. The big problem is to convert the required amplitude versus frequency response into an equation that describes the input impedance versus frequency response. Once in this form, the components can be "pried" loose from the equation one or two at a time by repeated division. The following is a very simple example of this process.

For a particular filter, it has been determined that

$$Z_{\text{in}} = \frac{2s^2 + 2s + 1}{2s^3 + 2s^2 + 1}$$

Using division and inversion (inversion changes the result from a series component to a shunt component and vice versa):

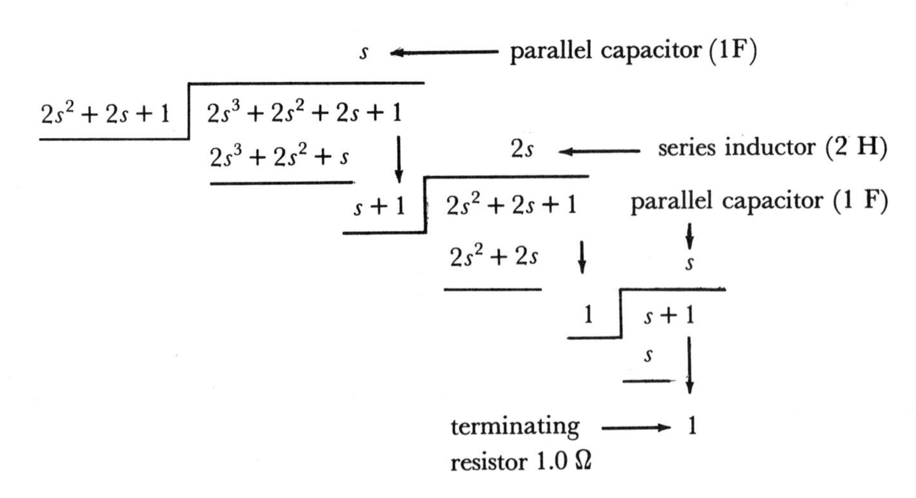

The results are a Butterworth filter with the form shown in Figure 4A-16:

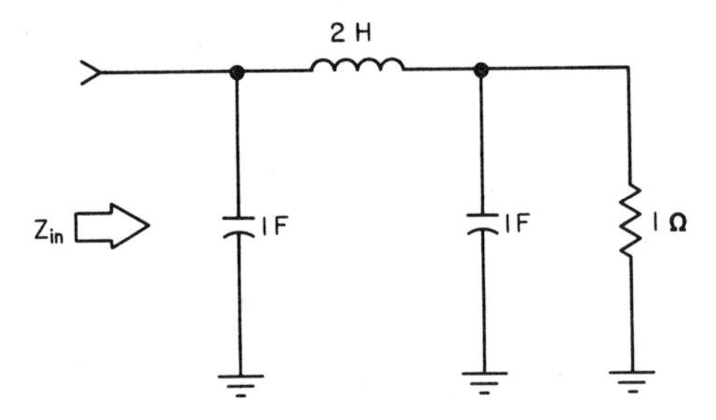

Figure 4A-16 Butterworth filter synthesized from an input impedance function.

chapter **5**

impedance matching

At high frequencies, gain and high power levels are not as easily obtained as they are at, say, audio frequencies. When a power gain of only 5 dB might be the most that can be squeezed out of a $50 transistor at 200 MHz, a loss of even 1 dB due to impedance mismatch cannot be tolerated. Since the output impedance of one stage is rarely the same as the input impedance of the next stage, some form of impedance-matching network will be required.

In addition to shifting impedance levels up and down, these networks may have to perform additional functions; wide bandwidths may be required or, instead, narrow bandwidths with good harmonic filtering may be called for. dc voltages may have to be passed for biasing purposes, or they may have to be blocked. The dissipative losses in the network must also be kept in mind; it would not make sense to build a matching network for a situation where the mismatch loss would have been only 2 dB if the newly designed network dissipated 3 dB of the applied power.

An example of matching circuits with additional requirements for filtering and dc passing and blocking is shown in Figure 5-1. This is a class C power amplifier for a transmitter. The input network must match the 50-Ω source resistance to the very low input resistance of the transistor and the bandwidth is relatively wide, about 60% of the center frequency. The output network steps the impedance up from the low level at the transistor to 50 Ω for the load. The harmonics that result from the class C operation must also be filtered out.

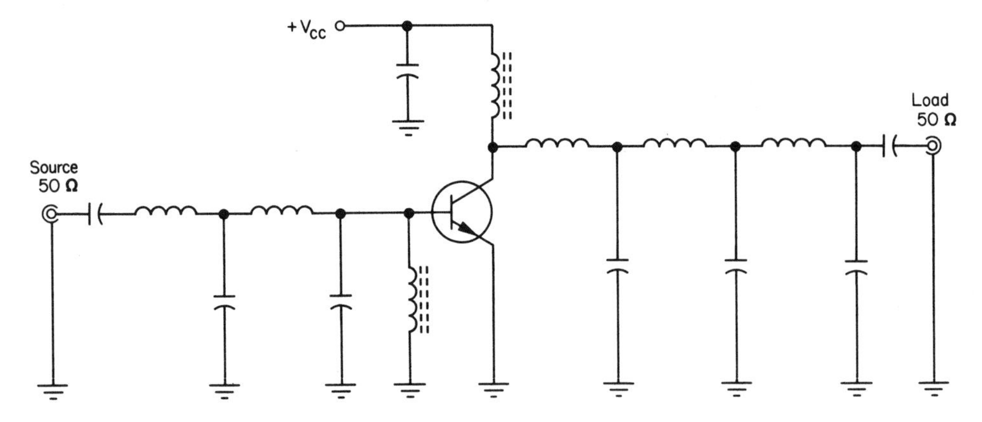

Figure 5-1 Application of impedance-matching networks to a broad-band, class C power amplifier.

In this chapter, then, we will start with the design of relatively narrow bandwidth matching circuits that resemble the filter circuits of Chapter 4. We will design these circuits first using equations that provide exact results, and then we will use a graphical method on the Smith chart that is not quite as accurate but is more descriptive.

5-1 TWO-ELEMENT L SECTIONS

Let us look at the simplest and most useful matching network, the two-element circuits of Figure 5-2, often referred to as L sections because of the component orientation.

The two elements can be put together in four different ways, the first two forming the low-pass filters of Figure 5-2 and the other two the high-pass filter

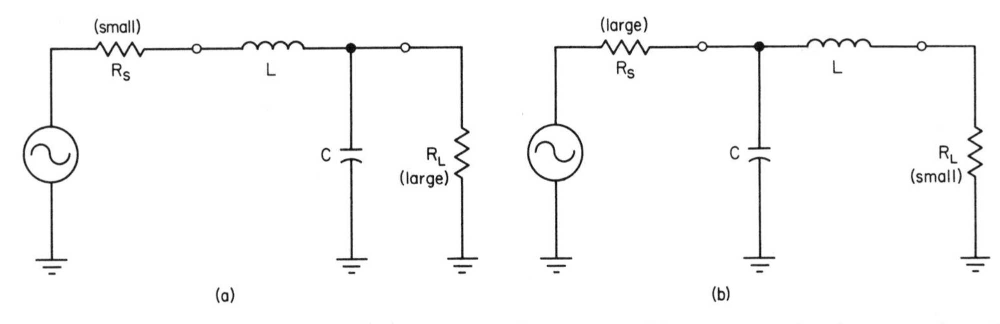

Figure 5-2 Two-element impedance-matching networks for transforming (a) a low source resistance up to a high load resistance, and (b) a high source resistance down to a low load resistance.

configurations of Figure 5-3. In either figure a low source resistance can be matched to a high load resistance with one network, or the opposite configuration can be used if the source resistance is higher than the load.

In Chapter 4 we saw that the two-element filter would provide maximum power transfer for an unequal value source and load resistance combination when the total loaded Q of the circuit was higher than 0.5. The match was perfect only at the very peak of the amplitude response curve, but in some cases, a fairly wide bandwidth was available where the loss was not too great. For higher values of total Q, the available bandwidth became increasingly more narrow. For an insertion loss of 0 dB at the peak frequency, the loaded Q of each arm of the filter had to be the same.

The secret of the design of any matching network lies in the ability to represent a series combination of components as an equivalent parallel combination, and vice versa. Each conversion is valid at only one frequency, where both the series and parallel circuits must have the same impedance. The

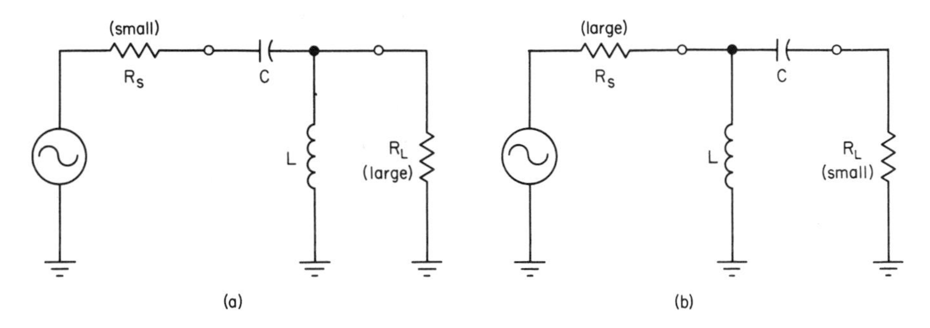

Figure 5-3 High-pass form of the two-element matching networks for transforming (a) a low source up to a high load resistance, and (b) transforming a high source down to a low load resistance.

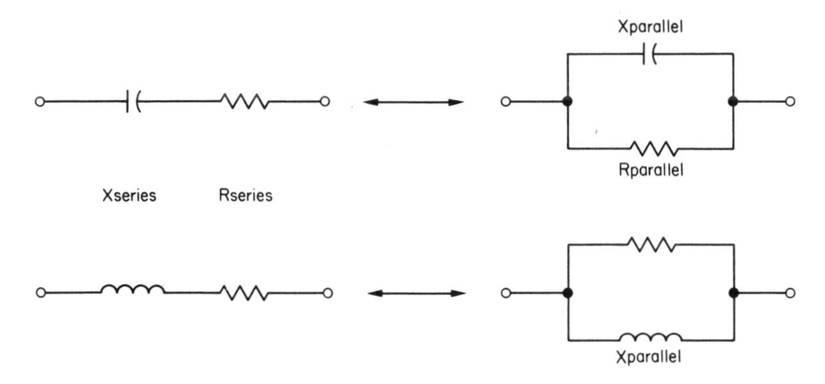

Figure 5-4

resistor of the series combination will always be smaller than the parallel resistor of the equivalent parallel combination (Figure 5-4).

The conversion can be performed using many different mathematical techniques, but we will use the idea that the Q of both the series and parallel circuit must be identical. For parallel components,

$$X_p = \frac{R_p}{Q} \tag{5-1}$$

For series components,

$$X_s = R_s \times Q \tag{5-2}$$

So

$$Xp = \left(1 + \frac{1}{Q^2}\right) X_S \tag{5-3}$$

$$R_p = (1 + Q^2) R_s \tag{5-4}$$

The Q used here is not the total Q of whatever network is being considered but rather the Q of just the two components being considered.

Now to apply this to an actual network. We have already seen in Chapter 4 that the two-element filter can be used for impedance matching at the frequency of its peak and is most efficient when the Q of the series arm is made the same as that of the parallel arm. The total Q of the network was then $\frac{1}{2}$ the Q of each arm.

Imagine that we have a signal source with a low internal resistance of 10Ω and that we want to transfer as much of the source power as possible to a high-value, 100-Ω load resistance. This situation is shown in Figure 5-5(*a*).

If the 10-Ω source were directly connected to the 100-Ω load, the mismatch would result in 4.8 dB less power reaching the load than the maximum available. Now place a capacitor across the load, as shown in Figure 5-5(b). At any single frequency we can then replace the parallel combination of the capacitor and load with a series equivalent, Figure 5-5(c). If the right value of parallel capacitor is used, the equivalent series load resistance will be 10 Ω. We still have not set up a condition for maximum power transfer because of the reactance of the series capacitance. However, if we add a series inductance, Figure 5-5(d), that has the same reactance as the series capacitor, the two reactances will cancel and so the 10-Ω source appears to be directly connected to the equivalent 10-Ω load resistance. Therefore, maximum power reaches the load. The total network is shown in Figure 5-5(e).

The value of this parallel capacitor can be found with the help of Equation (5-4). Since we know that the 100-Ω parallel load resistance must be transformed into an equivalent 10-Ω series resistance, then

$$1 + Q^2 = \frac{R_p}{R_s} = \frac{100}{10} = 10$$

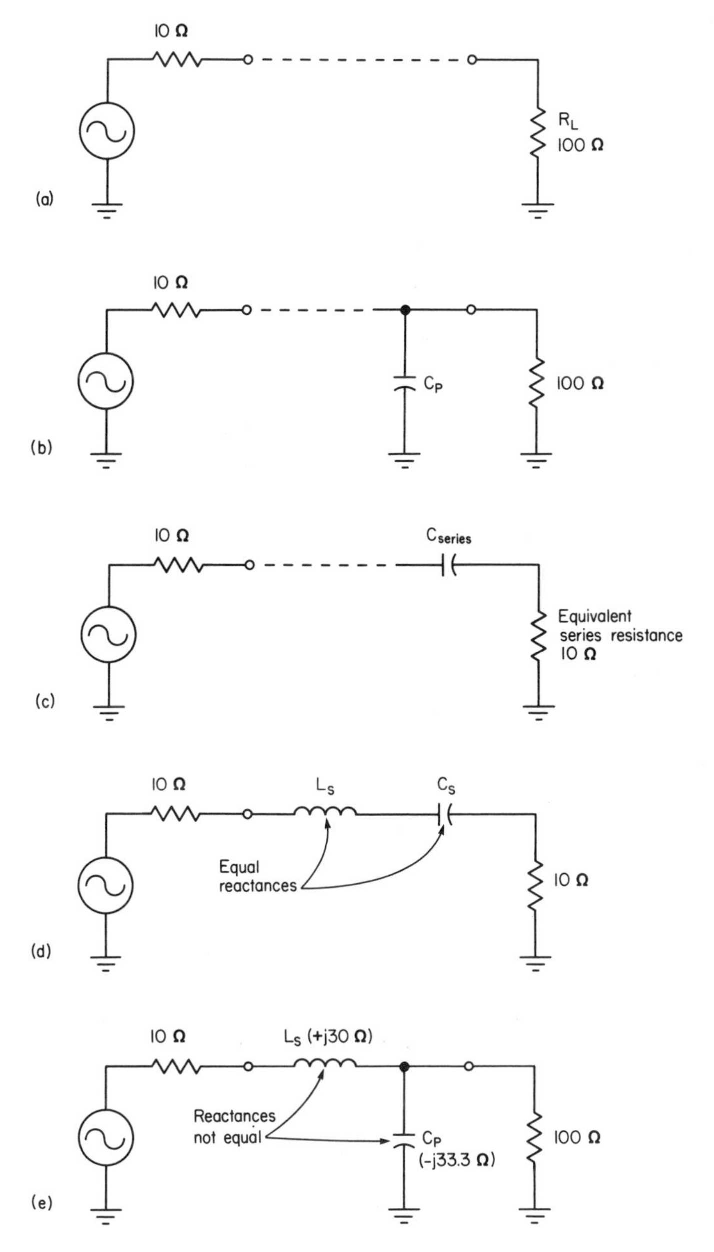

Figure 5-5 Step-by-step development of a two-element matching network to provide maximum power transfer from a 10-Ω source to a 100-Ω load resistance.

Therefore,

$$Q = \sqrt{10 - 1} = 3.0$$

for the parallel combination of R_L and C.

The reactance of the parallel capacitor is then

$$X_p = \frac{R_p}{Q} = \frac{100}{3.0} = -j33.3\ \Omega$$

The reactance of the series inductor can be found in either of two ways. One way would be to convert the capacitor and load resistor into a series equivalent circuit and then set the reactance of the inductor equal to the reactance of the equivalent series capacitor. For the other approach, recall that the Q of each half of the circuit must be the same; therefore, from Equation (5-2),

$$X_s = R_s \times Q = 10 \times 3.0 = +j30\ \Omega$$

The value of X_s will be the same for either approach. These reactances are shown in Figure 5-5(e); note that the parallel capacitor reactance is *not* the same as the series inductor reactance at the matching frequency.

If we were curious about the frequency response of this particular network, we could look back at Figure 4-5(a). Since the Q of each half of our matching network is the same, the total Q will be

$$Q_{\text{total}} = \frac{Q_1}{2} = \frac{Q_2}{2} \tag{5-5}$$

For our example, the total Q will be $3.0/2 = 1.5$, so we would look at this curve in Figure 4-5(a). The peak is 4.8 dB higher than the low-frequency level.

When designing a two-element matching network, we do not have a choice of the resulting bandwidth. As soon as the source and load resistance values are established, the Q of each arm of the network must be

$$Q_1 = Q_2 = \sqrt{\frac{R_{\text{larger}}}{R_{\text{smaller}}} - 1} \tag{5-6}$$

[This is just a rearranged form of Equation (5-4).]

The total network Q will be half this value,

$$Q_{\text{total}} = \frac{1}{2}\sqrt{\frac{R_{\text{larger}}}{R_{\text{smaller}}} - 1} \tag{5-7}$$

The bandwidth will be fixed:

$$BW \approx \frac{F_{\text{peak}}}{Q_{\text{total}}} \tag{5-6}$$

5-2 WIDER BANDWIDTH

For wider bandwidths a number of two-element circuits can be used together. Using two L sections as an example, we can match first from the lower value up to some intermediate value and then from this intermediate value up to the higher value, as shown in Figure 5-6. Since each circuit has a smaller difference between its effective source and load resistance, the total Q will be lower, and so the bandwidth wider. Maximum benefit is obtained when each L section has the same Q. This is achieved if

$$R_{\text{intermediate}} = \sqrt{R_{\text{smaller}} \times R_{\text{larger}}} \tag{5-8}$$

If the transformation is to be made in three steps, two intermediate values will be needed. Again, all three Q's should be the same, so

$$\frac{R_{\text{intermediate (1)}}}{R_{\text{smaller}}} = \frac{R_{\text{intermediate (2)}}}{R_{\text{intermediate (1)}}} = \frac{R_{\text{larger}}}{R_{\text{intermediate (2)}}} \tag{5-9}$$

The example shown in Figure 5-1 shows the input matching being made in two steps to obtain wider bandwidths. The output uses three L sections, both for wide bandwidth and for the additional harmonic filtering that can be obtained.

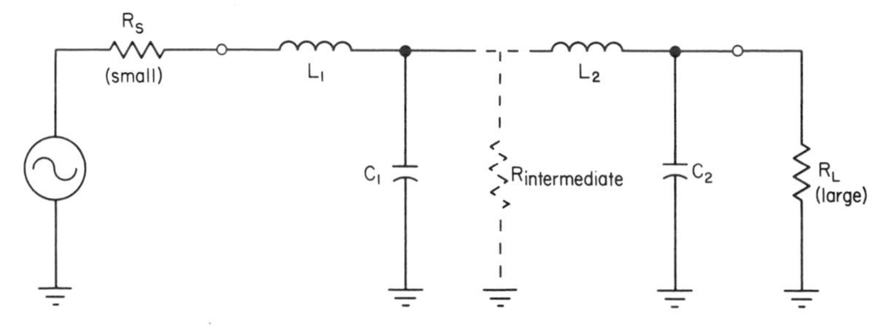

Figure 5-6 Impedance transformation using two sections to obtain wider bandwidths.

EXAMPLE 5-1

Design a matching network for 2.5 MHz, to go from a 10-Ω source to a 100-Ω load using two low-pass L sections to obtain wider bandwidths. Plot the approximate frequency response.

(The source and load resistances used here are intentionally the same values used in the previous discussion. A comparison of final bandwidths can then be made.)

Solution

The fictitious intermediate resistance will be

$$R_{\text{intermediate}} = \sqrt{R_S \times R_L} \qquad (5\text{-}8)$$

$$= \sqrt{10 \times 100} = 31.6\ \Omega$$

The Q of each arm in each L section will be

$$Q_1 = Q_2 = Q_3 = Q_4 = \sqrt{\frac{R_{\text{int}}}{R_{\text{source}}} - 1} \qquad (5\text{-}6)$$

$$= \sqrt{\frac{31.6}{10} - 1} = 1.47$$

The component reactances can be found:

$$X_{L_1} = Q_1 \times R_S$$

$$= 1.47 \times 10 = +j14.7\ \Omega$$

$$X_{C_1} = \frac{R_{\text{int}}}{Q_2}$$

$$= \frac{31.6}{1.47} = -j21.5\ \Omega$$

$$X_{L_2} = Q_3 \times R_{\text{int}}$$

$$= 1.47 \times 31.6 = +j46.5\ \Omega$$

$$X_{C_2} = \frac{R_{\text{load}}}{Q_4}$$

$$= \frac{100}{1.47} = -j68.0\ \Omega$$

The component values at 2.5 MHz are

$$L_1 = \frac{X_{L_1}}{2\pi f} = \frac{14.7}{2\pi 2.5 \times 10^6} = 0.936\ \mu\text{H}$$

$$C_1 = \frac{1}{2\pi f X_{C_1}} = \frac{1}{2\pi \times 2.5 \times 10^6 \times 21.5} = 2960\ \text{pF}$$

$$L_2 = \frac{X_{L_2}}{2\pi f} = \frac{46.5}{2\pi \times 2.5 \times 10^6} = 2.96\ \mu\text{H}$$

$$C_2 = \frac{1}{2\pi f X_{C_2}} = \frac{1}{2\pi \times 2.5 \times 10^6 \times 68.0} = 936\ \text{pF}$$

The final network is shown in Figure 5-7.

Now, what about the frequency response?

We can start by determining the total Q for each L section. The Q of each arm was found to be 1.47; therefore, the total Q will be half this, or 0.735. Using this value, we can look back at Figure 4-5 and estimate the response for *one* L section. But how do we go from this to the combined response for the two sections? We could simply double the attenuation produced by the one section; this is shown by the heavy-line response in Figure 5-8.

But the problem is that the response curves of Figure 4-5 assume constant terminating resistances at all frequencies. For our two sections, the load for the first section is the input impedance of the second section. At 2.5 MHz, this happens to be a proper resistive value of 31.6 Ω, our intermediate resistance. But at other frequencies, this impedance will change drastically and have reactive components. The true response was simulated on a computer and is included in Figure 5-8 as the dashed line, for comparison.

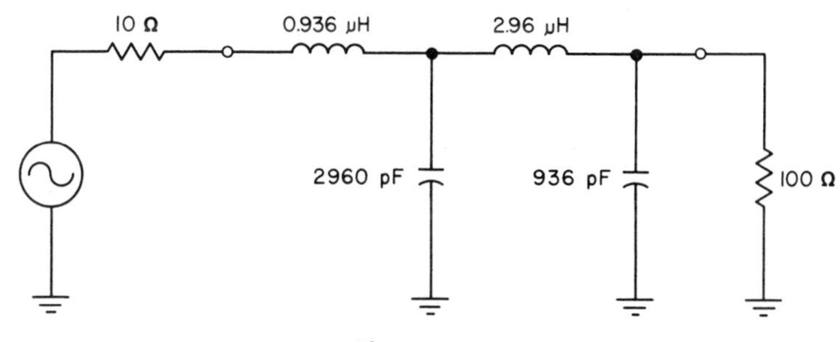

Figure 5-7

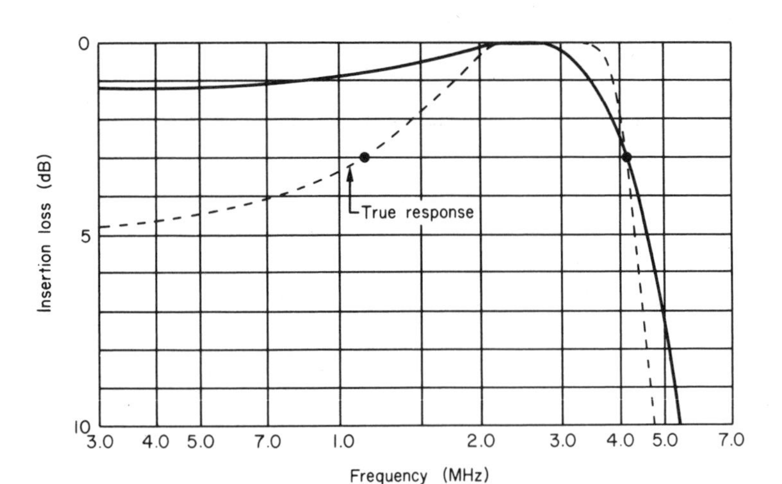

Figure 5-8

5-3 THREE-ELEMENT MATCHING

When three elements are used in a matching network, the designer is no longer limited to a single value of network Q as he or she was with the two-element circuit. For a given set of source and load resistances, he/she can now select any Q *higher* than that possible with the two-element L section. The Π arrangement for the three elements is shown in Figure 5-9. A corresponding T arrangement is also possible. The three-element Π section can match a source that is either higher or lower than the load resistance simply by altering the ratio of the two shunt components.

To understand the operation of the Π section, assume that it is cut into two parts, as indicated by the dashed lines in Figure 5-9. Each part then resembles an L section that is oriented to transform both the source and the load resistances down to some intermediate value that is lower than either one. This is illustrated on a resistance line drawing for the one situation where the source resistance is lower than the load resistance (Figure 5-10).

The intermediate value can be any value lower than the source resistance (or load resistance if it happens to be lower). The first portion of the Π network matches the source resistance down to this intermediate resistance; the relative values of the two resistances will determine the Q of this portion. The remainder of the network will match the intermediate resistance up to the higher load resistance. Since there is a greater jump involved, the Q of this section will be higher. The overall bandwidth of the circuit will depend on both Q's, but since the higher Q will dominate, it is usually used to describe the network and its bandwidth. Remember that this is more for convenience than accuracy.

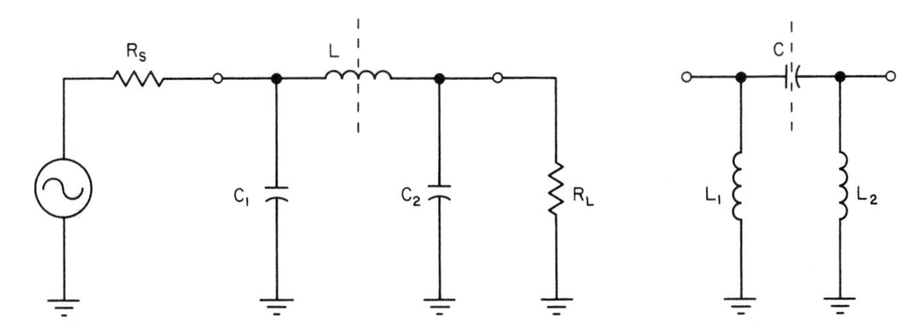

Figure 5-9 Three-element matching network arranged as a Π section; the low-pass form is shown in (a) and the high-pass form in (b).

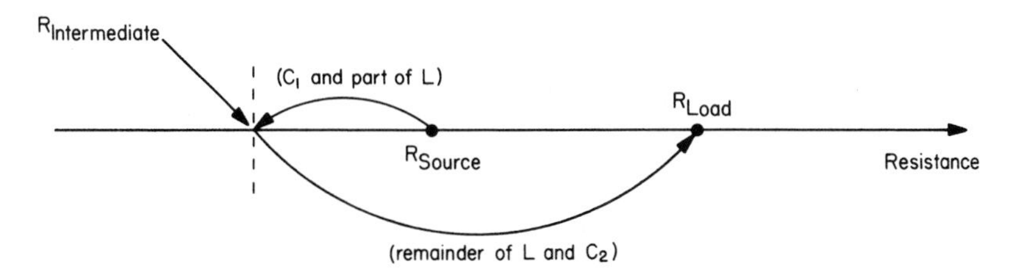

Figure 5-10

EXAMPLE 5-2

Design a three-element matching network for a source resistance of 10 Ω and a load resistance of 100 Ω. The best match is required at 3.75 MHz, and the Q_T of the higher portion will be 4.0.

Solution

First, let us check to see if the given value of Q_T will result in the intermediate resistance being lower than the source resistance.

$$Q_T = \frac{1}{2}\sqrt{\frac{R_{\text{larger}}}{R_{\text{smaller}}} - 1} \qquad (5\text{-}7)$$

$$R_{\text{intermediate (smaller)}} = \frac{R_{\text{load (larger)}}}{1+4Q_T^2} = \frac{100}{1+4\times 4^2} = 1.538\ \Omega$$

Since this is smaller than our 10-Ω source, we can proceed with the design. For the higher-Q section, a total Q of 4.0 requires each arm to have $Q_1 = Q_2 = 8.0$ [from Equation (5-5)].

For the final shunt capacitor,

$$X_{C_2} = \frac{R_{\text{load}}}{Q_2} = \frac{100}{8.0} = -j12.5\ \Omega$$

For the last part of the series inductor,

$$X_L = Q_1 \times R_{\text{intermediate}} = 8.0 \times 1.538 = +j12.30\ \Omega$$

For the first portion of the network, the Q of each arm will be

$$Q_1 = Q_2 = \sqrt{\frac{R_{\text{larger}}}{R_{\text{smaller}}} - 1} = \sqrt{\frac{10}{1.538} - 1} = 2.346 \qquad (5\text{-}6)$$

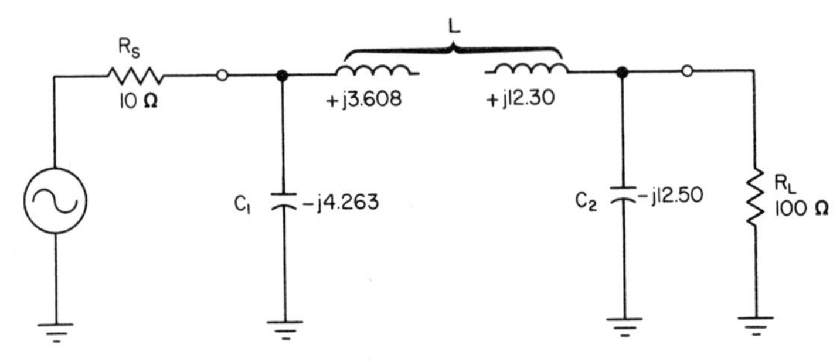

Figure 5-11

For the first shunt capacitor,

$$X_{C_1} = \frac{R_{\text{source}}}{Q_1} = \frac{10}{2.346} = -j4.263\ \Omega$$

For the other part of the series inductor,

$$X_L = Q_2 \times R_{\text{intermediate}} = 2.346 \times 1.538 = +j3.608\ \Omega$$

Our network now looks as shown in Figure 5-11. The component values at 3.75 MHz will be

$$C_1 = \frac{1}{2\pi X_C F_0} = \frac{1}{2\pi \times 4.263 \times 3.75 \times 10^6} = 9955\ \text{pF}$$

$$L = \frac{X_L}{2\pi F_0} = \frac{3.608 + 12.30}{2\pi \times 3.75 \times 10^6} = 0.675\ \mu\text{H}$$

$$C_2 = \frac{1}{2\pi X_C F_0} = \frac{1}{2\pi \times 12.50 \times 3.75 \times 10^6} = 3395\ \text{pF}$$

A summary of types of matching circuits is given in Figure 5-12.

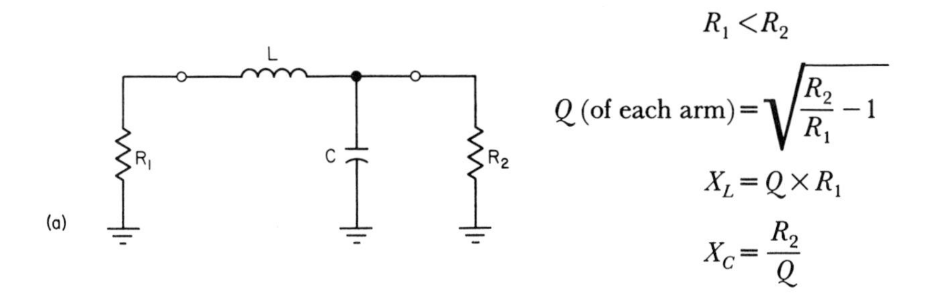

$$R_1 < R_2$$

$$Q\text{ (of each arm)} = \sqrt{\frac{R_2}{R_1} - 1}$$

$$X_L = Q \times R_1$$

$$X_C = \frac{R_2}{Q}$$

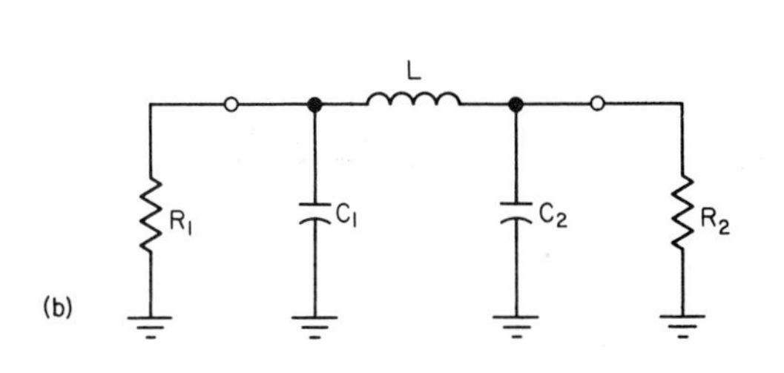

$$R_1 < R_2$$

$$Q = \frac{R_2}{X_{C_2}} \text{ (the end with the higher } Q\text{)}$$

$$X_{C_1} = \frac{R_1}{\sqrt{\frac{R_1}{R_2}(Q^2+1) - 1}}$$

$$X_L = \frac{QR_2}{Q^2+1}\left(1 + \frac{R_1}{QX_{C_1}}\right)$$

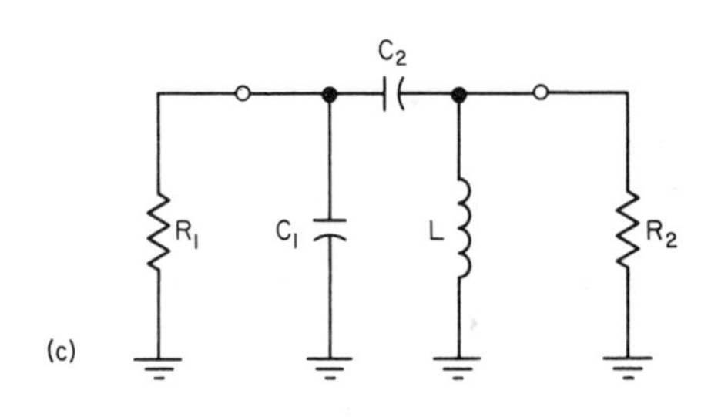

$$R_1 < R_2$$

$$Q = \frac{R_1}{X_{C_1}}$$

$$X_L = \frac{R_2}{\sqrt{\frac{R_2}{R_1}(Q^2+1) - 1}}$$

$$X_{C_2} = \frac{QR_1}{Q^2+1}\left(\frac{R_2}{QX_L} - 1\right)$$

Figure 5-12 Several types of matching circuits. The two already discussed are included. In all cases, a capacitor can be replaced with an inductor of the same reactance, and vice versa. The Q used here refers to the highest single Q in the network. The total circuit Q, which will determine the bandwidth, will be somewhat higher or lower, depending on the particular circuit.

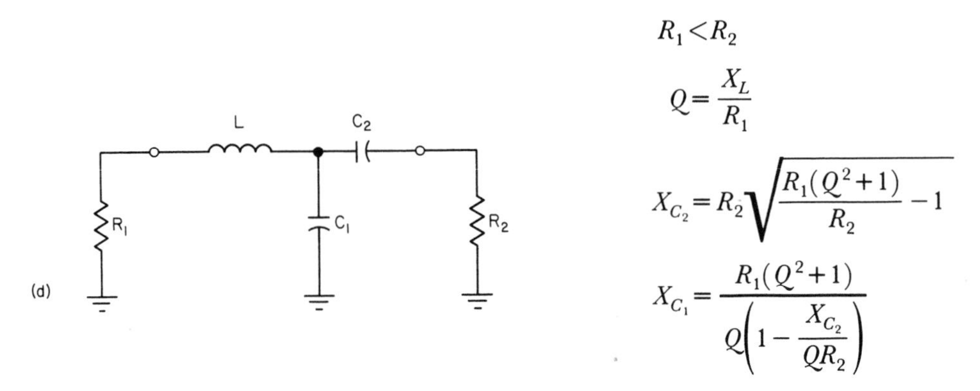

Figure 5-12 (Continued)

5-4 GRAPHICAL DESIGN OF MATCHING NETWORKS

The graphical design of matching networks on Smith charts is not as accurate as the mathematical techniques just discussed but is more visual. More complex designs can easily be made on the charts and the results will usually be accurate enough for the initial design work.

The Smith chart is available in two basic forms, an admittance chart and an impedance chart.[1] The two are shown in Figure 5-13. Any point on the impedance chart describes a series combination of resistance and reactance of the form $Z = R + jX$. A pure resistance is represented by a point on the central horizontal axis. Inductive impedances lie above this line and capacitive impedances below. The point marked in Figure 5-13(a) is a capacitive impedance of $1.0 - j0.5\ \Omega$. The arrow from this point indicates the direction in which the impedance would move if a *series* inductance were added to the impedance. The arrow is following a line of constant resistance since the addition of any series reactance does not change the original series resistance. The length of the arrow is determined by the amount of added inductive reactance.

[1]One popular form of the chart is "universal," in that the one chart is marked in both admittance and impedance, (see Appendix A).

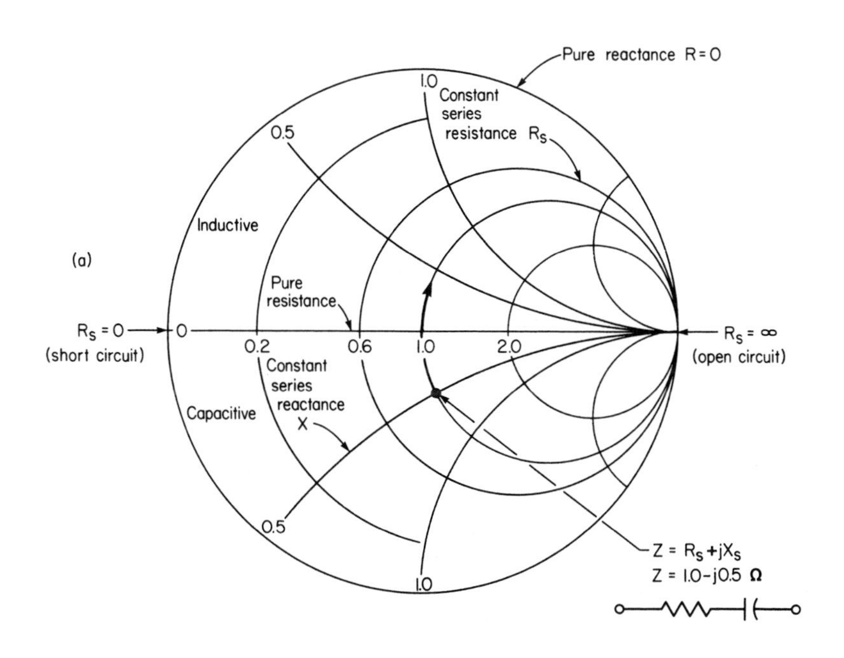

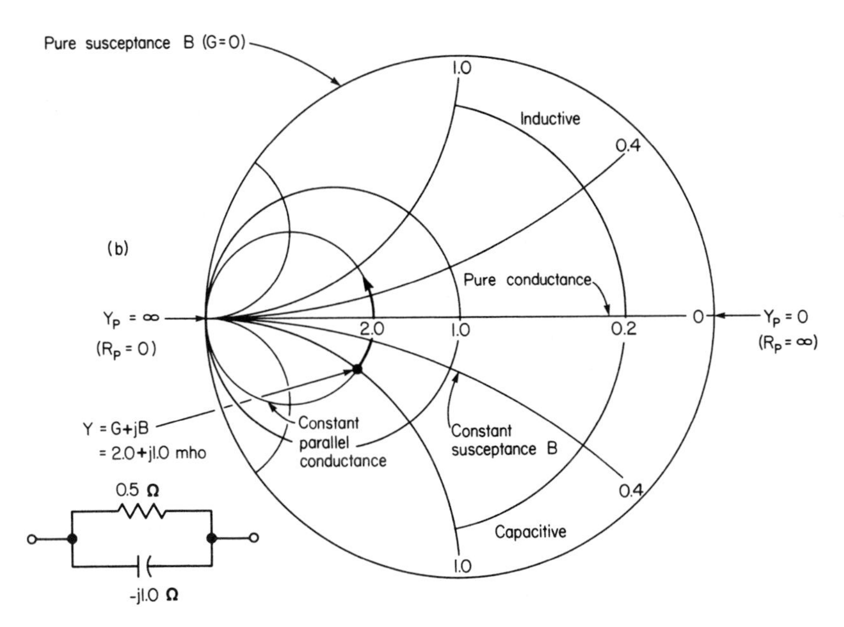

Figure 5-13 Impedance (a) and admittance (b) forms of the Smith chart.

In a similar manner, the admittance chart of Figure 5-13(b) describes a parallel combination of components of the form

$$Y = G + jB$$

where: Y = admittance

G = parallel conductance (resistive portion)

B = parallel susceptance (reactive portion)

[all are measured in mhos (℧) or siemens[2] (S)]

Note that the susceptance of a capacitor is $+jB$ and of an inductor is $-jB$. The signs are the opposite of those used for capacitive $(-jX)$ and inductive $(+jX)$ reactance.

The point marked on the admittance chart represents a 0.5-Ω resistor in parallel with a capacitive reactance of 1.0 Ω. The arrow indicates the direction that the admittance would move if an inductance were added in parallel. The arrow is following a line of constant conductance since the addition of any parallel reactance does not change the original parallel resistance.

We have already seen that matching networks can be easily designed if we have some means of converting from components in series to components in parallel (i.e., from impedance to admittance) and back again. Fortunately, if the impedance chart is positioned directly on top of the admittance chart and a pin pushed through both charts, the point on one chart is the series equivalent (impedance) of the parallel components (admittance) on the other chart.

Ladder-type impedance-matching networks of almost any complexity can therefore be designed by starting with a value of resistance that may represent the source (or load) and then adding series reactances on the impedance chart and shunt reactances on the admittance chart until the other value of load (or source) resistance is reached. A summary of which chart to use and which way to move is provided in Table 5-1.

The charts shown in Figure 5-13 are "normalized" charts in that the center point represents a pure resistance of 1 Ω. Other charts are available with 50 Ω or 75 Ω as the center point. The actual printing does not matter as long as you can work with convenience and accuracy. For example, if the normalized charts of Figure 5-13 are used, you will have a hard time locating points for a 25-Ω source resistance and 150-Ω load resistance. The points exist but they are crammed way over at the right-hand side. To make life easier, divide all

---

[2]The correct SI unit of conductance, as indicated, is the Siemen. However, Smith charts are not available yet that are printed in these units so the more traditional "mho" will be retained (1 mho = 1 Siemen).

TABLE 5-1

INSTRUCTIONS USED TO ADD SERIES AND PARALLEL COMPONENTS TO THE SMITH CHARTS.

TO ADD	*USE CHART*	*FOLLOW*	*DIRECTION*
Series L	Impedance	Constant R	CW
Series C	Impedance	Constant R	CCW
Series R	Impedance	Constant X	Toward higher R (open)
Parallel L	Admittance	Constant G	CCW
Parallel C	Admittance	Constant G	CW
Parallel R	Admittance	Constant B	Toward higher G (short)

impedances by some convenient number[3] so that the normalized source and load resistances are more conveniently located out in the middle of the chart. If we divide all values by 50, the source resistance point lies on the pure resistance line at $25/50 = 0.5$ and the load resistance lies at $150/50 = 3.0$.

EXAMPLE 5-3

For those readers not familiar with the Smith chart, two examples are provided here. The reader is encouraged to follow these through on a chart of his/her own.

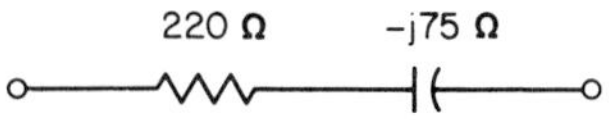

220 Ω -j75 Ω

Figure 5-14

(a) We are going to find the parallel equivalent components for the circuit of Figure 5-14. First, we normalize the impedance:

$$Z_n = \frac{220 - j75}{100} = 2.2 - j0.75\ \Omega$$

(The value 100 is arbitrarily chosen to locate the point close to the middle of the chart where the accuracy is the greatest. Any other value could have been used.) Now mark the point on the impedance chart A. Transfer the point to an admittance chart and read its value.

$$Y_n = 0.41 + j0.14 \text{ mhos}$$

[3]For discrete element matching networks, we can divide by whatever value is convenient. However, for lumped-element transmission-line applications, all values must be normalized so that the center point $(1 + j0)$ represents the characteristic impedance of the transmission line being used, normally 50 Ω or 75 Ω.

This admittance is now "unnormalized" by again dividing by the same value we used for the initial normalization.

$$Y = \frac{0.41 + j0.14}{100} = 0.0041 + j0.0014 \text{ mhos}$$

The parallel component values are then

$$R_p = \frac{1}{G_p} = \frac{1}{0.0041} = 243.9\ \Omega, \qquad X_p = \frac{1}{B_p} = \frac{1}{+j0.0014} = -j714.3\ \Omega$$

The parallel equivalent circuit looks as shown in Figure 5-16.

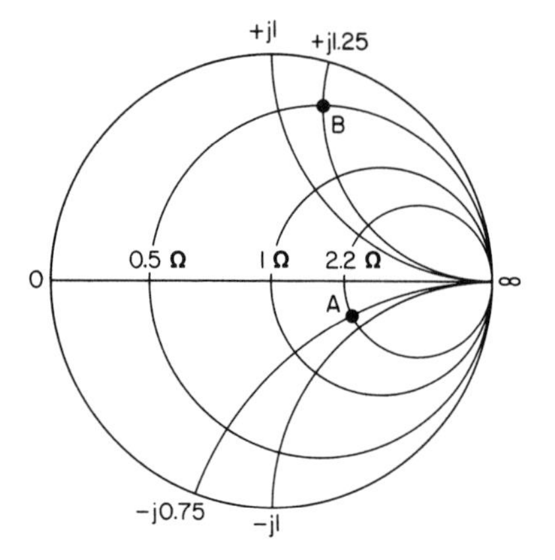

Figure 5-15

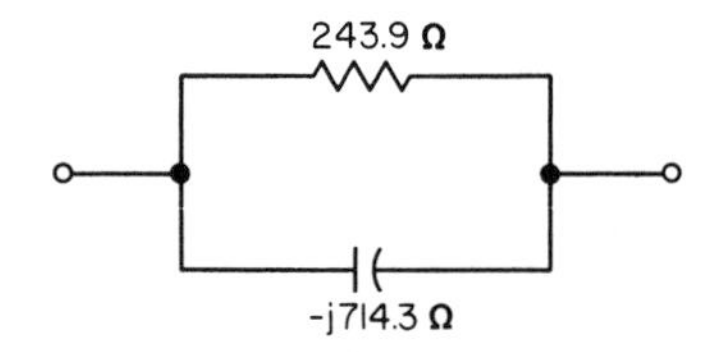

Figure 5-16

(b) In a similar manner, we can use the chart to convert from the series circuit shown in Figure 5-17 into the equivalent parallel circuit. The location of this impedance is marked *B* in Figure 5-15.

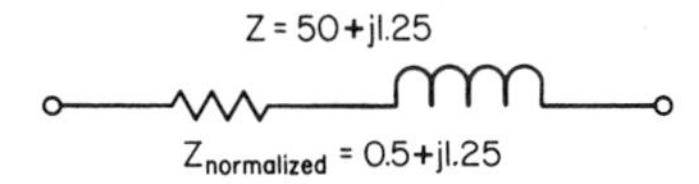

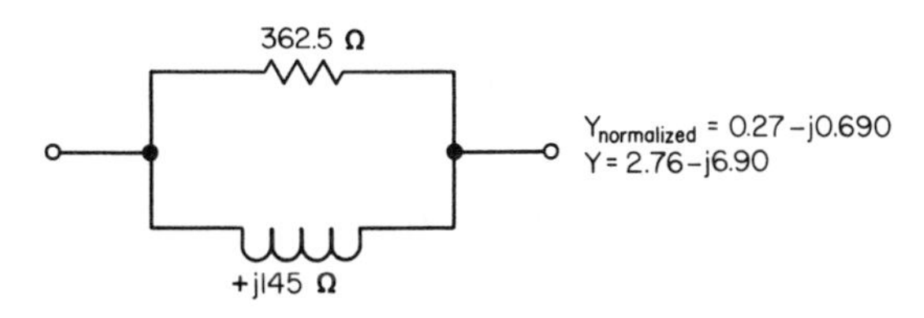

Figure 5-17

5-5 TWO-ELEMENT MATCHING ON THE CHART

The simplest use of the chart involves a two-element matching network. This is a design that could just as easily be handled mathematically but will illustrate the use of the chart. We will design a network to match a 30-Ω source resistance to an 80-Ω load resistance. For convenience, we will use a set of charts already scaled for 50 Ω (impedance chart) and 20 mmhos (admittance chart) at the center point. This will save us the trouble of having to normalize and unnormalize values. The network can have two forms, as shown in Figure 5-18. The first is a high-pass filter, the second is a low-pass filter.

The first step in using the chart is to locate the source and load points on each chart (Figure 5-19). The source resistance is marked at A (30 Ω) on the impedance chart and at B (33.3 mmhos) on the admittance chart. The load resistance is marked at C (80 Ω) on the impedance chart and at D (12.5 mmhos) on the admittance chart. Now all we have to do is draw a path from

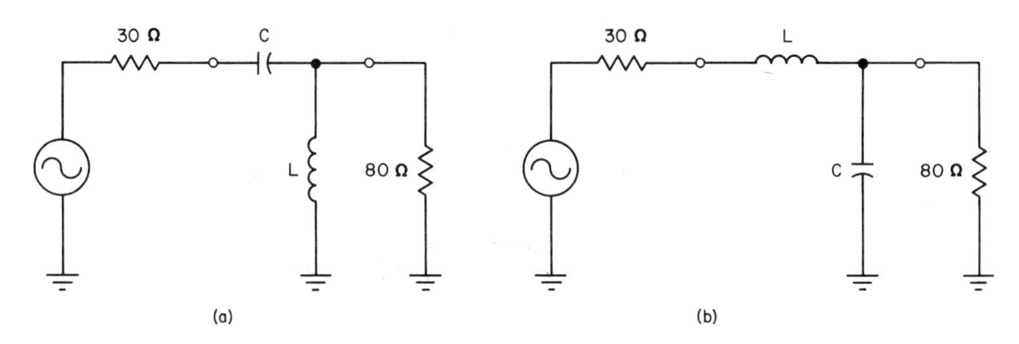

Figure 5-18 Two forms of a two-element matching network.

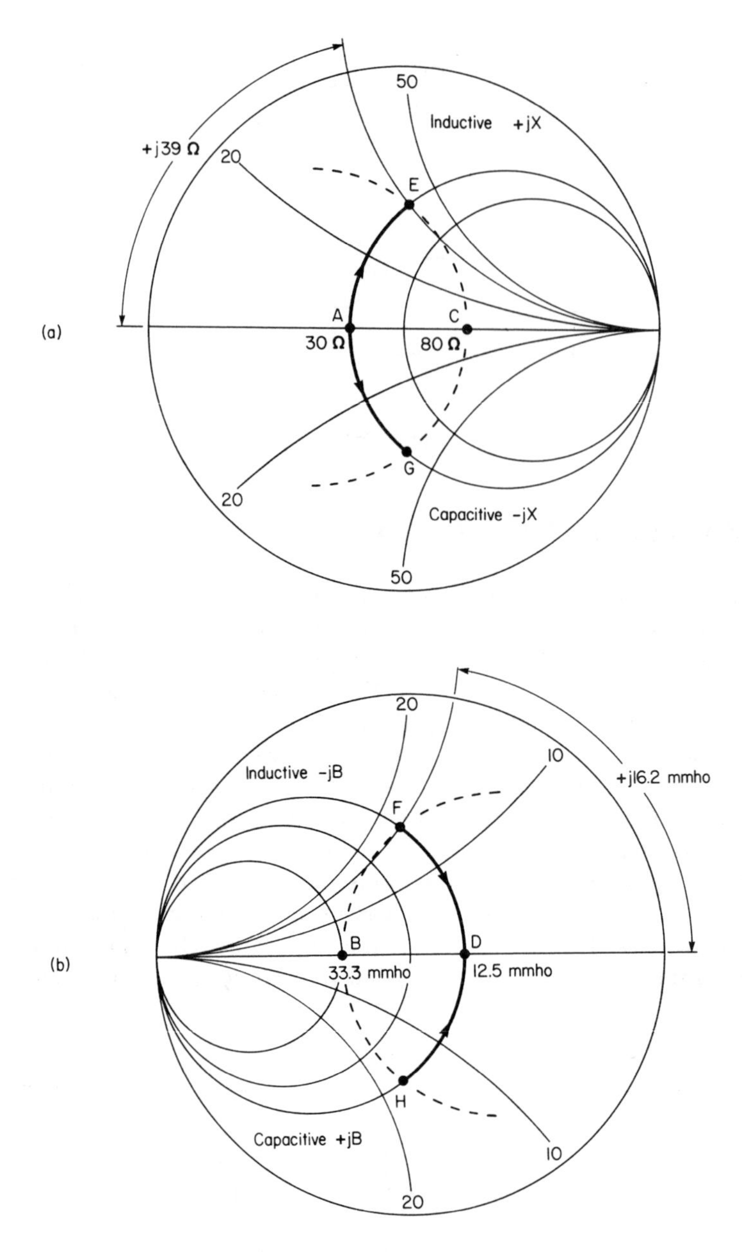

Figure 5-19 Fifty-ohm impedance chart (a) and 20-mmho admittance chart (b) used to design a two-element matching network.

the source to the load along lines of constant resistance and conductance, using the instructions of Table 5-1. The starting point can be at either end; for our example, we will start at the 30-Ω source end. Since the first component will be a series element, we will begin with the impedance chart. Draw a long curved line up from point *A* along the constant-resistance curve. At the moment, this line has no particular length; it just represents a series inductance. By using a shunt component we must move from some point on this line down to the 80-Ω load point, the starting point is marked as *E* but at the present we do not know where *E* is. Since the second component is a parallel component, it will be drawn on the admittance chart. Draw a long curved line down from the inductive susceptance side of the chart toward 12.5 mmhos. The starting point of the line is not known, but the end point is *D*. By overlaying the two charts, a point of intersection of the two curved lines should appear; this is marked *E* on the impedance chart and *F* on the admittance chart.

The length of the line *AE* is the reactance of the series inductor (39 Ω), and the length of *FD* is the susceptance of the shunt capacitor (16.2 mmhos or 61.5 Ω). The final circuit will have the form of Figure 5-18(b). The other network of (a) could be designed on the charts by moving in the other direction *A* to *G* and then *H* to *D*. The series components of both matching networks, (a) and (b), will have equal reactances, as will the two parallel components. This happens because both the source and load are purely resistive, and so the matching networks drawn on the Smith chart are symmetrical about the horizontal axis.

5-6 MATCHING COMPLEX LOADS

Our second design on the charts will match a resistive source (50 Ω) to a complex load (25 Ω in parallel with $-j12.5$ Ω). This is a problem that could also be handled mathematically but is usually easier with the charts.

The resistive source can be easily marked on the chart, but the complex point requires a little thought first. Maximum power will be transferred when a source is conjugately matched to a load. If a load is purely resistive, the source should appear to be resistive with the same value. If the load is complex, as in this case, the source should appear to have the same resistance but the opposite reactance. The two reactances will then cancel (resonance), leaving only the identical source and load resistances (Figure 5-20).

For the Smith chart design, we must consider which end of the network we are starting from. If we start at the load end, we mark the complex load point on the chart and work from there toward an input impedance of 50 Ω, purely resistive. If we start at the source end, we mark the 50-Ω point on the chart. From here we work toward an output impedance that would provide the conjugate match for the load; for our example, then, the output must be inductive. The design on the chart of Figure 5-21 is worked from the source to

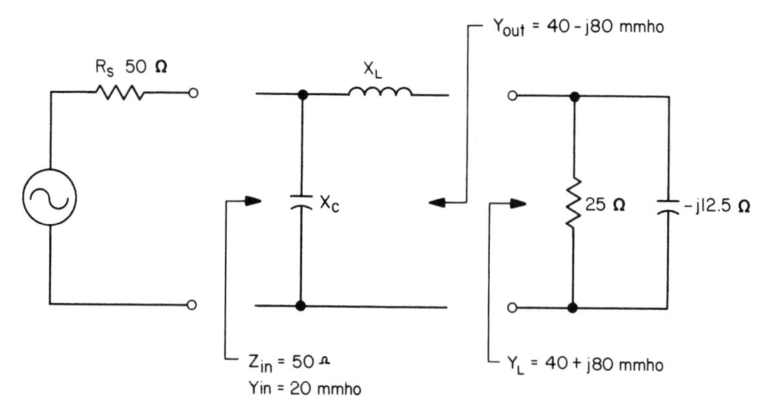

Figure 5-20 Two-element network used to match a resistive source to a complex load.

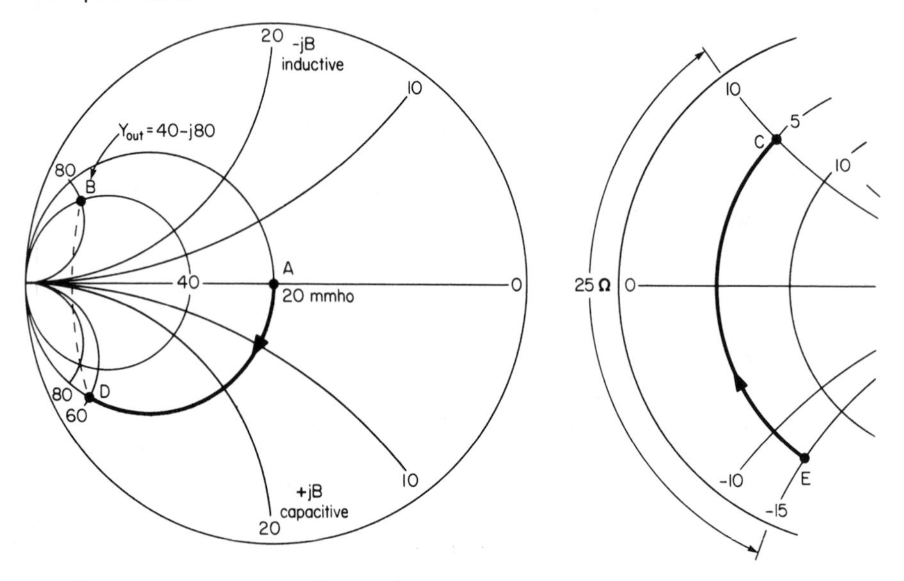

Figure 5-21 Design of the two-element matching network from a resistive source to a complex load.

the conjugate of the load. The 20-mmho admittance chart is shown along with a portion of the 50-Ω impedance chart. The source admittance is marked at 20 mmhos and the conjugate of the load admittance is marked at B:

$$\text{conjugate } Y_L = \frac{1}{25} - j\frac{1}{12.5} = 40 - j80 \text{ mmhos} \quad \text{(inductive)}$$

Starting at 20 mmhos (A), we draw a line down along the constant-conductance line. The length of this curve is unknown at the moment. On the impedance chart we can locate the conjugate of the load impedance (C) by transferring the point (B) from the admittance chart. We must approach this point using a series inductor, so we draw a line from the lower portion of the chart up to point C. The starting point of this curve is also unknown for the moment. By lining up the two charts, the intersection of the two curved lines appears at D on the admittance chart and E on the impedance chart. The admittance of the shunt capacitor is given by the length of AD.

$$Y_C = 0 \quad \text{to} \quad +j60 \text{ mmhos}$$

$$X_C = \frac{1}{+j60} = -j16.7\ \Omega$$

The reactance of the series inductor is found from the impedance chart and is the length of EC.

$$X_L = -j15 \quad \text{to} \quad +j10 = +j25\ \Omega$$

One point should be noted. If the high-pass form of the matching network was needed instead, the series reactance and the parallel reactance would not be numerically the same as the values just obtained. The reason is that the two designs would not be symmetrical on the chart, owing to the complex load.

5-7 THREE-ELEMENT MATCHING ON THE CHART

For a three-element matching design, we are free to select a value of loaded Q as long as it is above a minimum value. Lines of constant Q will form arcs of circles, as shown in Figure 5-22. Note that all arcs pass through the ends of the pure resistance axis on the chart. A third point on each arc can be found by locating any point on the chart with the required Q. Three such points are shown in Figure 5-22: $20+j40$ has $Q=2$; $50+j50$ has $Q=1$; and $6-j30$ has $Q=5$. We will use this arc of constant Q to design a three-element matching network for a 50-Ω source and a 150-Ω load. The minimum Q (if two elements were used) for these values is 1.414, so we will try a design for $Q=3.0$.

Start by drawing a $Q=3.0$ curve on the upper and lower halves of the chart; the point $20\pm j60$ and the two ends of the resistance axis were used for the chart in Figure 5-23. Begin with the higher-resistance end (150 Ω) since this is the point with the higher Q. On the admittance chart, draw a curve downward until it hits the constant-Q curve. This is the line AB, shown dashed in Figure 5-23, that represents a shunt capacitor.

$$Y_{AB} = +j20 \text{ mmhos}$$

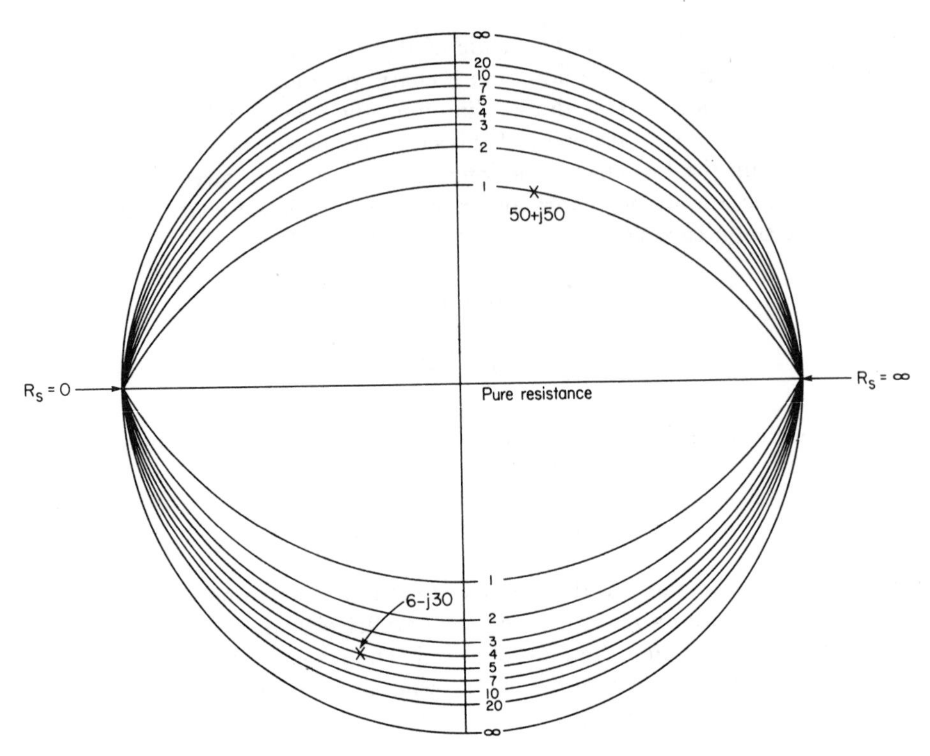

Figure 5-22 Lines of constant Q for the Smith chart.

Therefore,

$$X_c = \frac{1}{20} = -j50\ \Omega$$

The action now switches to the impedance chart, so we can add a series inductance. For this we draw a line from B following a curve of constant resistance (happens to be 15 Ω) and extending into the upper half of the chart; the location of the point D is not known yet. Back on the admittance chart, draw a line along a constant-conductance path extending upward from 20 mmhos (50 Ω). The location of D is now found as the point of intersection of the curve from B and the curve from C. The length of BD is $45 + 22.5 = +j67.5\ \Omega$ and is the reactance of the series inductor. The length of CD on the admittance chart is 33.3 mmhos, which represents a shunt capacitor of $-j30\ \Omega$. The final circuit is shown in Figure 5-24. A corresponding high-pass T form could also be designed by drawing the lines toward the opposite half of the chart. The lines would have the same length and the components the same value reactances (but opposite sign) since the 50-Ω and 150-Ω points were located exactly on the pure resistance line, the line of symmetry between the top and the bottom of the chart.

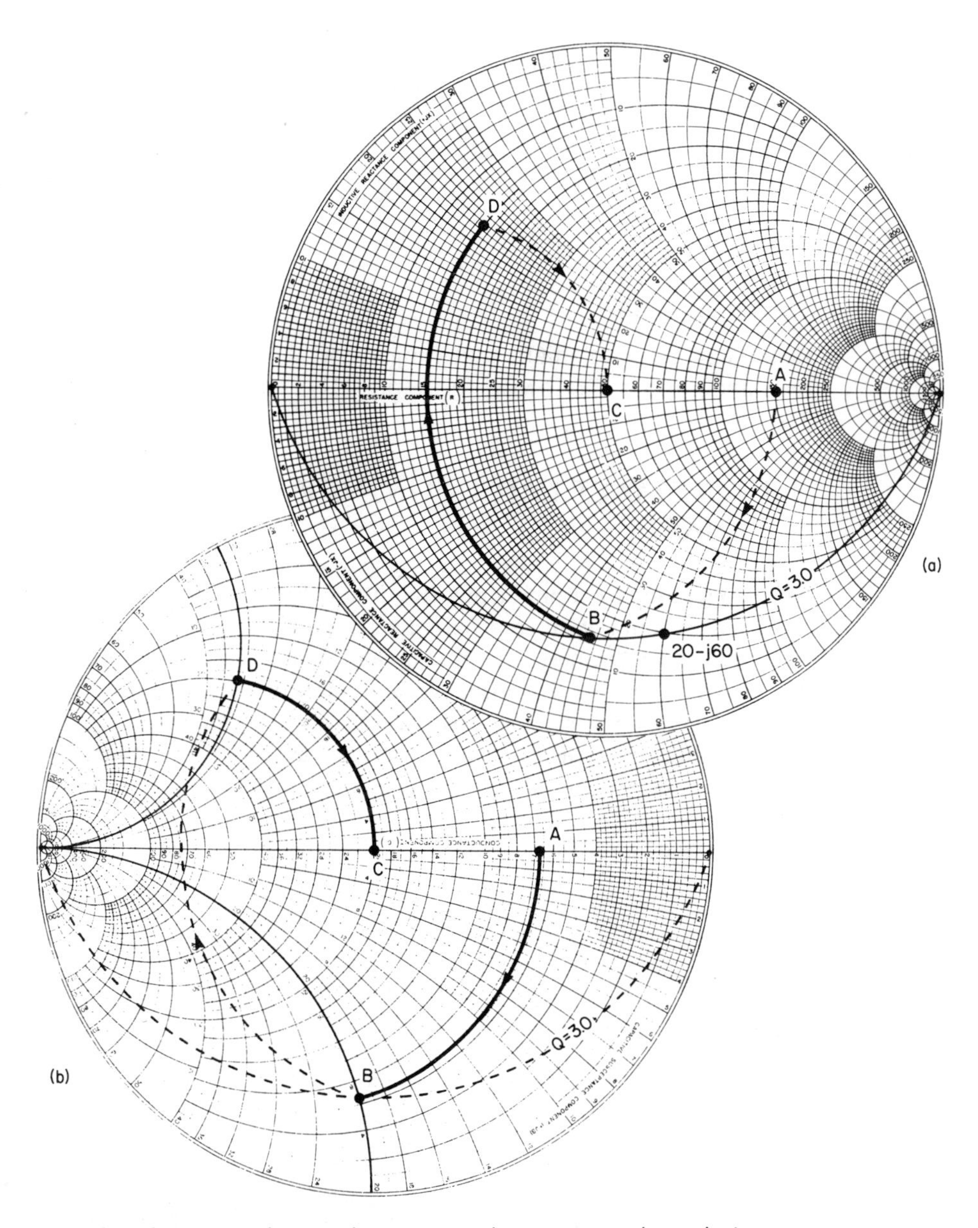

Figure 5-23 Fifty-ohm impedance chart (a), and twenty mmho admittance chart (b) used for the design of a three-element matching network to match a 50-Ω source to a 150-Ω load.

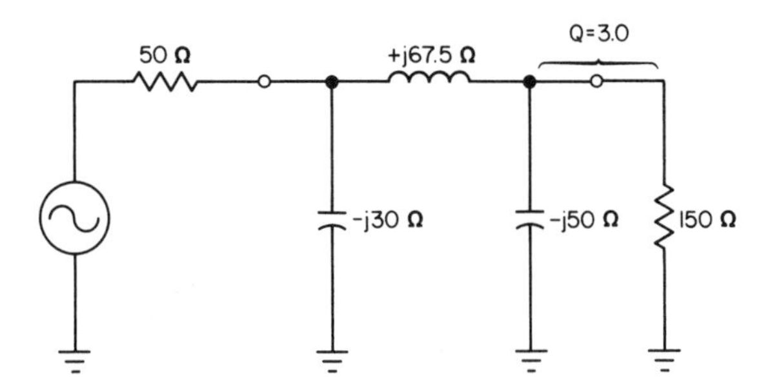

Figure 5-24 Three-element matching network designed for a 50-Ω source resistance and a 150-Ω load resistance. The maximum Q of the circuit is 3.0.

5-8 *TOLERANCE CALCULATIONS WITH THE CHART*

When designing for a production run, it may be necessary to find what range of impedances could occur in a particular circuit just from the normal variations of component tolerances. The Smith chart can be used in this situation to find the limits of the final impedance and indicate whether tighter component tolerances or even adjustable components might be required.

To illustrate this procedure, we will determine the limits of the input impedance of the circuit shown in Figure 5-25 if each component has a tolerance of $\pm 10\%$.

The median values necessary to make the input impedance look like 25 Ω will be

$$X_C = -j57.7\ \Omega$$

$$X_L = +j43.3\ \Omega$$

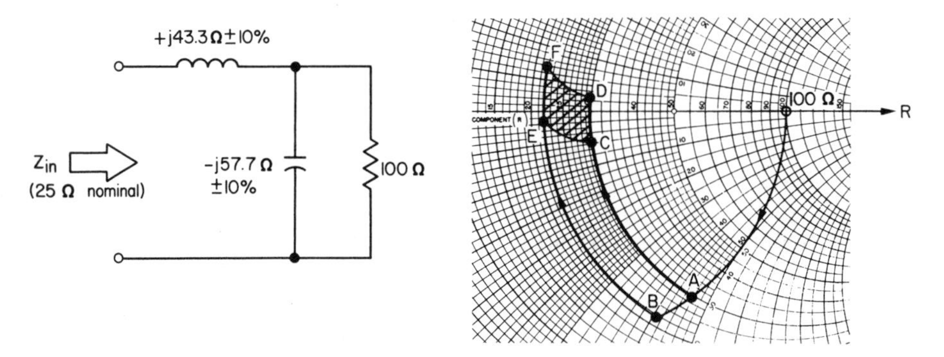

Figure 5-25 Input impedance variation for a $\pm$ 10% change in component value.

To find the changes that result from a $\pm 10\%$ variation in the capacitor and inductor values, we start first on an admittance chart (not shown). The load admittance of $1/100\ \Omega = 10$ mmhos would be located, and then a curve of constant conductance would be followed downward a distance proportional to the two extremes of the shunt capacitor's admittance. These two extremes would be

$$Y_{\text{min}} = \frac{1}{-j57.7\ \Omega + 10\%} = \frac{1}{-j63.47} = +j15.75 \text{ mmhos}$$

$$Y_{\text{max}} = \frac{1}{-j57.7\ \Omega - 10\%} = \frac{1}{-j51.93} = +j19.26 \text{ mmhos}$$

These would correspond to the two points marked A and B on the 50-Ω impedance chart of Figure 5-25.

The next step is made on the impedance chart. The two extremes of the reactance of the series inductor would be

$$X_{\text{min}} = +j43.3 - 10\% = +j38.97\ \Omega$$

$$X_{\text{max}} = +j43.3 + 10\% = +j47.63\ \Omega$$

Starting from each point, A and B, constant-resistance lines are followed back up toward the resistance axis. Each line will have two points marked, corresponding to the maximum and minimum series reactances. Four points, C, D, E, and F, have now been located that surround the nominal input impedance point of 25 Ω. These form the corners of the input impedance variations. The sides are all curves, and several more points may be necessary to draw these curves if extreme accuracy is required.

5-9 BROAD-BAND TRANSFORMERS

For very wide bandwidth matching, as required in cable TV amplifiers, a transformer must be used. In its simplest form, a transformer consists of two separate windings on some type of magnetic core (Figure 5-26). The core provides a path for the magnetic field that couples the two windings and also increases the winding inductance, which, along with the source and load resistances, determines the lower cutoff frequency. The impedance ratio will depend on the square of the turns ratio as long as the core does a good job of coupling the one winding to the other.

$$R_{\text{in}} = \left(\frac{n_1}{n_2}\right)^2 \times R_L \tag{5-10}$$

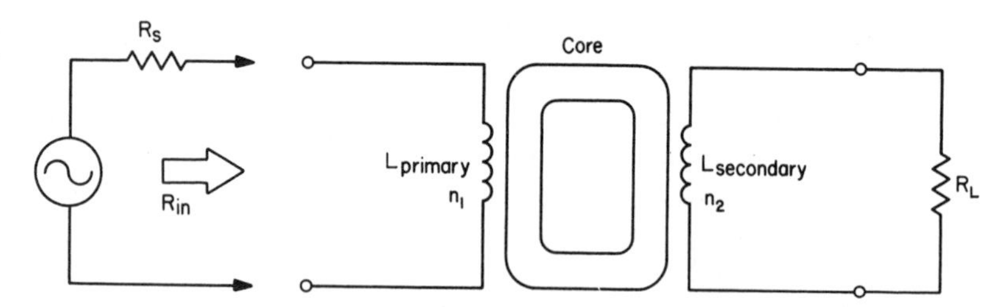

Figure 5-26 Basic transformer used for wide-band impedance matching.

In a practical transformer, a number of additional factors arise that affect the efficiency and frequency response of the device. First, not all of the magnetic field from the primary will connect with the secondary. The portion that does not couple forms a primary and secondary *leakage inductance.* The portion of the field that does couple forms the *mutual inductance.* If a proper core is not used for the frequency range desired, the permeability will fall off at the higher frequencies and the core losses will increase. As a result, less of the magnetic field will couple with the secondary. Finally, when the primary and secondary are wound, a capacitance will be created between adjacent turns of wire, causing further losses. The equivalent circuit of a practical transformer is shown in Figure 5-27.

The frequency response that would be obtained with such a practical transformer is shown in Figure 5-28. If it is assumed that the transformer is matching the source to the load so that R_{in} is the same as R_{source}, a typical situation, then the lower 3-dB frequency will be

$$f_1 = \frac{R_{source}}{4\pi L_p} \tag{5-11}$$

The primary inductance (L_p) is easily controlled by selecting the right core size and permeability (Section 2-2.6) and winding on the correct number

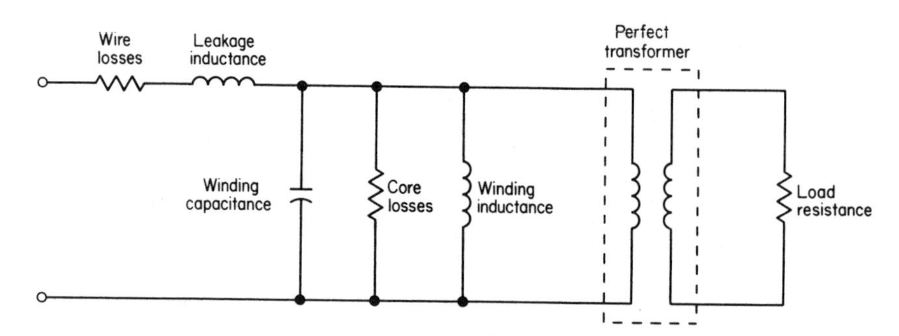

Figure 5-27 Equivalent circuit of a practical transformer, showing wire and core losses, leakage inductance, winding inductance, and winding capacitance.

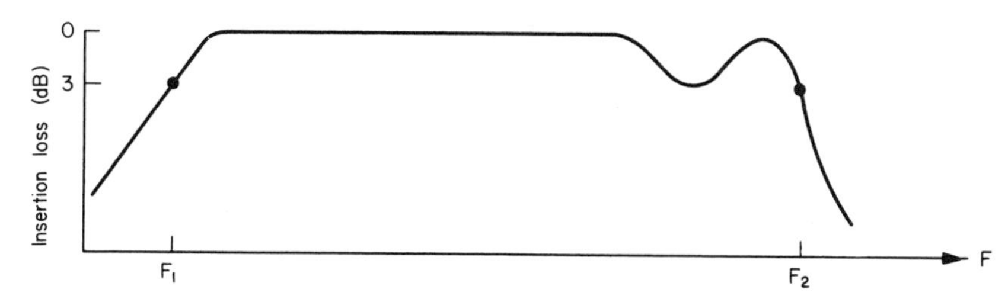

Figure 5-28 Frequency response of a practical wide-band transformer.

of turns of wire. The low-frequency response can therefore be easily controlled. The slope will be 6 dB/octave, since only one component is involved.

The high-frequency response is much more difficult to control. The important factors here are the leakage inductance and the winding capacitance, and neither of these are easily predicted. To further complicate matters, the permeability of the core will also be decreasing at the high-frequency end and will alter the leakage inductance. In any case, resonance is likely to occur between the leakage inductance and the winding capacitance, and so a peak is likely to appear near the upper cutoff frequency. The load and source resistance will control the Q of this resonance and so affect the size of the peak. The high-frequency cutoff slope will be 12 dB/octave, since two reactive components are involved.

5-10 TRANSMISSION-LINE TRANSFORMERS

Since the high-frequency response of a transformer is limited by the leakage inductance and the winding capacitance, they must be controlled if wide bandwidths are to be obtained. A major improvement can be made if each turn of the primary is kept very close to one of the secondary turns and the next turn of the primary is spaced farther along the core. The primary and secondary turns now form a transmission line and the leakage inductance and winding capacitance become part of that line's impedance and so improve the coupling rather than interfere with it.

The construction of a transmission-line transformer is shown in Figure 5-29. Two wires of equal length are wound on the core in such a way that they stay close together. Often this is accomplished by twisting the two together initially, but use of two wires in a common strip of insulation is also possible. After winding, the beginning of the one wire and the end of the other wire are soldered together (1 and 4 in the diagram). The result is a step-up transformer with a 2:1 turns ratio and a 4:1 impedance ratio. Additional windings may also be added to make larger ratios or to form balanced to unbalanced transformers, as shown in Figure 5-30.

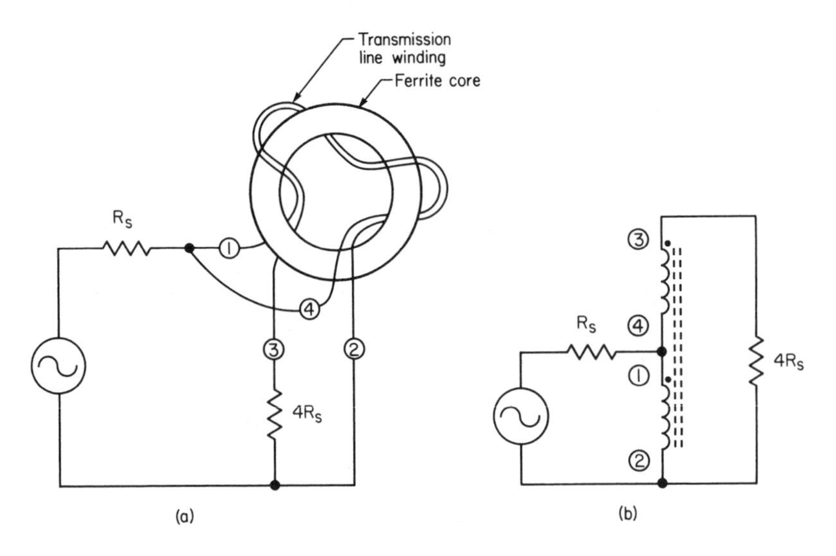

Figure 5-29 Wide-band transmission-line transformer wound on a toroid. The physical form of the windings are shown in (a) and the equivalent electrical diagram in (b).

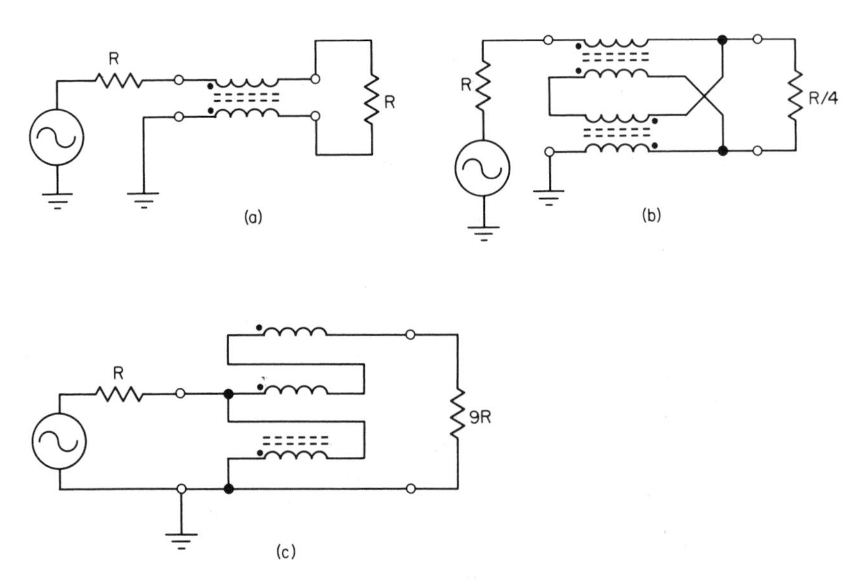

Figure 5-30 Additional toroidal transformers: (a) 1:1 unbalanced to balanced; (b) 4:1 unbalanced to balanced; (c) 1:9 unbalanced to unbalanced.

At low frequencies, the transformer behaves much the same as an ordinary transformer with a center-tapped winding, so the winding inductance and the core properties are important. At higher frequencies, the transmission-line properties are more important and the core less important. This is very convenient, as bandwidths can now be obtained that are much wider than the frequency range of the core material used.

To design a transmission-line transformer:

1 Select the longest length of winding that can be used, as this determines the high-frequency cutoff point. The response will drop 1.0 dB when the winding length becomes equal to $\lambda/4$. For dielectrics other than air, the line will appear electrically longer than it physically is. Therefore, it is best to use a shorter line length, so the length (after twisting) should be no longer than

$$l = \frac{3750}{f_2\,(\text{MHz})} \qquad (\text{cm}) \qquad (5\text{-}12)$$

where: f_2 = high-frequency 1.0-dB point

2 Find the winding inductance needed for the desired low-frequency point. The response of the transformer will drop 1.0 dB when the winding reactance is $= 2.86 \times R_{\text{source}}$. The inductance formula used for ferrite cores is

$$L = N^2 A_L \times 10^{-3} \qquad (\mu\text{H}) \qquad (5\text{-}13)$$

where: A_L = inductance index provided by the core manufacturer
N = number of turns

3 The characteristic impedance of the transmission line should be set as close as possible to the following value if widest bandwidth and flattest passband are required:

$$Z_{\text{line}} = \sqrt{Z_{\text{source}} \times Z_{\text{load}}} \qquad (5\text{-}14)$$

To maintain this impedance uniformally along the length of the winding, the two wires may have to be twisted together; this will reduce the impedance somewhat.

Two enamel-coated No. 22 wires lightly twisted will have an approximate impedance of 55 Ω. The use of No. 20 wire will reduce this to 45 Ω and No. 18 will have a lower impedance, of about 35 Ω. For still lower values, a number of wires can be twisted together to form a larger-diameter, flexible conductor, and then two such conductors can be twisted to form the transmission line. Flat strips of thin flexible copper can also be used if two strips are insulated from each other.

A big advantage of the transmission-line type of transformer is that the core handles the power transfer from the primary to the secondary only at the very low frequency end of the total passband. For the higher frequencies, it is the transmission line that links more and more of the primary flux to the secondary, until, at the very high end of the passband, the transformer plays no part. The relative importance of the two transfer mechanisms is shown in Figure 5-31.

The benefits of the transformer at the higher frequencies are:

1 Overall losses are very low, since ferrite losses have a reduced effect.
2 Less dependency on the temperature and magnetic nonlinearities of the core.
3 Very high powers can be handled with a small transformer, since magnetic saturation is less likely to occur.

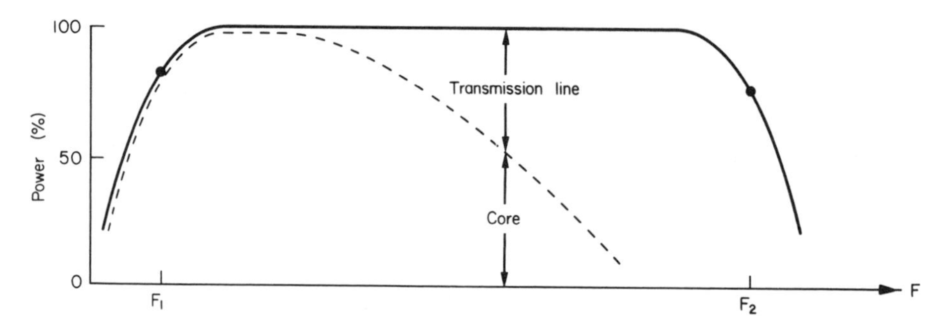

Figure 5-31 Power transferred from the primary to the secondary of the transmission-line transformer is handled partly by the magnetic core and partly by the transmission line.

5-11 TRANSMISSION-LINE MATCHING

For transmitter circuitry operating above 100 MHz, matching circuits can be made by constructing "microstrip-line" networks on a printed circuit board. As shown in Figure 5-32, this is a type of transmission line consisting of one wide conductor on one side of the board and a ground plane made by leaving all the copper on the other side of the board. The characteristic impedance of the line will depend on the width of the upper conductor and the thickness and dielectric constant of the circuit board. Since the copper forms a transmission line, sharp corners in the path must be avoided or reflections will be set up on the line. Corners should either have a large radius to them or, as shown, be mitered.

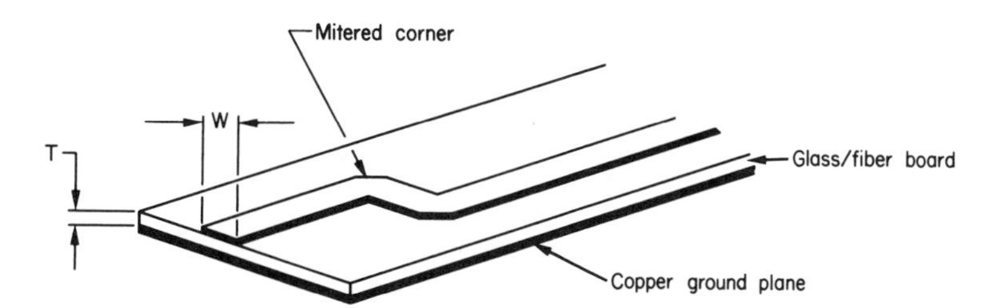

Figure 5-32 Microstrip transmission line formed on a double-sided printed circuit board. When corners are necessary, they should be mitered as shown.

Part of the electromagnetic wave travels in the air above the line and part travels in the board material itself. Typical boards will have a dielectric constant between 2 and 10 times that of the air, so the two parts of the wave will tend to travel at different speeds along the line. Calculations of exact impedance and phase velocity are therefore difficult. If the wave stayed completely within the board material, the characteristic impedance would be

$$Z_0 = 377\frac{T}{W\sqrt{\varepsilon_r}} \tag{5-15}$$

where: T = thickness of the board's dielectric
W = conductor width
ε_r = relative dielectric constant of the board

Glass-epoxy circuit boards have a relative dielectric constant of 4.8 and can be used to about 300 MHz before losses become serious. Teflon-loaded glass boards have lower losses and can be used to several GHz. Its relative dielectric constant is 2.5.

The characteristic impedance can be calculated with greater accuracy (better than 5%) with the following equation:

$$Z_0 = 377\frac{T}{W\sqrt{\varepsilon_r}}\,\frac{1}{\left[1+1.735\varepsilon_r^{-0.0724}(W/T)^{-0.836}\right]} \tag{5-16}$$

The wavelength, if the wave stayed in the board, would be

$$\lambda_{strip} = \frac{\lambda_0}{\sqrt{\varepsilon_r}} \tag{5-17}$$

where: λ_0 = free-space wavelength

A more realistic value would be

$$\lambda_{\text{strip}} = \frac{\lambda_0}{\sqrt{\varepsilon_r}} \sqrt{\frac{\varepsilon_r}{1 + 0.63(\varepsilon_r - 1)(W/T)^{0.123}}} \tag{5-18}$$

When a transmission line is used for matching, it is possible to move to any point on a Smith chart that lies on a circle passing through the load impedance and having a center equal to the characteristic impedance of the line. This is shown in Figure 5-33, and it should be noted that the circle must be centered at the midpoint of the chart. This is different from the constant-resistance and -conductance circles used with the previous Smith chart matching networks. One of the first problems, then, is to find what characteristic line impedance is needed to match two impedances. If the two impedances are pure resistances, then

$$Z_0 = \sqrt{R_1 R_2} \tag{5-19}$$

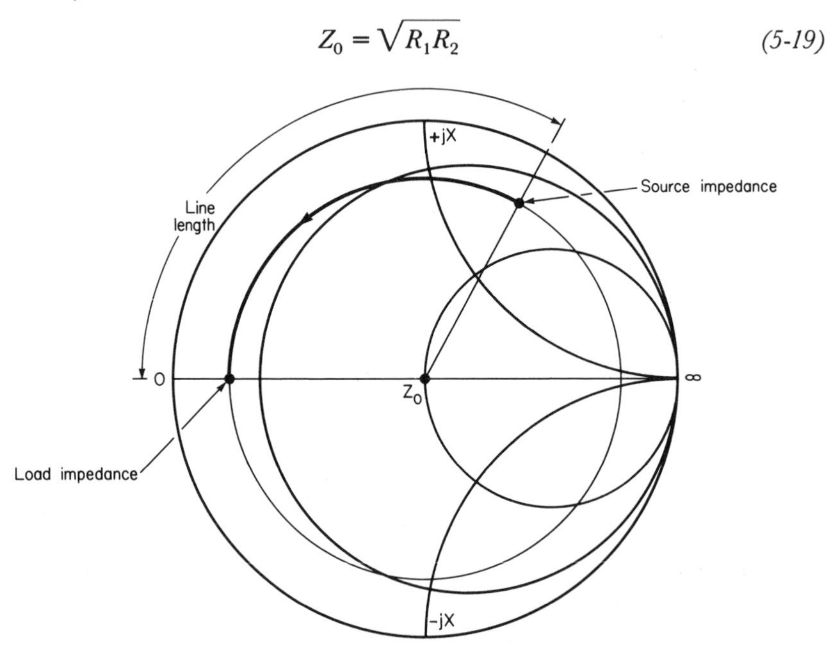

Figure 5-33 Transmission-line matching will move along circles of constant VSWR centered at the midpoint of the chart. This point must also be the characteristic impedance of the line used.

If one of the points to be matched is complex, the calculation is a little more involved. For one real impedance of R_1 and a complex impedance of $R_2 + jX$, the characteristic line impedance will be

$$Z_0 = \sqrt{R_1 R_2} \times \sqrt{1 - \frac{X^2}{R_2(R_1 - R_2)}} \tag{5-20}$$

The length of the circle on the chart will represent the fraction of one wavelength needed for the line. One complete revolution around the chart represents a half-wavelength line and would result in a return to the impedance we started with. Most matching problems will then take less than a half-wavelength. The direction of movement around the chart depends on which end of the line the source and load are placed. If we move from the load impedance toward the source, the circle is drawn clockwise on the chart. Movement from the source toward the load is drawn counterclockwise.

Because the required line impedance is rarely a standard value, the normalized chart is the easiest chart to use. An example follows.

EXAMPLE 5-4

Find the characteristic impedance and length of a microstrip transmission line needed to match a 50-Ω source to the complex input impedance of a transistor $Z_{in} = 20 - j10$ (Figure 5-34).

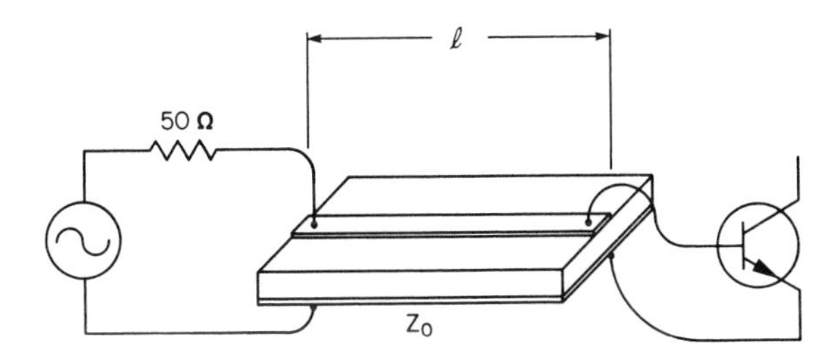

Figure 5-34

Solution

The first step is to find the characteristic impedance that would be required:

$$R_1 = 50\ \Omega \qquad X = -10\ \Omega$$

$$R_2 = 20\ \Omega$$

$$Z_0 = \sqrt{R_1 R_2} \cdot \sqrt{1 - \frac{X^2}{R_2(R_1 - R_2)}}$$

$$= \sqrt{50 \times 20} \cdot \sqrt{1 - \frac{(-10)^2}{20(50-20)}}$$

$$= 28.87\ \Omega$$

Before plotting on the normalized chart, we have to first normalize all the impedance values using the characteristic line impedance:

$$R_n = \frac{50}{28.87} = 1.732, \qquad Z_n = \frac{20 - j10}{28.87} = 0.693 - j0.346$$

The Smith chart is shown in Figure 5-35 with the line impedance at the center and the source and load impedances equidistant from the center. A circle can then be drawn clockwise from the load to the source. The total angle involved is 241°, so the length of the microstrip line will be

$$l = \frac{241°}{360°}\frac{\lambda_{\text{strip}}}{2} = 0.335\,\lambda_{\text{strip}}$$

If 1.6-mm glass fiberboard ($\varepsilon_r = 4.8$) were used for this microstrip line, the width of the line would be 6.4 mm and the length would be $0.51\lambda_0$. The actual physical length would depend on the frequency:

$$\lambda_0(\text{cm}) = \frac{30{,}000}{f(\text{MHz})} \tag{5-21}$$

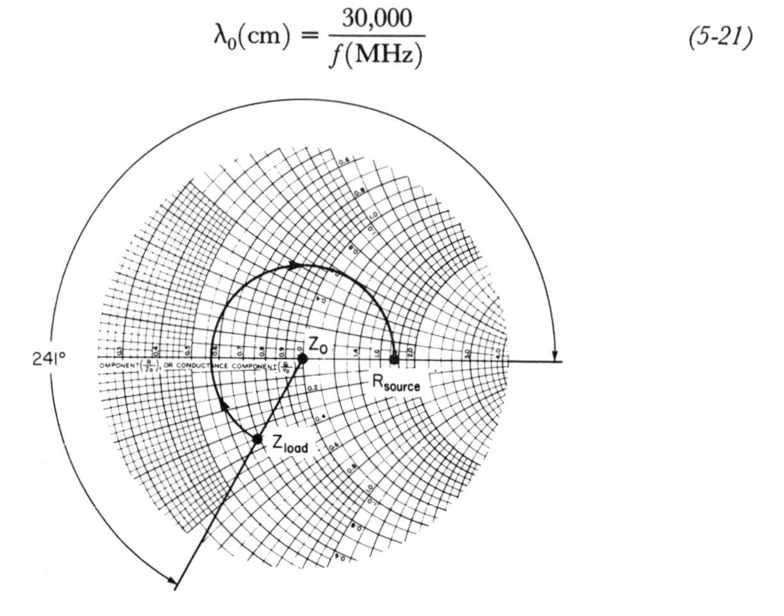

Figure 5-35 Portion of a normalized Smith chart used to design a transmission-line matching network. The characteristic impedance used for normalizing is 28.87 Ω.

QUESTIONS

1 Design a two-element high-pass matching network for a 25 Ω-source and an 85-Ω load at a center frequency of 3.0 MHz. What will the approximate 3- dB bandwidth be?

2 Design a four-element low-pass matching network for a 35-Ω source and a 75-Ω load. The center frequency is 5.0 MHz.

3 Design a three-element high-pass matching network to work between a 15-Ω source and a 75-Ω load at 7.3 MHz. The highest Q in the network will be 3.0.

4 Design a matching network similar to that of Figure 5-12(c) to match 15 Ω to 200 Ω at 1.65 MHz with a 3-dB bandwidth of 100 kHz. (Note that the Q to be used with the formulas given in Figure 5-12(c) must be twice the Q calculated to give the desired bandwidth.)

5 Using a Smith chart, match 10 Ω to 50 Ω at 75.0 MHz using a low-pass filter form.

6 Using a Smith chart, design a three-element matching network at 15.0 MHz. The load resistance is 50 Ω and the complex source impedance consists of 10 Ω in parallel with $+j6.0$ Ω. The highest Q in the network should be 2.0.

7 Design a microstrip line for a 50-Ω source and a load impedance of $12+j5.0$ Ω. The operating frequency is 200 MHz and the circuit board to be used is Teflon loaded with a thickness of 2.0 mm.

REFERENCES

Kraus and Allen. Aug. 16, 1973 Designing toroidal transformers. *Electronics.*

Motorola Semiconductor Products Inc. 1975. *13-Watt Microstrip Amplifier.* Motorola Application Note AN-728, Phoenix, Arizona.

RCA Solid-State Division. 1971. *RCA Designer's Handbook—S.P-52 Solid State Power Circuits.* Somerville, N.J.

chapter 6

transistors at high frequency

As the operating frequency climbs, transistor characteristics change dramatically and circuit design techniques must be altered accordingly. Above 1 MHz, depending on the transistor, the input and output impedances will decrease and become increasingly more reactive. The voltage, current, and power gains will decrease and there will be a greater tendency for signals at the output to feedback to the input through internal capacitance. This leads to loss of gain, input impedance changes, and a strong tendency to oscillate.

Transistors are often required to amplify very small high-frequency signals, and any internal noise will interfere with this task. Proper transistor selection and biasing will minimize the problem.

Although bipolar and field-effect transistors are very different in construction and in low-frequency performance, their differences tend to decrease as the frequency increases. In this chapter, we will examine the construction of both types of transistors and develop an equivalent circuit for them.

6-1 BIPOLAR TRANSISTOR CONSTRUCTION

All modern transistors are constructed using some variation of the diffusion process. The three different layers are then developed vertically in the silicon (or germanium or gallium arsenide) material, as shown in Figure 6-1. The active portion of the transistor lies immediately beneath the emitter region,

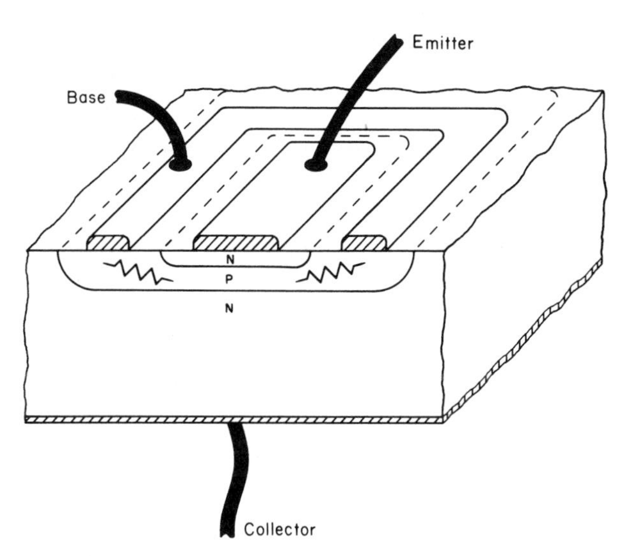

Figure 6-1 Cross section of a diffused *NPN* transistor.

where the *p*-type base region is the thinnest. The surrounding area serves only to bring the base region up to the surface to make contact with the metallization. The collector contact is made to the bottom of the silicon chip to provide high thermal conductivity, for proper heat flow.

The cross section in Figure 6-1 shows small resistances from the base terminal down to the active base region. These are unavoidably due to the resistance of the semiconductor material between the metal contact and the active base. When the equivalent circuit is described, the active base region will be given the letter (b') while the actual base lead will be called (b). The resistance that separates the two points is called the *base spreading resistance* and will be given the symbol ($r_{bb'}$).

6-2 EQUIVALENT CIRCUIT

Many different equivalent circuits can be used to characterize a bipolar transistor, but the hybrid Π is the best description of what actually exists inside the transistor, and its values are reasonably independent of frequency. It is, however, a little difficult to use for everyday design work, so later we will show the relation of the hybrid Π to the more useful h parameters.

The equivalent circuit is shown in Figure 6-2. The values are small-signal values that will change if the bias current or voltages change but are otherwise fixed values. The circuit can therefore be used to calculate gains and impedances over a wide range of frequencies.

Each of the components of the equivalent circuit is now described and an indication of typical values and how they change with bias voltage and current

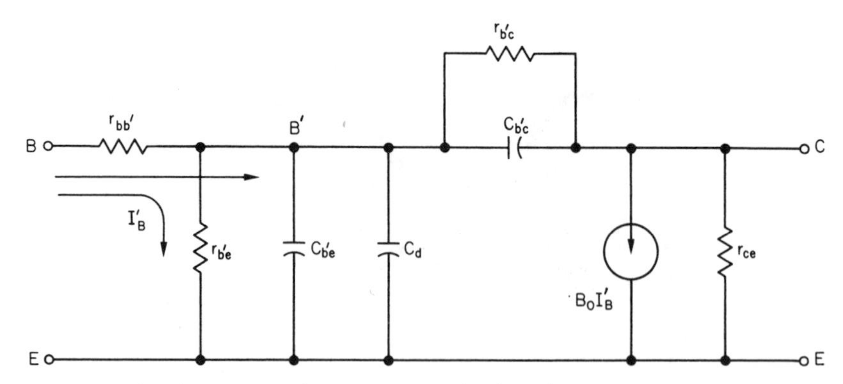

Figure 6-2 Hybrid-Π equivalent circuit of a bipolar transistor.

is given:

1 $r_{bb'}$: *base spreading resistance.* Its origin has already been given. Its value must be kept as low as possible if high gain and low noise are to be obtained at high frequencies. For small-signal transistors, it will be approximately 5–100 Ω. For larger-signal transistors having larger chip sizes, $r_{bb'}$ will be even lower. The value will increase as collector voltage increases. Collector current does not alter the value.

2 $r_{b'e}$: *equivalent ac resistance of the forward-biased base-emitter junction.* Its value decreases as the emitter bias current increases, and increases with the ac current gain (β_0). Typical values would be 500–2000 Ω when the transistor is biased at 1.0-mA emitter current.

3 $C_{b'e}$ *and* C_d: *emitter-base junction capacitance* and *emitter diffusion capacitance*, respectively. The diffusion capacitance is used as an analog for the slow movement of charges in the base region, and its value is proportional to emitter-bias current. The two capacitances are usually lumped together under the name *emitter capacitance* (C_e) and would typically have a total value of 100 pF for a small-signal transistor biased at 1.0-mA emitter current.

4 $r_{b'c}$: *feedback resistance from collector to base.* Its typical value is 2–5 MΩ and is considered to be insignificant for most radio-frequency work.

5 $C_{b'c}$ (or C_c): *collector-to-base feedback component*, in this case, a small capacitance. It results from the reverse-biased collector-base junction and its value decreases as the collector voltage increases. Typical values might be 0.5–5 pF for a small-signal transistor.

6 r_{ce}: *output resistance of the transistor.* Decreases with increase in collector current. Typical values could be 10–100 kΩ.

The hybrid-Π equivalent circuit just presented may look rather cumbersome but it does accurately describe the transistor's operation, and the values are relatively independent of frequency. Fortunately, for most applications, even at low RF frequencies, many simplifications can be made:

1 The two emitter capacitances have already been reduced to one value.
2 The collector feedback capacitance will have a lower reactance than $r_{b'c}$, so the resistor can be ignored.
3 Most load resistances are considerably less than the output resistance r_{ce} and it, too, can be ignored.

The simplified hybrid-Π circuit that will be used from now on is shown in Figure 6-3.

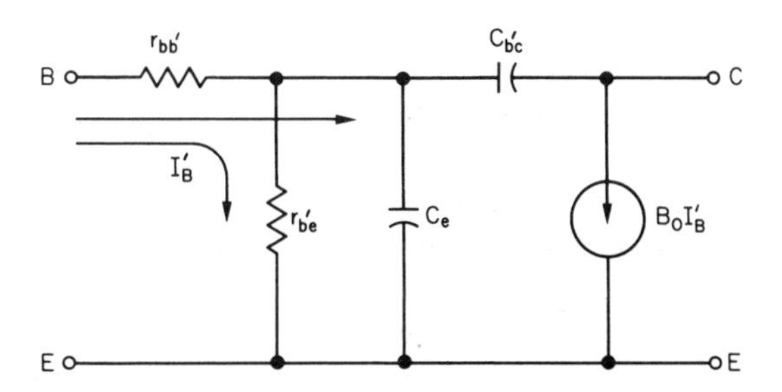

Figure 6-3 Simplified hybrid-Π equivalent circuit useful for frequencies above 100 kHz.

6-3 SHORT-CIRCUIT CURRENT GAIN

At low frequencies the base current all flows through $r_{bb'}$ and $r_{b'e}$. The collector current is simply β_0 times the base current. As frequency increases, the capacitances have an increased shunting effect, and less current flows through $r_{b'e}$. It is only this current that is amplified by β_0.

Both the emitter capacitance (C_e) and the collector capacitance (C_c) shunt some of this base current; the amount that flows through C_c depends on its Miller equivalent, and this in turn depends on the collector load resistance. To simplify matters, the current gain with the output short-circuited will be discussed first. The equivalent circuit for this purpose is shown in Figure 6-4. Since $r_{bb'}$ is a series component, it has no effect on the current gain of the transistor. It does, however, have a significant effect on the voltage gain and therefore also on the power gain, as we will see later.

At a certain high frequency (usually several MHz), the reactance of the two capacitances will have become low enough that equal amounts of the input current will flow through $r_{b'e}$ and through the capacitances (70.7% through each at different phase angles). The output current will then be down to 0.707 (-3 dB) of its low-frequency value. This frequency is called the β cutoff frequency (f_B). Above this frequency the current gain continues to drop at a

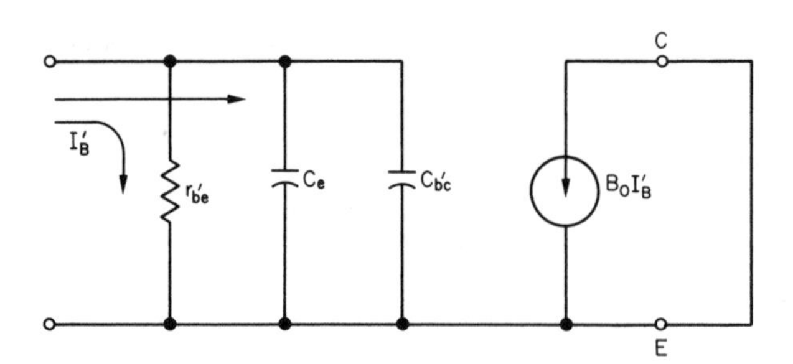

Figure 6-4 Equivalent circuit for high-frequency current gain, with the output short-circuited.

constant rate of −6dB/octave (Figure 6-5). At a very high frequency (usually several hundred MHz) the current gain will be reduced to 1.0, this being called the *short-circuit current-gain bandwidth product*. For any frequency between f_β and f_T, the product of frequency and current gain is a constant.

The actual value of f_T will change with the different transistor types and will often have values beyond 1000 MHz. For any particular transistor, a large variation will be noted, depending on the collector current and voltage used for biasing. The typical results for one transistor are shown in Figure 6-6.

$$f_T = \beta_0 \times f_\beta \tag{6-1}$$

$$f_\beta = \frac{1}{2\pi r_{b'e}(C_e + C_{b'c})} \tag{6-2}$$

$$f_T = \frac{\beta_0}{2\pi r_{b'e}(C_e + C_{b'c})} \tag{6-3}$$

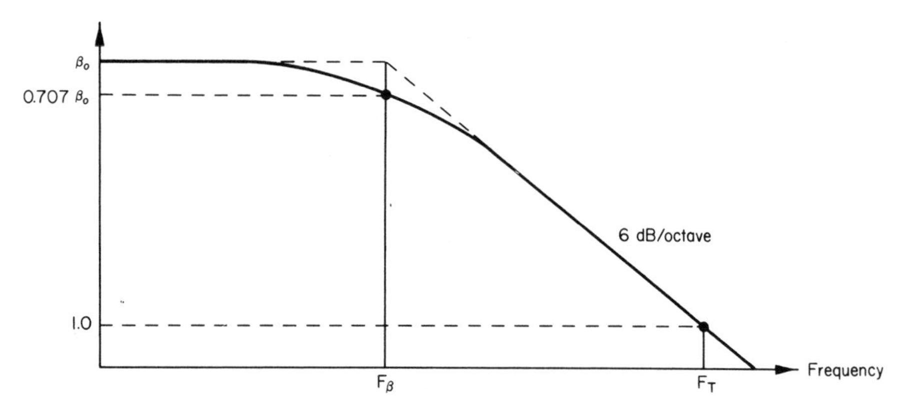

Figure 6-5 Short-circuit current gain versus frequency.

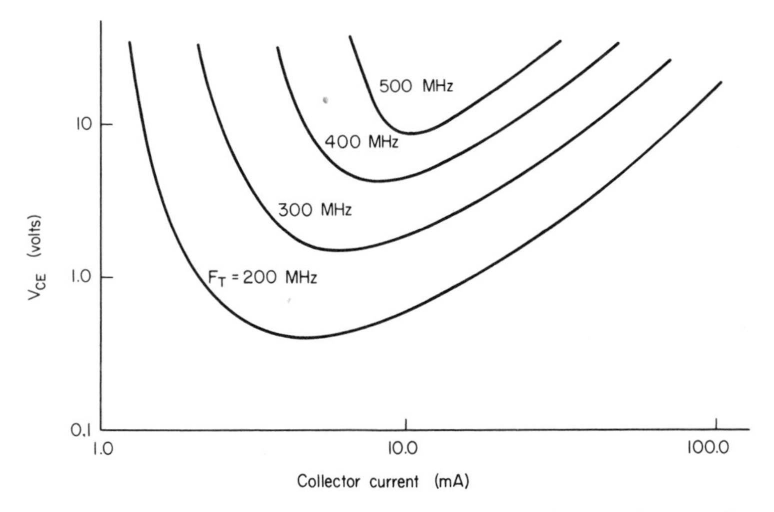

Figure 6-6 Typical variation of gain-bandwidth product with I_c and V_{CE}.

As the frequency increases, the phase angle between the base and collector currents will increase, a result of the extra current flowing through C_e. At the β cutoff frequency, the angle will be 45° and, as f_T is approached, the angle will approach 90°.

For circuits operating at frequencies above f_β, then, it is important to select the correct bias point if maximum gain is to be obtained. For most RF work, this results in bias currents higher than those normally used in small-signal audio work.

6-4 VOLTAGE GAIN

When a load resistor is included, the transistor is now capable of some voltage gain (A_v); the appropriate equivalent circuit is shown in Figure 6-7. The capacitance $C_c(1+A_v')$ is the equivalent Miller capacitance that replaces the collector-to-base capacitance.

We have to watch very carefully what voltage gain we are talking about. Since the collector capacitance was connected[1] from the collector back to the internal base (b'), the voltage gain we need is the gain from (b') to collector.

[1]For the "overlay" type of transistor, the collector capacitance, is made up of two parts, one from the collector to the internal base (b') and the other from collector to the external base terminal (b). A careful analysis is more difficult in this case.

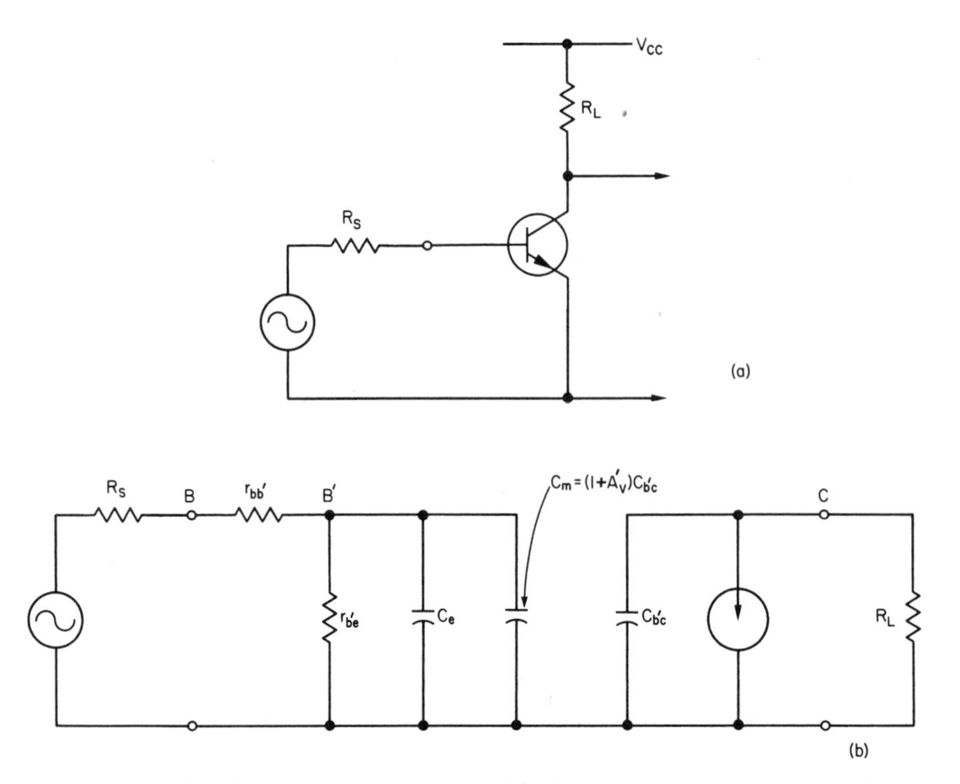

Figure 6-7 (a) Simple transistor amplifier (b) its high-frequency equivalent circuit.

Fortunately, this gain is fairly independent of frequency and is given by

$$A_v' = \frac{V_{ce}}{V_{b'e}} = \frac{\beta_0 R_L}{r_{b'e}} \tag{6-4}$$

The voltage gain of the transistor that the designer is more interested in is the gain from the external base terminal to the external collector terminal:

$$A_v = \frac{V_{ce}}{V_{be}} \tag{6-5}$$

This gain will drop off at the higher frequencies; the difference is caused by the increasingly larger voltage drop across that "nasty" resistor $r_{bb'}$.

Before we actually calculate the terminal voltage gain, one more detail should be pointed out. We have already examined the current gain, but that was under short-circuit conditions at the output. Now, with a load resistor there instead, a much larger capacitance has appeared across the internal base junction. The current gain of the *amplifier* will therefore be lower than the calculated short-circuit current gain of the *transistor*. This helps to point out the unfortunate fact that, because of the internal feedback capacitance $C_{b'c}$, many of the parameters, such as Z_{in} and A_i, are dependent on what happens at the output, in other words, the load impedance.

Let us return to our voltage gain. The actual calculation will be a little easier to follow if we use some actual component values instead of developing theoretical equations.

EXAMPLE 6-1

For a transistor having the following parameters, calculate the current and voltage gains at a frequency of 7.5 MHz.

$$r_{bb'} = 50\ \Omega \qquad C_e = 150\ \text{pF}$$
$$r_{b'e} = 1000\ \Omega \qquad C_{b'c} = 5\ \text{pF}$$
$$R_L = 2000\ \Omega \qquad \beta_0 = 100$$

The complete equivalent circuit for the amplifier is shown in Figure 6-8.

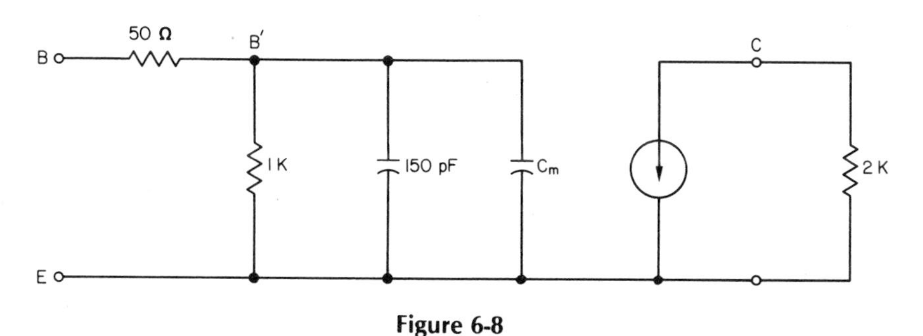

Figure 6-8

Solution

The internal voltage gain is

$$A'_v = \frac{\beta_0 R_L}{r_{b'e}} = \frac{100 \times 2000}{1000} = 200$$

The value of the Miller capacitance can then be found:

$$\begin{aligned} C_m &= C_{b'c}(1 + A'_v) \\ &= 5(1 + 200) \\ &= 1005\ \text{pF} \end{aligned}$$

The total capacitance across the emitter is

$$C_T = C_e + C_m$$
$$= 150 + 1005$$
$$= 1155 \text{ pF}$$

The reactance of this capacitor at 7.5 MHz is

$$X_C = \frac{1}{2\pi f C}$$
$$= \frac{1}{2\pi \times 7.5 \times 10^6 \times 1155 \times 10^{-12}}$$
$$= -j18.37\ \Omega$$

The current gain at this frequency will depend on the ratio of the capacitor $(C_e + C_m)$ current to the resistor $(r_{b'e})$ current. Only the resistor current will be amplified by β_0.

$$A_i = \beta_0 \times \frac{I_{\text{resistor}}}{I_{\text{total}}}$$
$$= \beta_0 \times \frac{(-jX_c)}{r_{b'e} - jX_c}$$
$$= 100 \times \frac{(j18.37)}{1000 - j18.37}$$
$$= 1.84 \angle -89^\circ$$

The internal voltage gain has already been found. The external voltage gain can be found by taking the loss across $r_{bb'}$ into account. The input attenuation circuit will look like Figure 6-9 if we ignore the 1000 Ω resistor. The attenuation of this circuit will be

$$\frac{Vb'e}{Vbe} = \frac{-j18.37}{50 - j18.37}$$
$$= 0.345 \angle -69.8^\circ$$

The overall voltage gain of the amplifier will therefore be

$$A_v = \text{input attenuation} \times \text{internal gain}$$
$$= 0.345 \angle -69.8^\circ \times 200$$
$$= 69 \angle -69.8^\circ \text{ (36.8 dB)}$$

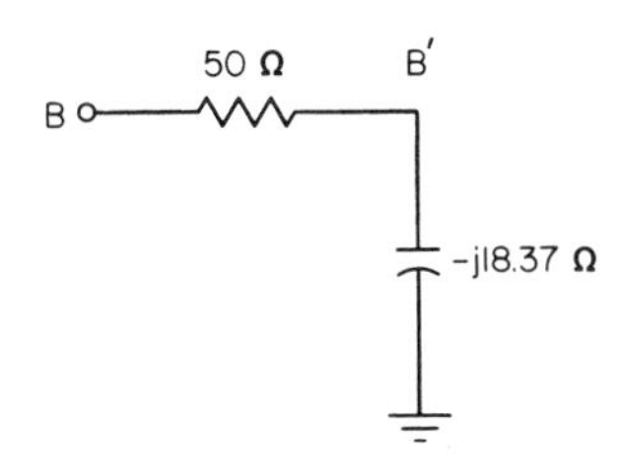

Figure 6-9

From this example, we see that it was the current gain that suffered the most. The voltage gain is reduced by a much smaller amount. The two most serious components in the equivalent circuit are the collector capacitance ($C_{b'c}$) and the base spreading resistance ($r_{bb'}$). Any manufacturing steps that reduce either of these will result in a transistor capable of operating at higher frequencies. The product of the two components is often used as a figure of merit for different small-signal transistors and is occasionally included on data sheets. For the values used in Example 6-1, the *base collector time constant* would be 50 $\Omega \times 5$ pF or 250 ps (a rather poor transistor—a good "microwave" transistor might have a time constant as low as 2.0 ps).

6-5 INPUT IMPEDANCE VERSUS FREQUENCY

At low frequencies, the input impedance of the transistor is purely resistive. At higher frequencies the impedance starts to decrease and become increasingly more reactive. This is caused by the shunting capacitance across the base-emitter junction.

Figure 6-10 shows a typical equivalent input circuit of a small-signal transistor. At low frequencies, the input resistance would be 1100 Ω. At very high frequencies, the capacitance reactance becomes extremely low, and the input resistance approaches 100 Ω. The impedance variation at intermediate frequencies will be a complex value that varies as shown. At very high frequencies, the inductance of the base and emitter lead wires may be significant and so would be included as shown.

Notice that only the emitter capacitance (C_e) was shown in Figure 6-10. This diagram would then have to represent the input impedance with the output short-circuited; otherwise, the Miller capacitance would need to be included. Since the load resistance value is up to the designer, the manufacturer could not possibly publish input impedances for all possible conditions. Short-circuit parameters are therefore published, and the designer must keep in mind that his/her in-circuit impedances will be different (usually lower), depending on the load impedance used. Data sheets will often present this short-circuit impedance in terms of the two rectangular components, each

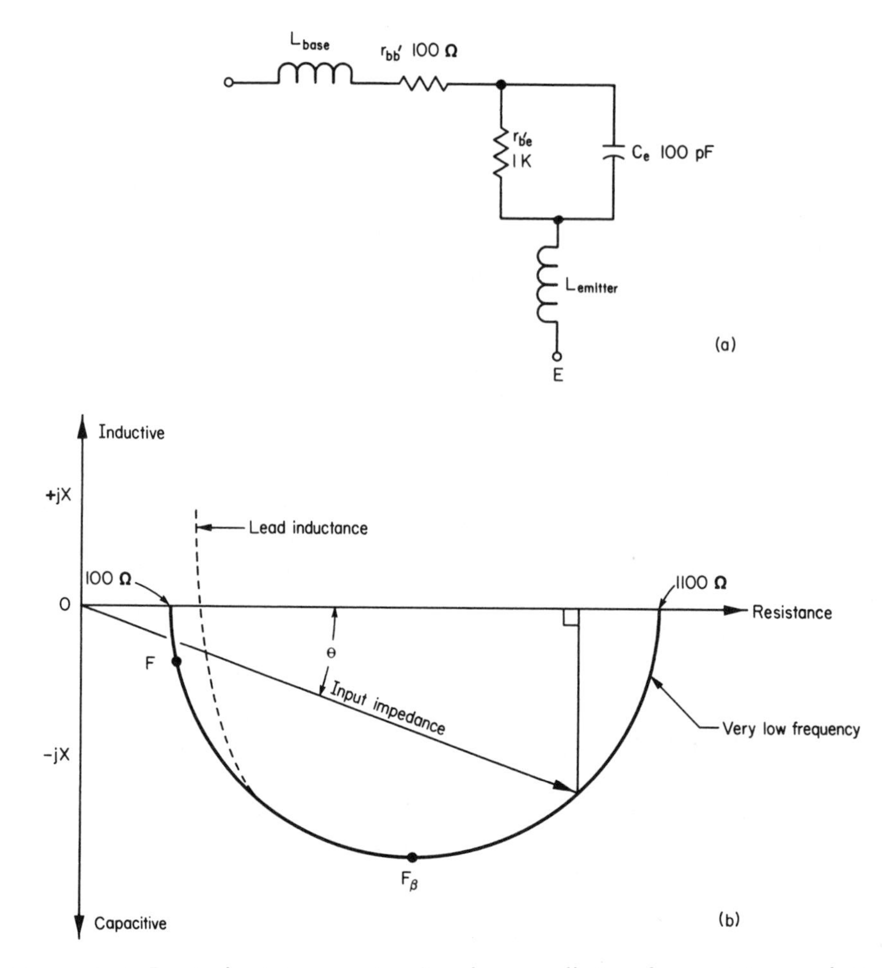

Figure 6-10 Equivalent input circuit of a small-signal transistor and its impedance variation with frequency. The output of the transistor is short-circuited.

plotted on a graph as a function of frequency. An even more common presentation is the two rectangular admittance components, $Y_{ie} = G_{ie} + jB_{ie}$. An example of this is shown in Figure 6-12(a).

6-6 OUTPUT IMPEDANCE VERSUS FREQUENCY

As with the input impedance, the output impedance is usually presented on data sheets as a complex value measured with the input short-circuited (base bias is still applied, but a good bypass capacitor is placed base to emitter). The source impedance used by the designer will lower the output impedance somewhat. The components of the hybrid circuit that affect the output impedance are shown in Figure 6-11 together with a signal generator used to measure the impedance. A source resistance is shown connected across the base and emitter terminals; for the short-circuit measurements, its value would be zero. The output impedance will be equal to the collector-to-emitter voltage applied by the signal generator, divided by the resulting collector current. At first it may look as if the only path for current from the generator is through the small capacitor $C_{b'c}$. This would suggest that the output should appear mainly as a capacitive reactance, decreasing proportional to frequency. But look at where the current goes after $C_{b'c}$; it has three paths. One path is through C_e, the second is through $r_{b'e}$ and the third is through $(r_{bb'} + R_s)$; the relative amounts will depend on the relative impedances of the three paths. The current from the collector passing through $r_{b'e}$ will act in the same manner as any base input current, and so cause a collector current to flow. By a circuitous route, then, a change in collector voltage produced by the signal generator has caused a proportional change in collector current, and so a much lower output impedance is the result.

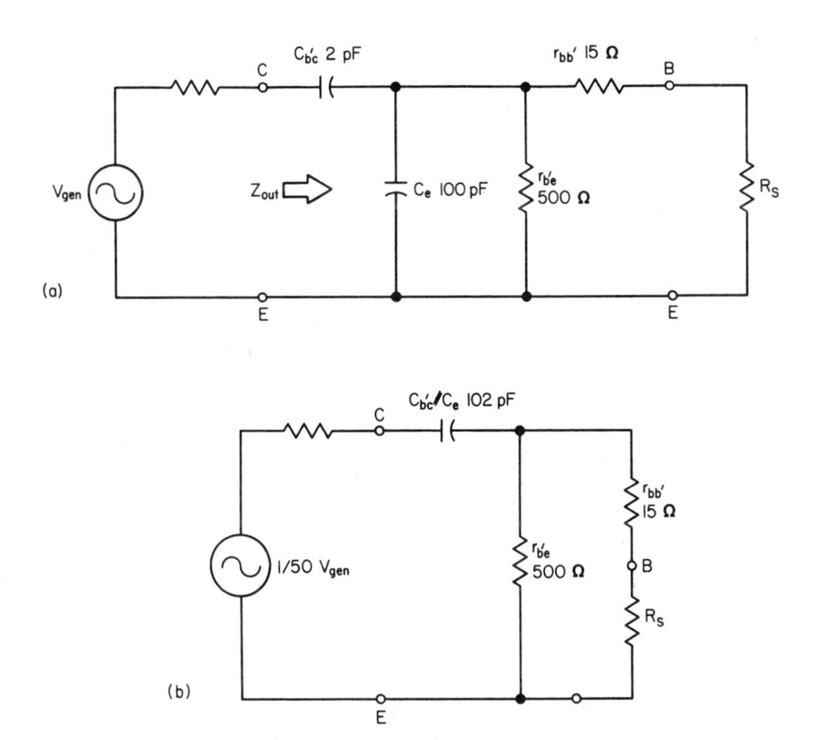

Figure 6-11 Components affecting the output impedance of a transistor. The full set is shown in (a) and a simplified version using a Thévenin equivalent circuit is shown in (b). Typical values are included.

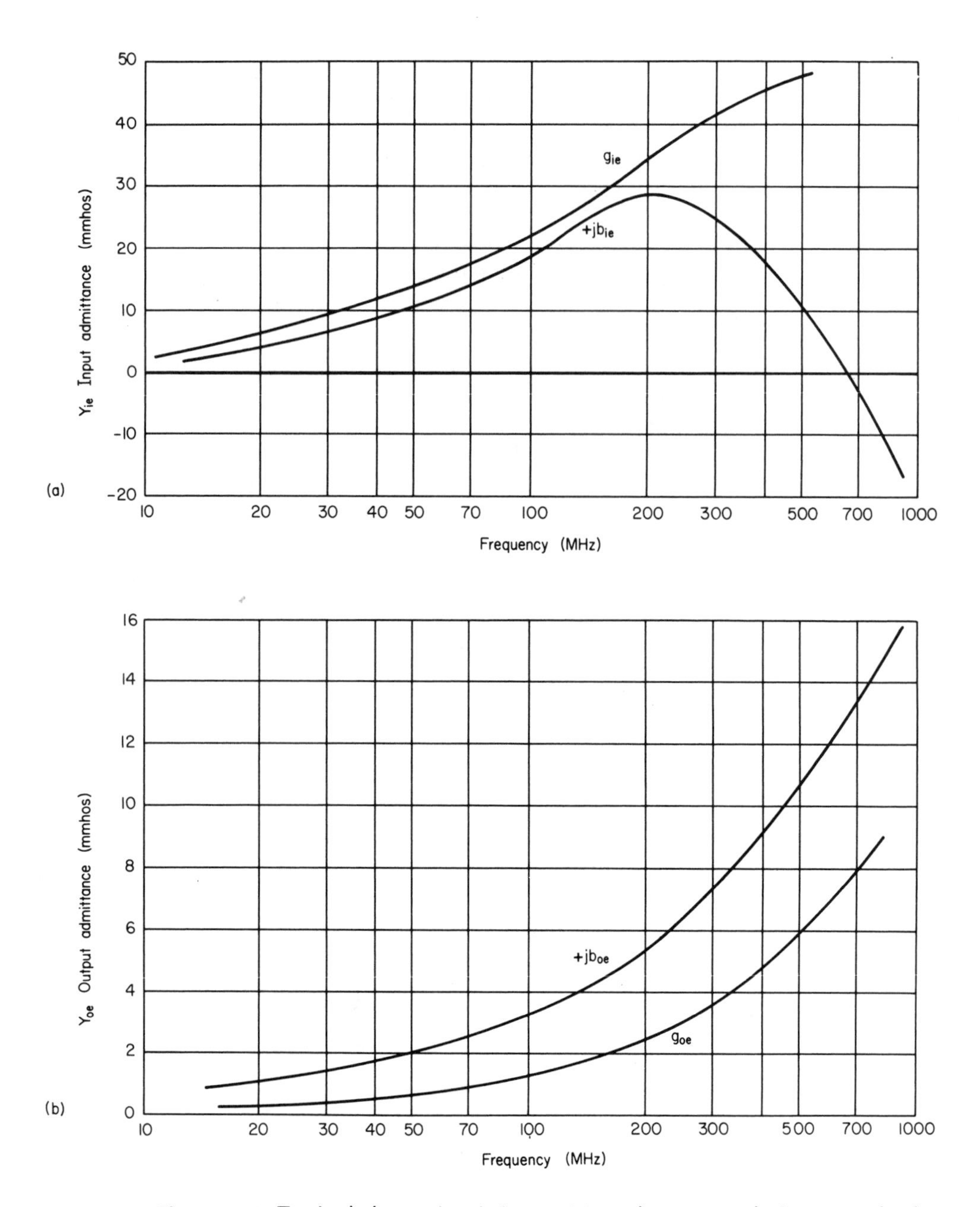

Figure 6-12 Typical short-circuit input (a) and output admittances (b) for a high-frequency transistor. These values would apply to only one particular set of bias conditions, such as $V_{ce} = 10.0$ V and $I_c = 8.0$ mA.

The amount of current fed back from the collector to $r_{b'e}$ will increase with frequency as the reactances of the capacitors decrease. The amount will also increase as the source resistance is increased, since this will lessen the shunting effect of the $R_s + r_{bb'}$ path. Any current passing through $r_{b'e}$ will be amplified by β_0, the low-frequency current gain of the transistor. The output impedance will therefore decrease as the frequency increases and also as the source resistance is increased. Figure 6-12(b) shows a typical variation of the output admittance of a transistor with its input short-circuited.

6-7 INSERTION VOLTAGE GAIN

So far we have discussed the voltage gain of the transistor amplifier in terms of the collector voltage divided by the base voltage. Defined this way, the gain will stay relatively flat over a wide frequency range and then start to fall off. This gain will be down 3 dB at the frequency

$$f = \frac{1}{2\pi r_{bb'}(Ce + C_{\text{Miller}})} \tag{6-6}$$

Of more practical interest is the variation of voltage gain obtained by operating the transistor from a constant-internal-voltage signal generator. As the input impedance of the transistor drops with increasing frequency, the voltage across the base-to-emitter terminals of the transistor will also drop. The results will be similar to the insertion losses obtained for the filters of Chapters 3 and 4. Using the equivalent circuit of Figure 6-13(b), the insertion voltage gain will drop 3 dB at

$$f = \frac{1}{2\pi(R_s + r_{bb'})(C_e + C_{\text{Miller}})} \tag{6-7}$$

This cutoff frequency can be kept high by keeping the source resistance low. The resulting power gain will suffer as a result of the mismatch between source and input resistance, but the bandwidth will be wider.

6-8 POWER GAIN

Power gain or *real gain* is often discussed at high frequencies to emphasize the difference between active and passive circuits. A passive network may have a voltage gain or a current gain but not both at the same time. A transformer is a

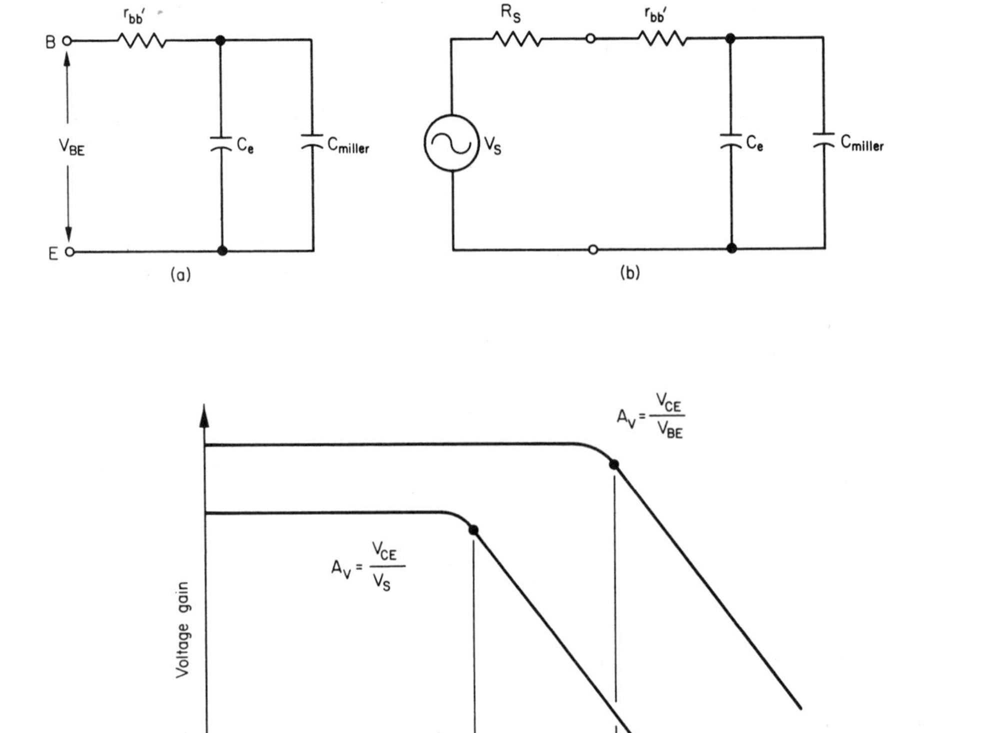

Figure 6-13 Equivalent circuit used for transistor terminal voltage gain (a) and for insertion voltage gain (b). The resulting difference in cutoff frequencies is shown in (c).

good example; a higher voltage is always obtained at the expense of lower currents, the overall power gain remains at 1.0 or even less if losses occur in the transformer. The majority of audio-frequency work involves only minor changes in impedance level and very few impedance-changing devices. Voltage gain is a meaningful term under such conditions. At radio frequencies, however, impedance matching is common and impedance levels throughout a circuit change tremendously. The only true indication of how good a job a transistor is doing is to calculate its power gain. If the power gain drops below 1.0 (0 dB), that transistor might as well be replaced by a passive impedance matching circuit.

Power gain cannot simply be calculated by multiplying the current gain times the voltage gain, since, as we saw in Example 6-1, both have different phase angles. The gain calculation must therefore include this angle.

$$A_{power} = A_v \times A_i \times \cos(\theta_v - \theta_i) \quad (6\text{-}8)$$

where θ_v = phase angle for the voltage gain.
θ_i = phase angle for the current gain.

EXAMPLE 6-2

Let us now return to the transistor we used in Example 6-1, and calculate the power gain of the device at 7.5 MHz.

Solution

We had already found that

$$A_i = 1.84 \quad \angle -89^\circ$$

$$A_v = 69 \quad \angle -69.8^\circ$$

The power gain is then

$$\begin{aligned} A_p &= A_v \times A_i \times \cos(\theta_v - \theta_i) \\ &= 69 \times 1.84 \times \cos(-69.8 + 89^\circ) \\ &= 119.9 \end{aligned}$$

In decibels this would be

$$10 \times \log 119.9 = 20.8 \text{ dB}$$

6-9 MAXIMUM POWER GAIN

The *maximum power gain* that a transistor can produce at any frequency is a rather involved topic. It is well known that maximum power will be transferred into and out of the device if the source and load impedances are conjugately matched to the transistor's input and output impedances. One of the problems is finding what these input and output impedances actually are, since the use of source and load impedances alters the conditions used for the measurement of the manufacturer's short-circuit Y_i and Y_o parameters. The values can be calculated using equations given in Chapter 7.

The next problem is whether this gain will actually be achieved. The published values are typical values, meaning that the actual values of a transistor sitting in front of you could be higher or lower. Also, even though the transistor is capable of the calculated maximum gain, there is a good chance that the circuit may oscillate, thanks to the internal feedback provided by $C_{b'c}$. This would then lead to attempts at counteracting or neutralizing the feedback path.

6-10 NEUTRALIZED POWER GAIN

If the internal feedback capacitor ($C_{b'c}$) is counteracted, the transistor has been "neutralized" (the techniques will be described in Chapter 7). If all internal feedback, resistive, and capacitive, paths are counteracted, the transistor has been *unilateralized*. Under this condition, the effect of the load on the input admittance and of the source on the output admittance will have been eliminated. The published figures should then be typical of the actual admittances of the transistor in the circuit. If all goes well, the amplifier should also be free of oscillations. The maximum power gain obtained under unilateralized conditions is referred to as *maximum available gain* and will be higher than the previous power gain due to the change in impedances and the removal of the negative feedback. The calculation is also much simpler.

$$\text{MAG} = \frac{\beta^2 \cdot g_{ie}}{4 \cdot g_{oe}} \tag{6-9}$$

where g_{ie} and g_{oe} are the real parts of the input and output admittances

The MAG of a typical transistor is shown in Figure 6-14. Also included is the typical power gain that will be obtained with the same transistor when not unilateralized, assuming that it does not oscillate. The point where the MAG curve drops below the 0-dB line is labeled $f_{\max}$. This is called the *maximum frequency of oscillation* and theoretically indicates the highest frequency at which that transistor will oscillate. All the output power from the transistor would have to be fed back to the input to maintain the oscillation, leaving nothing for losses or external use. This frequency can be calculated with

$$f_{\max} = \sqrt{\frac{f_T}{8\pi r_{bb'} C_{b'c}}} \tag{6-10}$$

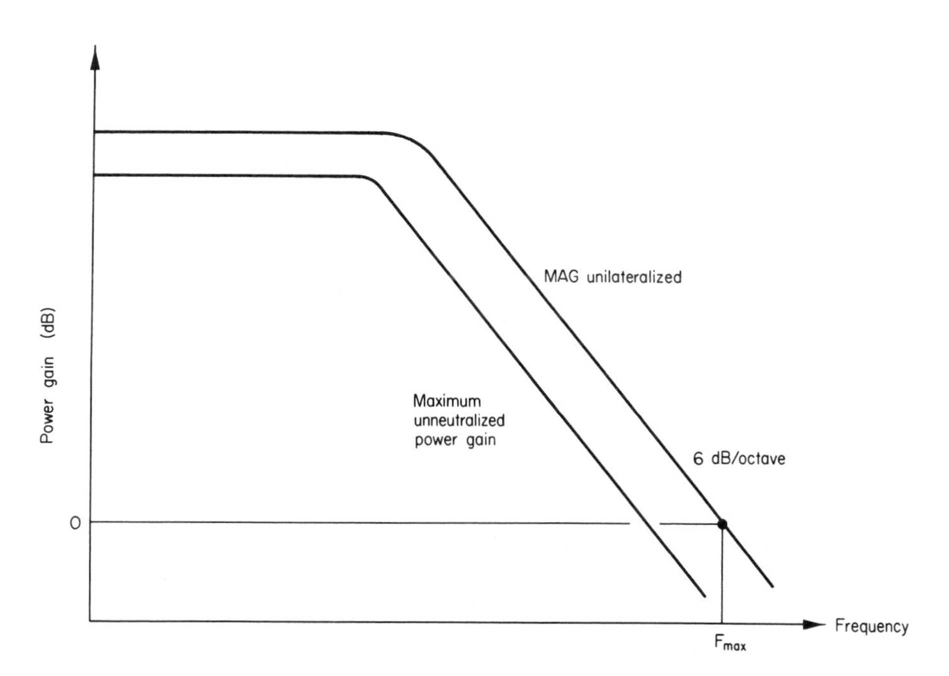

Figure 6-14 Typical variation of amplifier power gain with frequency.

6-11 FIELD-EFFECT TRANSISTORS

Field-effect transistors, both J-FET and MOS-FET types, find numerous applications in communication circuits. The mention of FET may conjure up visions of fantastically high input impedances, hundreds of megohms or more. Such is not the case at high frequencies, where the input resistance drops rapidly with the inverse square of the frequency and develops a highly reactive component. While the input impedance is still much higher than an equivalent bipolar transistor at the same frequency, this is not enough to justify its use, as the difference can easily be made up with a matching circuit. The power gain of FETs is usually lower than can be provided by a bipolar transistor, so this does not make them very attractive. What is important is the nature of their transfer characteristic. The drain current (Figure 6-15) is very closely given by

$$I_D = I_{DSS}\left(\frac{V_{GS}}{V_P} - 1\right)^2 \qquad (6\text{-}11)$$

where: I_D = drain current (mA)
I_{DSS} = saturation drain current with zero gate-to-source bias voltage (mA)
V_{GS} = gate-to-source voltage
V_P = pinchoff voltage

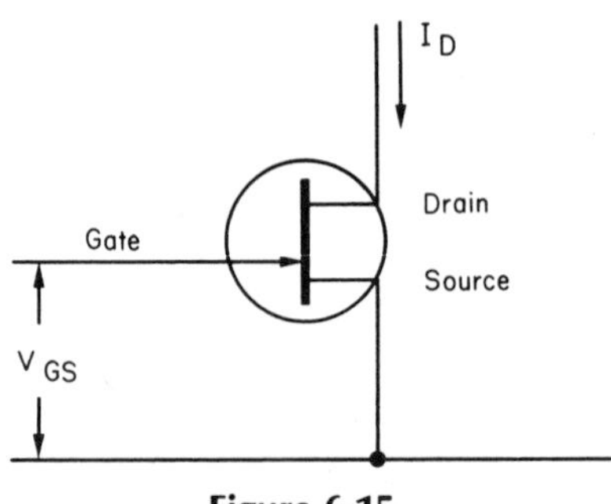

Figure 6-15

The benefits of such a square-law relationship were described in Section 1-5; no in-band mixing products result, and cross-modulation is nonexistent. Field-effect transistors are therefore used where this square-law relationship is important—RF amplifiers, mixers, and gain-controlled stages. The fact that FETs can also have very low noise figures makes them the logical choice for many applications.

The equivalent circuit useful for both J-FETs and MOS-FETs is shown in Figure 6-16 together with a cross section of a typical depletion mode, *n*-channel MOS-FET.

The origin of the components shown in the equivalent circuit and some typical values follow.

1 R_g, R_s, R_d: *small resistances* created by the semiconductor material itself; typically, 2–40 Ω.

2 R_{ds}: *effective resistance* of the channel for the bias conditions used; value will increase rapidly with increasing drain-to-source voltages near the saturation voltage; typically, 500–50 kΩ

3 C_{gd}: *feedback capacitance* caused by the reverse-biased drain-to-source junction; 0.05–5 pF

4 C_{gs}: *input capacitance* that results from the gate-to-source junction; typically, 0.1–10 pF.

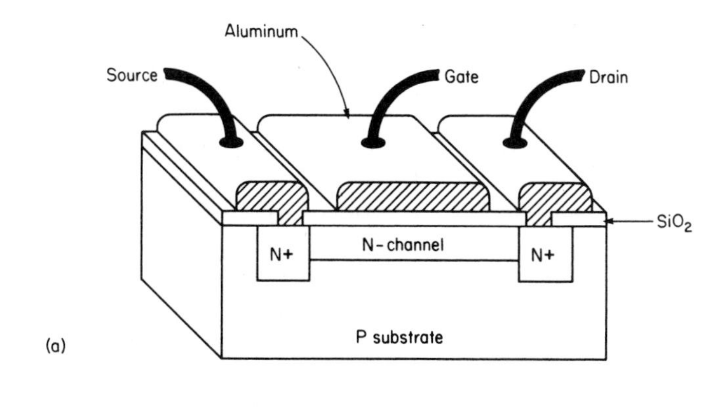

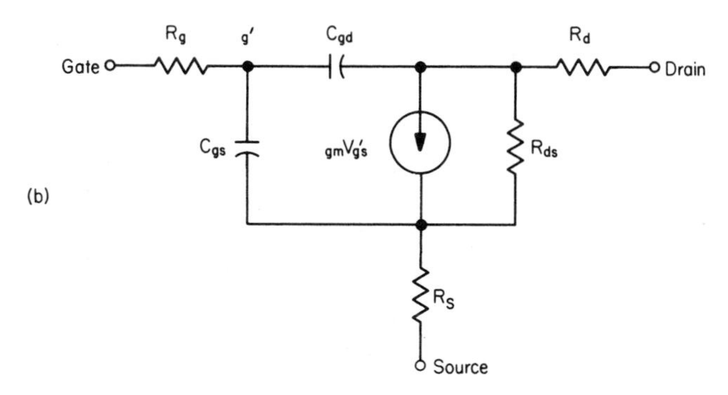

Figure 6-16 (a) Cross section of a depletion mode, *n*-channel MOS-FET, (b) an equivalent circuit that can be used for either MOS-FET or J-FET devices.

5 $g_m\left(=\dfrac{\Delta I_D}{\Delta V_{GS}}\right)$: *transconductance* that provides the forward gain of the FET; this is the value that varies with the square of the gate bias voltage; typically, maximum values are 4000–20,000 μmhos.

6-12 FET *Y* PARAMETERS

If the FET is described by the four admittance parameters in the same manner as the bipolar transistor, all the equations for voltage and power gain will still apply. A typical set of *Y* parameters for a MOS-FET capable of UHF operation is shown in Figure 6-17.

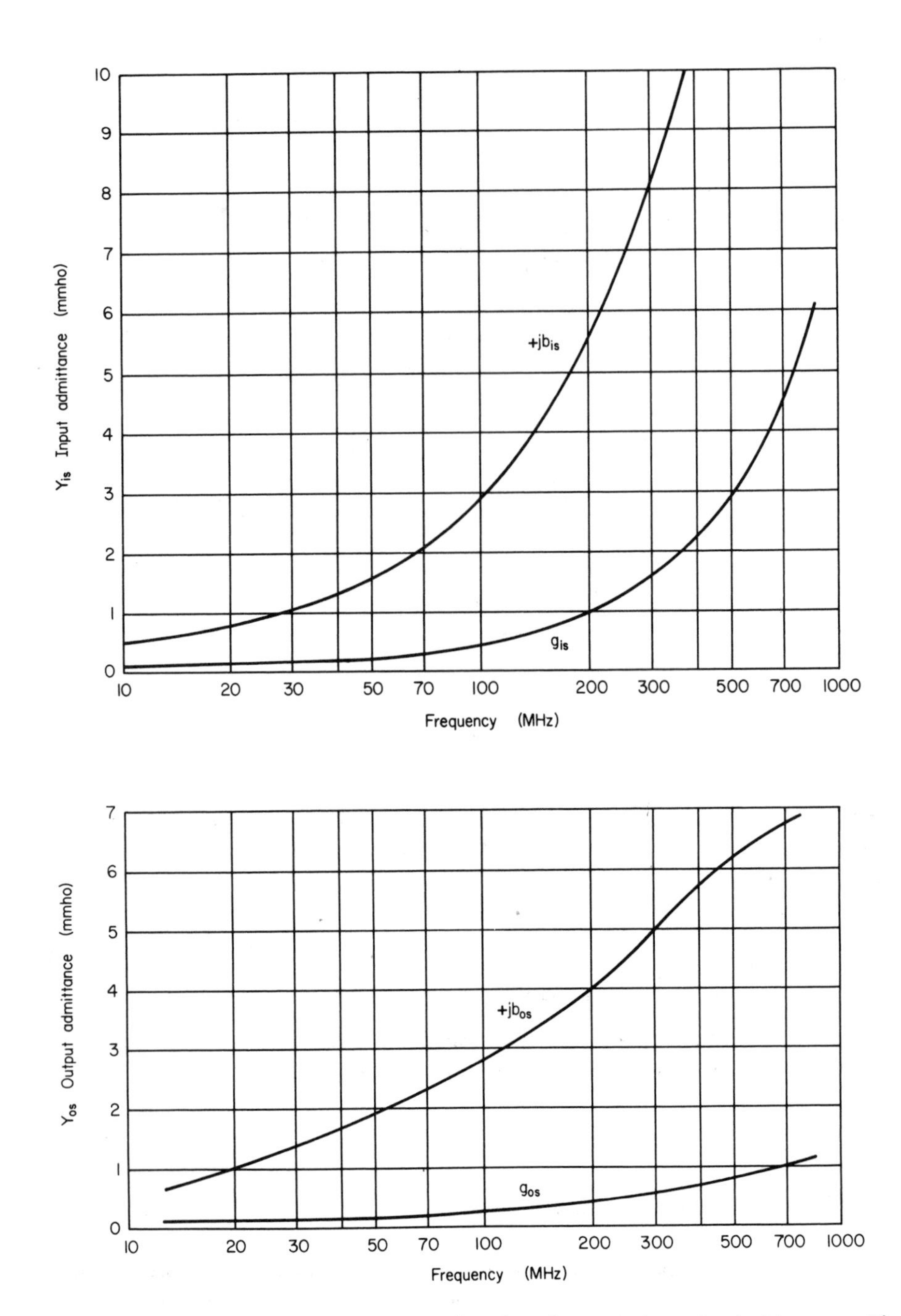

Figure 6-17 Four *Y*-parameter curves for a MOS-FET, showing the variation of admittances with frequency.

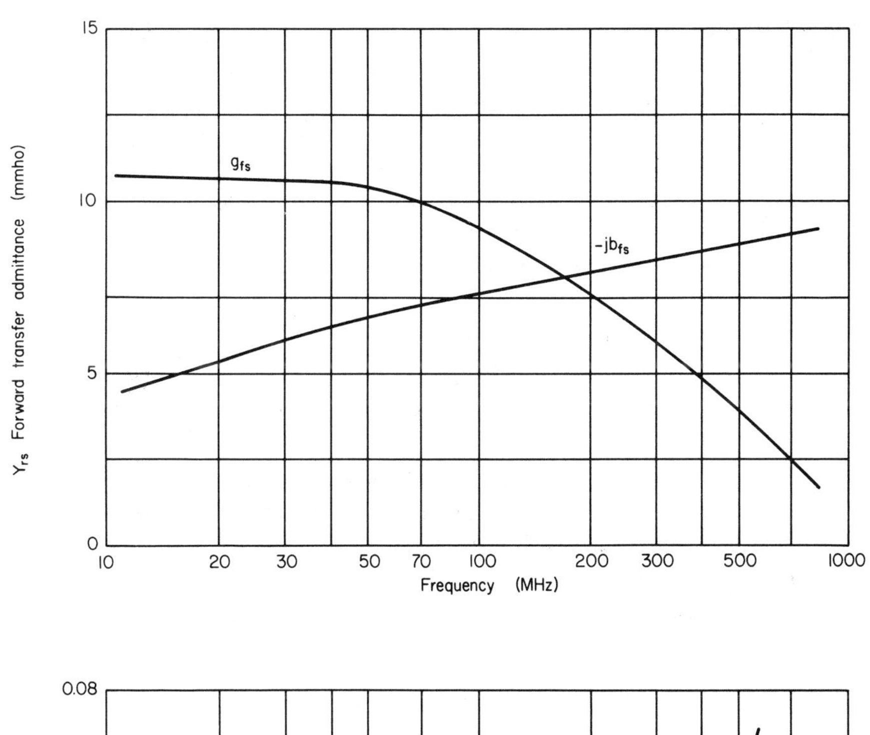

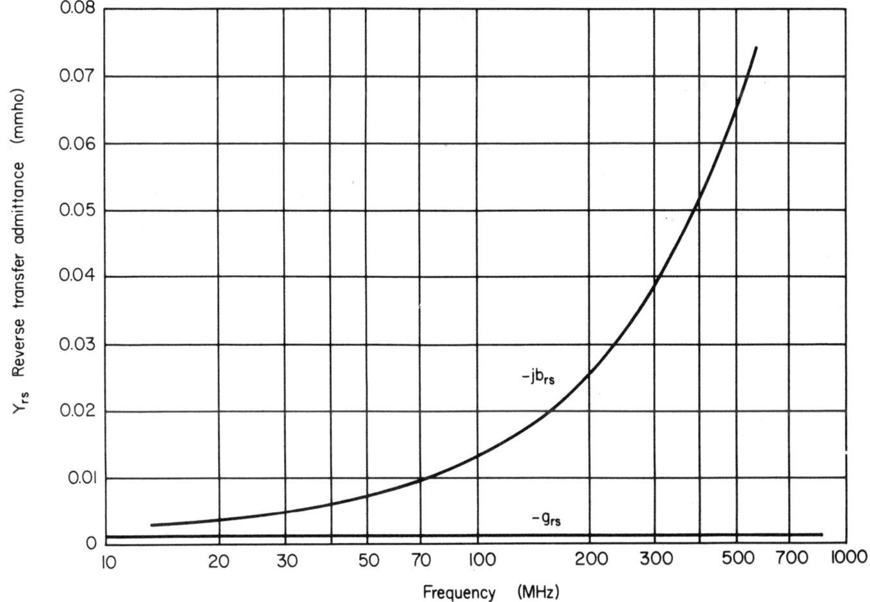

Figure 6-17 (cont'd)

QUESTIONS

1 A transistor has the following hybrid-Π parameters: $r_{b'e}=800\Omega$, $r_{bb'}=20\Omega$, $\beta_0=85$, $C_e=110$ pF and $C_{b'c}=2.5$ pF. Calculate f_β and f_T for this transistor and then find the short-circuit current gain at 25 MHz.

2 Using the same parameters as given in Question 1 and with a load resistance of 560 Ω, find the voltage gain (base to collector) and current gain at 25 MHz. Both values should have a magnitude and a phase angle.

3 Calculate the power gain (dB) of the transistor from Question 2.

4 Calculate the input impedance of the transistor amplifier from Question 2 and also the input impedance of the same transistor with its output shorted.

5 Write a computer program to calculate the values for Z_{in}, A_v, A_i, and A_p for the amplifier of Question 2. Cover the frequency range from 100 kHz to 100 MHz in 20 logarithmic steps.

6 Calculate the maximum available gain (MAG) at 200 MHz for the transistor with the characteristics shown in Figure 6-12 and with f_T of 650 MHz.

REFERENCES

Millman and Halkias, 1972. Integrated electronics. New York: McGraw Hill.

RCA Solid-State Division. 1971. *RCA Designer's Handbook SP-52, Solid-State Power Circuits*. Somerville, N. J.

chapter 7

small signal amplifiers

Small-signal amplifiers are usually considered to be linear amplifiers operating in class A conditions. They might have very wide bandwidths such as in TV video and cable distribution amplifiers and oscilloscope vertical amplifiers. They could, on the other hand, have narrow bandwidths, as in RF or IF stages for communications receivers. The gain of the amplifiers might be set to one value, or it might be variable through the use of automatic gain control. In most cases, the noise generated by the transistor should be kept as low as possible and the distortion of the signal due to transistor nonlinearities must be minimized.

In addition to normal design considerations of gain, bandwidth, and so on, the designer must also keep in mind the possibility that the amplifier may oscillate. The final circuit might then include various neutralization techniques to prevent this possibility.

7-1 MANUFACTURERS' DATA

The choice of transistor for a particular application will depend on a number of variables, including frequency, supply voltages, signal and noise levels, and the type of gain control required. The semiconductor manufacturers are the best source of information and will recommend several transistors for any given application.

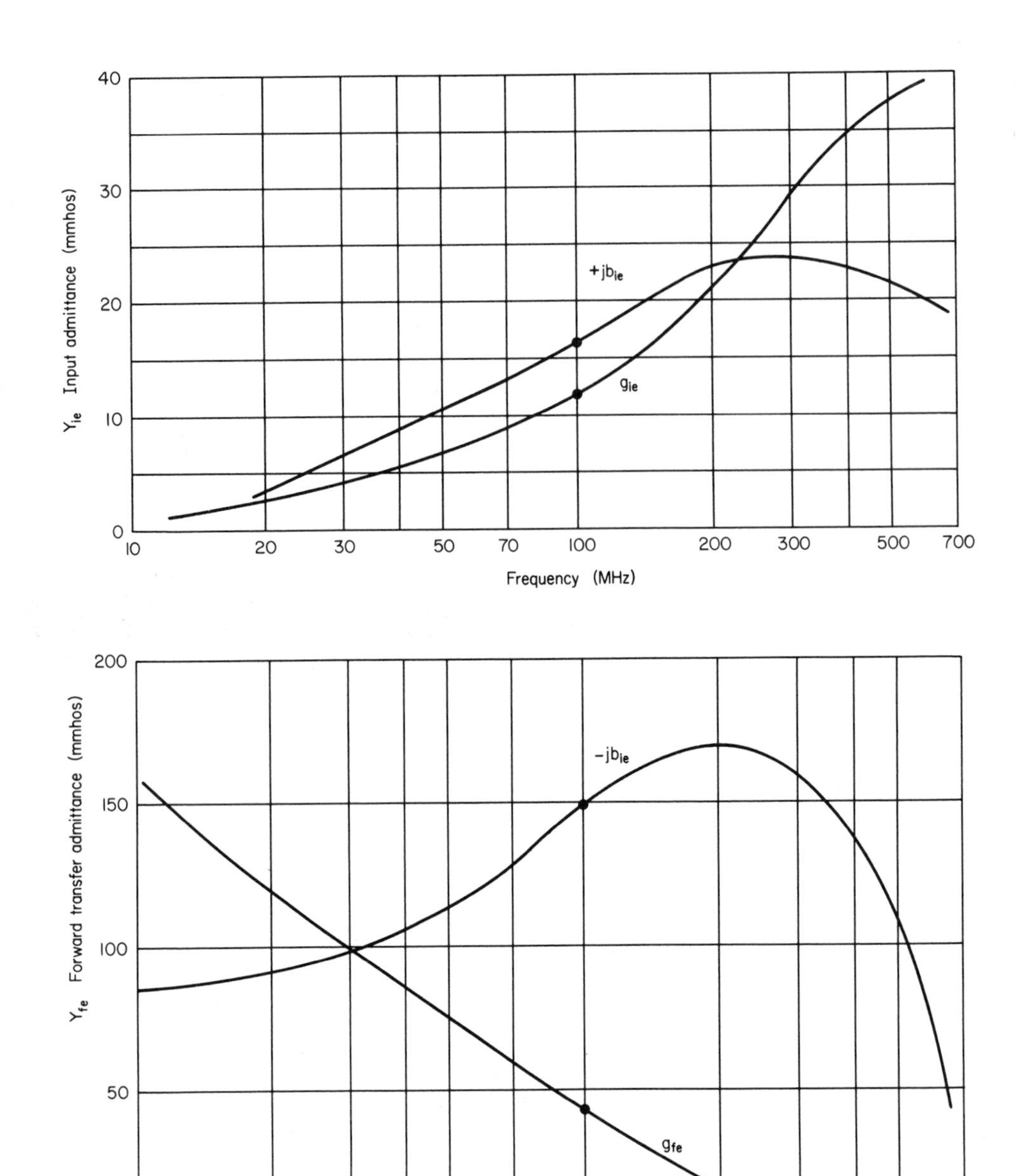

Figure 7-1 Admittance parameters versus frequency for a small-signal bipolar transistor. $V_{CE}=8.0$ V, $I_C=5.0$ mA.

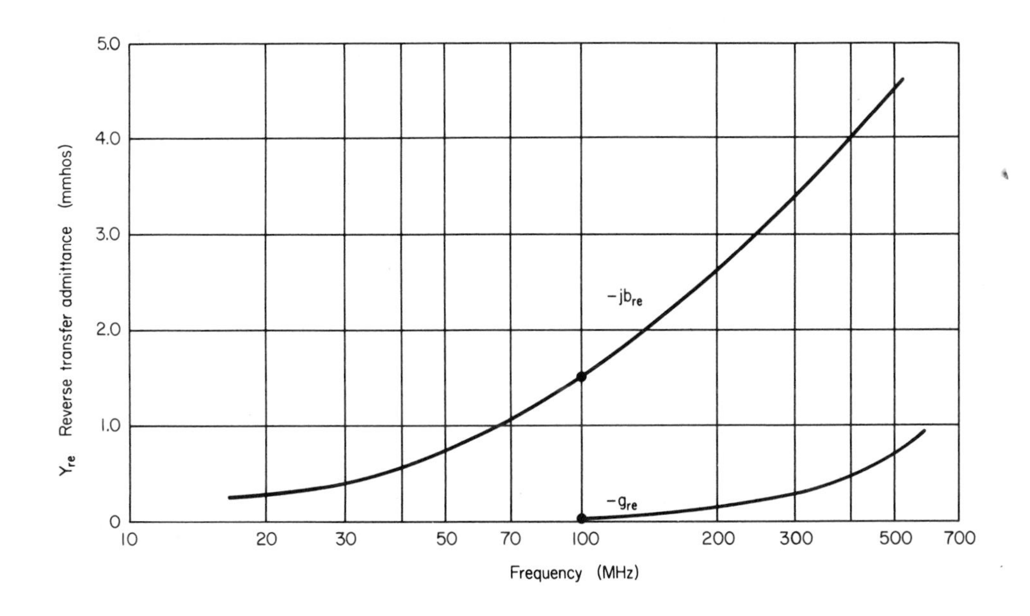

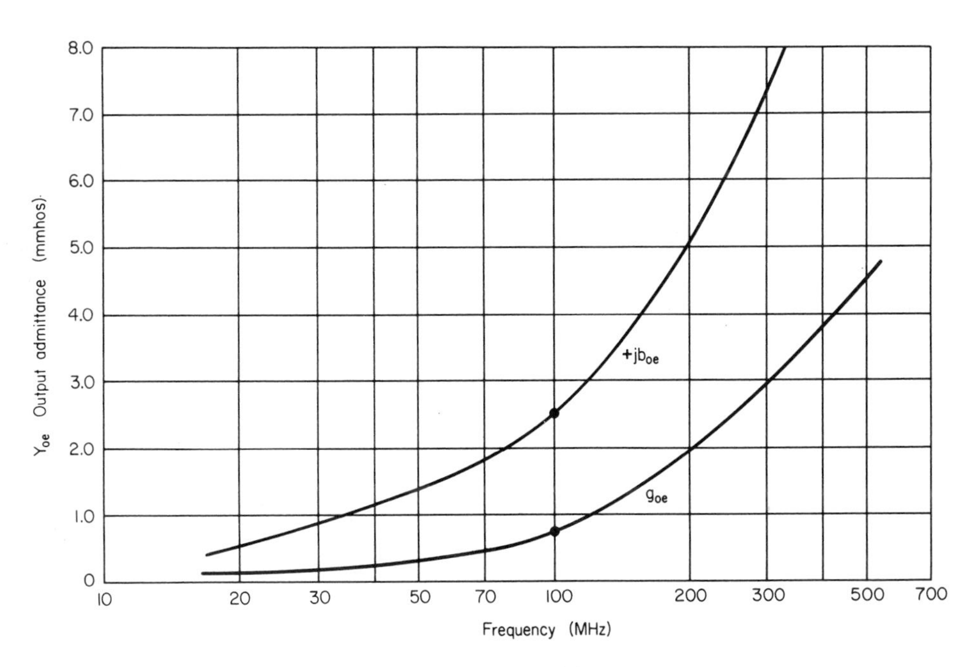

Figure 7-1 (cont'd)

Once the device is selected, the data sheet containing the admittance (Y) parameters must be obtained and an optimum bias current and voltage selected. Again, the manufacturer will have done most of the hard work and will recommend an operating point and provide corresponding curves of Y-parameter variation with frequency. Occasionally, bias conditions must be changed from the recommended values for low-noise operation or perhaps higher signal levels. The admittance parameters will then probably change, making the designer's job more challenging, as a little educated guessing at the new parameters must be made. The admittances most sensitive to bias current changes are the input admittance (Y_i) and the forward transfer admittance (Y_f).

The parameters[1] to be used for the following discussion (Figure 7-1) represent a small-signal bipolar transistor with a current gain–bandwidth product of 500 MHz. The RF design procedures are identical for bipolar and field-effect transistors once the admittance parameters are obtained; the biasing, of course, would be different. Remember that data sheets provide only "typical" parameters, so any one transistor could be better or worse. FETs are more likely to have a wider parameter spread than bipolar transistors.

7-2 AMPLIFIER STABILITY

As already indicated, the chances of starting to design an amplifier and ending up with an oscillator are very good, thanks to the internal collector-to-base capacitance ($C_{b'c}$) in the bipolar transistor or the corresponding C_{dg} in the FET.

The likelihood of the circuit oscillating can be predicted beforehand, so this is the best first step to take. If oscillations appear likely, changes can be made in the design to prevent it. There are two different stability factors that can be calculated. The first is the *Linvill stability factor*:

$$C_L = \frac{|Y_{re} \cdot Y_{fe}|}{2 \cdot g_{ie} \cdot g_{oe} - \text{Re}(Y_{re} \cdot Y_{fe})} \qquad (7\text{-}1)$$

where: $|\ |$ = absolute value

g_{ie} and g_{oe} = real part (conductances) of the corresponding admittances

Re( ) = real part of the result

This factor checks the possibility of the transistor oscillating under any conditions of source or load impedance, resistive or otherwise. If C_L is 1.0 or greater,

[1] These parameters actually represent a combination of data sheets from three separate transistors. They have been modified slightly but are still very typical of a real transistor.

the transistor is potentially unstable and may oscillate if the right combination of source and load are accidentally used. If C_L is less than 1.0, the transistor will be unconditionally stable regardless of the terminating impedances (it could still oscillate if very poor circuit layout is used, with lots of stray capacitance and magnetic coupling of inductors).

The second stability factor comes closer to the real truth. It will test the stability of the circuit when the actual design values of source and load are used. This is *Stern's stability factor*:

$$K_s = \frac{2(g_{ie} + G_s)(g_{oe} + G_L)}{|Y_{re} \cdot Y_{fe}| + \mathrm{Re}(Y_{re} \cdot Y_{fe})} \tag{7-2}$$

where: G_S = source conductance
G_L = load conductance

The significance of this factor is the opposite of Linvill's factor. The circuit will be unconditionally stable if K_S is greater than 1.0 (as long as the selected values of G_S and G_L are maintained) and will be potentially unstable and may oscillate if K_S is less than 1.0. Keep in mind, though, that the data-sheet values are *typical* values, that different values of G_S and G_L may appear in a circuit during tuning, and that extra feedback is likely to exist in a circuit as a result of the proximity of input and output circuits. A calculated value of 1.05 should not then be taken as an indication of complete stability in a production run of 1000 amplifiers.

For the moment, let us assume that whatever amplifier we are going to design will be stable. Later we will examine methods of taming an unstable circuit.

7-3 INPUT AND OUTPUT ADMITTANCES

We are assuming, for the moment, that the internal capacitance is not neutralized. The input and output admittances of the transistor will therefore depend on the actual value of load and source resistances used. The new values can be found from:

$$Y_{in} = Y_{ie} - \frac{Y_{re} \cdot Y_{fe}}{Y_{oe} + Y_L} \tag{7-3}$$

$$Y_{\text{out}} = Y_{oe} - \frac{Y_{re} \cdot Y_{fe}}{Y_{ie} + Y_S} \tag{7-4}$$

where: Y_S and Y_L = source and load admittances, respectively

The optimum value of source and load admittances that will provide the maximum gain without neutralization are given by

$$G_{\text{source}} = \frac{1}{2 \cdot g_{oe}} \sqrt{[2 \cdot g_{ie} \cdot g_{oe} - \text{Re}(Y_{re} \cdot Y_{fe})]^2 - |Y_{re} \cdot Y_{fe}|^2} \tag{7-5}$$

$$G_{\text{load}} = \frac{1}{2 \cdot g_{ie}} \sqrt{[2 \cdot g_{ie} \cdot g_{oe} - \text{Re}(Y_{re} \cdot Y_{fe})]^2 - |Y_{re} \cdot Y_{fe}|^2} \tag{7-6}$$

$$B_{\text{source}} = -jb_{ie} + \frac{\text{Im}(Y_{re} \cdot Y_{fe})}{2 \cdot g_{oe}} \tag{7-7}$$

$$B_{\text{load}} = -jb_{oe} + \frac{\text{Im}(Y_{re} \cdot Y_{fe})}{2 \cdot g_{ie}} \tag{7-8}$$

where: Im( ) = imaginary part of the result

7-4 AMPLIFIER DESIGN

Let us now design an amplifier. We will try for a center frequency of 100 MHz and a bandwidth at 10 MHz and a power gain of at least 20 dB. The transistor whose admittance parameters were presented in Figure 7-1 will be used and the biasing will be set at $I_C = 5.0$ mA and $V_{CE} = 8.0$ v, as recommended by the data sheet. The general circuit will appear as shown in Figure 7-2. Bias resistor values have already been included.

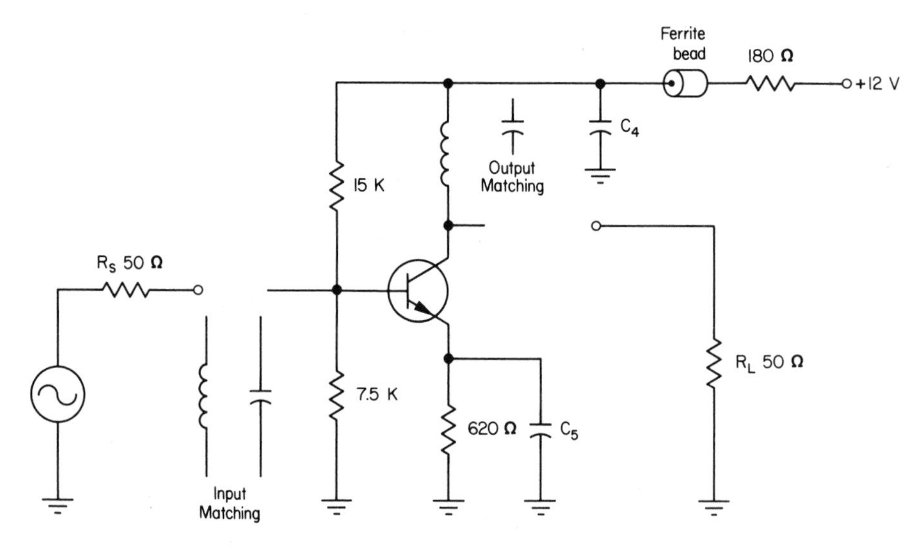

Figure 7-2 Schematic diagram of 100-MHz amplifier, including bias resistors.

The base bias circuit will absorb some of the input signal and will add some thermal noise, but since the input resistance of the transistor will be so low in comparison, the degradation of performance will hardly be noticeable. The collector voltage is supplied with some extra filtering by the 180-Ω resistor and ferrite bead and C_4.

The remainder of the design now concentrates on calculating values for impedance transformation and bandwidth setting components. The source and load resistances will be matched to the input and output of the transistor so that maximum gain can be obtained.

From Figure 7-1, the four admittance parameters at 100 MHz are

$$Y_{ie} = 12 + j16 \text{ mmhos}$$
$$Y_{fe} = 40 - j150 \text{ mmhos}$$
$$Y_{re} = 0 - j1.5 \text{ mmhos}$$
$$Y_{oe} = 0.75 + j2.5 \text{ mmhos}$$

7-4.1 Source and Load Admittances

For maximum power transfer, the source admittance must be

Real part:

$$G_s = \frac{1}{2 \cdot g_{oe}} \sqrt{[2 \cdot g_{ie} \cdot g_{oe} - \text{Re}(Y_{re} \cdot Y_{fe})]^2 - |Y_{re} \cdot Y_{fe}|^2}$$

$$= \frac{1}{2(0.75)} \sqrt{[2(12)(0.75) - \text{Re}((-j1.5)(40 - j150))]^2 - |(-j1.5)(40 - j150)|^2}$$

$$= 71.1 \text{ mmhos}$$

Imaginary part:

$$B_s = -jb_{ie} + \frac{\text{Im}(Y_{re} \cdot Y_{fe})}{2 \cdot g_{oe}}$$

$$= -j16 + \frac{\text{Im}[(-j1.5)(40 - j150)]}{2(0.75)}$$

$$= -j56.0 \text{ mmhos}$$

The source admittance must therefore be $71.1 - j56$ mmhos to transfer maximum power to the transistor input impedance of $71.1 + j56$ mmhos. Compare this value to the short-circuit input admittance Y_{ie}. The difference is rather large, about 6:1. Similarly, at the collector side, the load admittance should be:

Real part:

$$G_L = \frac{1}{2 \cdot g_{ie}}\sqrt{[2 \cdot g_{ie} \cdot g_{oe} - \text{Re}(Y_{re} \cdot Y_{fe})]^2 - |Y_{re} \cdot Y_{fe}|^2}$$

$$= \frac{1}{2(12)}\sqrt{(\text{this portion is the same as the } G_s \text{ calculation})}$$

$$= 4.44 \text{ mmhos}$$

Imaginary part:

$$B_L = -jb_{oe} + \frac{\text{Im}(Y_{re} \cdot Y_{fe})}{2 \cdot g_{ie}}$$

$$= -j2.5 + \frac{\text{Im}[(-j1.5)(40 - j150)]}{2(12)}$$

$$= -j5.0 \text{ mmhos}$$

The load admittance must therefore be $4.44 - j5.0$ mmhos and the output admittance of the transistor will be $4.44 + j5.0$ mmhos—again a very significant increase in admittance from the short-circuit output value Y_{oe}.

7-4.2 Stability Factor

Using these calculated source and load admittances, we can now find the value of Stern's stability factor to see if the amplifier is going to be stable.

$$K_s = \frac{2(g_{ie} + G_s)(g_{oe} + G_L)}{(Y_{re} \cdot Y_{fe}) + \text{Re}(Y_{re} \cdot Y_{fe})}$$

$$= \frac{2(12 + 71.1)(0.75 + 4.44)}{(-j1.5)(40 - j150) + \text{Re}[(-j1.5)(40 - j150)]}$$

$$= 1.90$$

The amplifier should therefore be stable for the source and load calculated, since the value of K_s is greater than 1.0. We can then proceed with the original circuit diagram of Figure 7-2, since no additional feedback circuits for neutralization are necessary.

7-5 MATCHING CIRCUIT DESIGN

Now that we know the input and output admittances of the transistor and the required source and load, we can design the input and output circuits. The transistor is drawn as a block in Figure 7-3 and the admittances are each represented as a parallel combination of a resistance and a capacitive reactance. The output network will be designed as a transformer to match from 225 Ω down to 50 Ω. The input network will be designed as a tapped capacitor to match the low transistor input resistance of 14.1 Ω up to the 50-Ω generator.

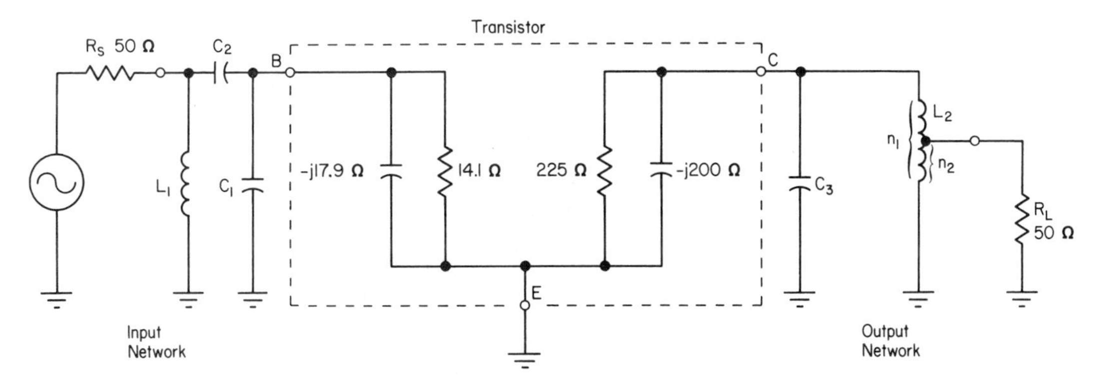

Figure 7-3 Input-and output-matching networks for the 100-MHz amplifier.

The required bandwidth between the 3 dB points is 10 MHz, and this would imply a loaded Q of

$$Q_L = \frac{f_0}{\text{BW}} = \frac{100}{10} = 10$$

But we have two tuned circuits and they cannot both have a Q_L of 10, or the bandwidth of the final amplifier would be too narrow. We must therefore select two lower Q_L values that will result in the required 10-MHz bandwidth. If we use a loaded Q of 5.0 for the input-matching network and $Q_L = 7.5$ for the output circuit, the overall bandwidth will be very close to 10 MHz. The lower Q is used on the input side to maintain a higher ratio of unloaded to loaded Q for those components, thereby minimizing insertion loss and the generation of thermal noise. Different requirements might make a higher Q_L at the input circuit necessary for better rejection of unwanted signals before the nonlinearities of the transistor were encountered.

For the input circuit (referring to Figure 7-3), the component reactances can be found using the formulas of Figure 5-12(c).

$$Q = 5.0$$

$$X_{C_1} = \frac{R_{\text{transistor}}}{Q} = \frac{14.1}{5} = -j2.82\ \Omega$$

(this will include the input reactance of the transistor).

$$X_{L_1} = \frac{R_{\text{source}}}{\sqrt{\frac{R_{\text{source}}}{R_{\text{transistor}}}(Q^2+1)-1}}$$

$$= \frac{50}{\sqrt{\frac{50}{14.1}(5^2+1)-1}}$$

$$= +j5.24\ \Omega$$

$$X_{C_2} = \frac{Q \cdot R_{\text{transistor}}}{Q^2+1}\left(\frac{R_{\text{source}}}{QX_L} - 1\right)$$

$$= \frac{5 \times 14.1}{5^2+1}\left(\frac{50}{5 \times 5.24} - 1\right)$$

$$= -j2.47\ \Omega$$

Because the input of the transistor also contains a shunt capacitive reactance of $-j17.9\ \Omega$, the final reactance of C_1 will have to be raised to $-j3.35\ \Omega$.

For the output circuit, the turns ratio of L_2 will be

$$\frac{n_1}{n_2} = \sqrt{\frac{225}{50}} = 2.12$$

The inductive reactance can be found from the Q_L (remember that the 50-Ω load is transformed up to 225 Ω):

$$X_{L_2} = \frac{R}{2Q_L} = \frac{225}{2 \times 7.5} = +j15\ \Omega$$

The total resonating capacitance will have a reactance of $-j15\ \Omega$, but part of this ($-j200\ \Omega$) is inside the transistor. The outside capacitor reactance would then be

$$C_3 = \frac{200 \times 15}{200 - 15} = -j16.2\ \Omega$$

The final component values at 100 MHz are

$$L_1 = 8.34 \text{ nH}$$
$$C_1 = 564 \text{ pF}$$
$$C_2 = 475 \text{ pF}$$
$$L_2 = 23.9 \text{ nH}$$
$$C_3 = 98.2 \text{ pF}$$

The two bypass capacitors, C_4 and C_5, were selected as 0.001 μF, as their low calculated reactance (1.6 Ω at 100 MHz) will provide adequate filtering and also because they stand a good chance of being self-resonant and therefore providing even better performance. The final schematic of this amplifier is shown in Figure 7-4.

The best test of a design is to build it and see. The amplifier in Figure 7-4 was constructed using all the values as calculated, and a transistor with characteristics very close to those assumed was used. The amplifier did not oscillate, fortunately, and produced the gain and frequency characteristics shown in Figure 7-5.

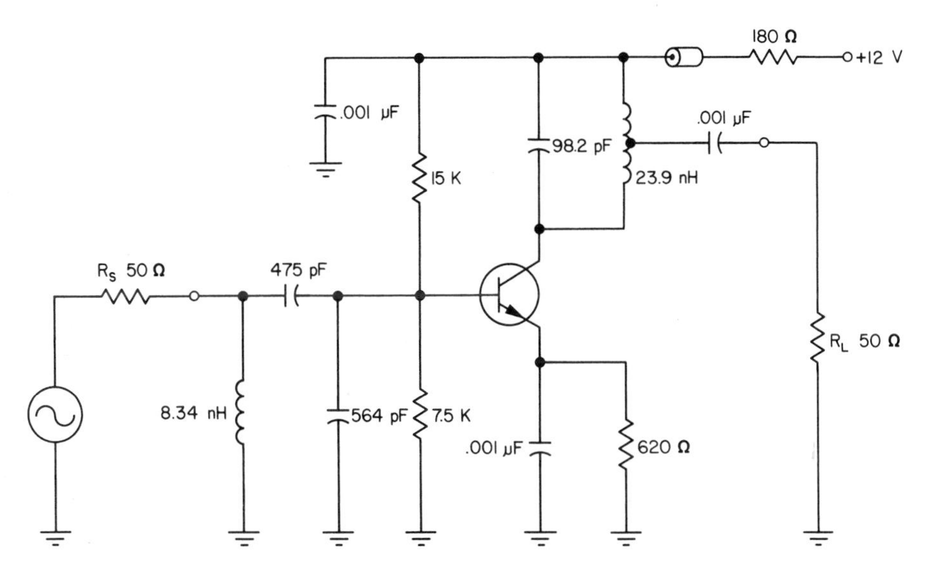

Figure 7-4 Schematic of 100-MHz amplifier.

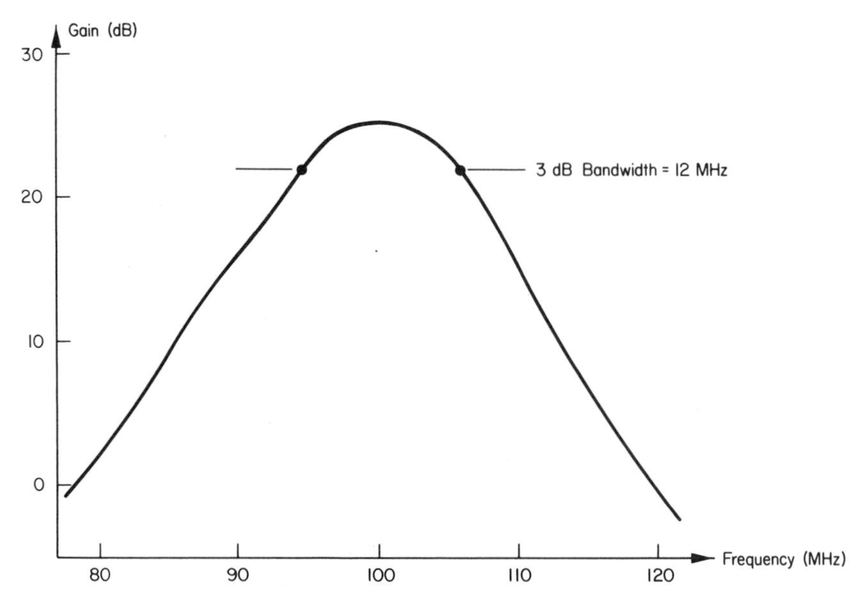

Figure 7-5 Frequency response and gain of transistor amplifier of Figure 7-4.

7-6 AMPLIFIER INSTABILITY

For the circuit just designed we found that oscillations would not occur as long as the calculated values of source and load resistance were used. What causes this tendency to oscillate and how can it be controlled?

We know that in audio amplifiers operating with resistive sources and loads, oscillations do not occur as easily, at least not with a single stage. The feedback path still exists through $C_{b'c}$ but no oscillations occur. The reason is that, to make a circuit oscillate, two conditions must exist. First, enough power must be fed back to the input so that the oscillations can sustain themselves. This means that the forward gain of the transistor times the losses back through the feedback path must produce a result greater than 1.0. The other requirement is that the overall phase shift through the transistor and the feedback path must be 360°. The amplified signal will then be back in step with the original input and is able to reinforce it. At low audio frequencies, the phase shift within a transistor is 180°. The feedback through the resistor $r_{b'c}$ will not produce any phase shift, so oscillations cannot occur. At higher frequencies, $C_{b'c}$ becomes more dominant, so the feedback will be shifted 90°, for a total of 270° —still no oscillations. With resistive terminations, then, oscillations are not likely to occur.

What is needed is a little outside help with the phase shift. (Remember, we are not trying to build oscillators yet, only to understand why amplifiers sometimes oscillate.) The additional phase shift that could produce unwanted oscillations can be created by the tuned circuits at the input and output of a tuned amplifier, as shown in Figure 7-6. At some frequency just below the resonance point, each tuned circuit will be inductive. The resulting $L, C_{b'c}, L$ filter can easily produce the extra shift necessary to make the circuit potentially unstable and prone to oscillate. Therefore, with any tuned amplifier, the unwanted oscillations will usually occur at a lower frequency than the tuned center frequency.

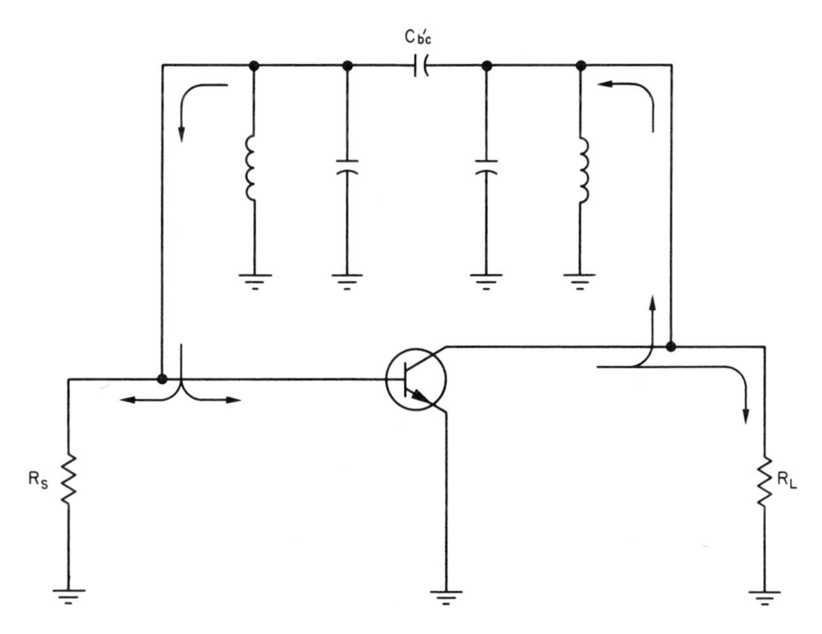

Figure 7-6 Feedback loop formed by a tuned amplifier and the internal feedback capacitance of the transistor.

7-7 NEUTRALIZATION TECHNIQUES

For a number of reasons, a designer may decide to neutralize the internal feedback capacitance of a transistor. The most obvious reason would be to avoid the tendency for the amplifier to oscillate; but another worthwhile benefit would be to obtain the increased power gain that results from the removal of the negative feedback. Neutralization will also raise the input and output impedances to the approximate values corresponding to the short-circuit admittance parameters. This makes matching and tuned circuit design easier and also eases alignment of tuned circuits, since the input impedance will no longer be a function of the load impedance.

Neutralization involves creating a second feedback path in such a way that the signal flowing back through it will be out of phase with and cancel the signal coming back through $C_{b'c}$. This can be done in a number of ways, as shown in Figure 7-7. The first example (a) uses an inductance that will parallel-resonate with the internal capacitance, thereby canceling the reactive feedback. The equivalent parallel resistance of the coil at resonance will leave a resistive feedback path that may result in a very minor loss of gain. With reasonable attention to coil losses, this will not be a problem; in any case, the tendency of the amplifier to oscillate will have been reduced. The series capacitor is simply for dc blocking. This is a narrow-band neutralization technique that allows some capacitive feedback to occur at higher frequencies and some inductive feedback at lower frequencies. Adjustment of an amplifier with this type of neutralization is fairly difficult.

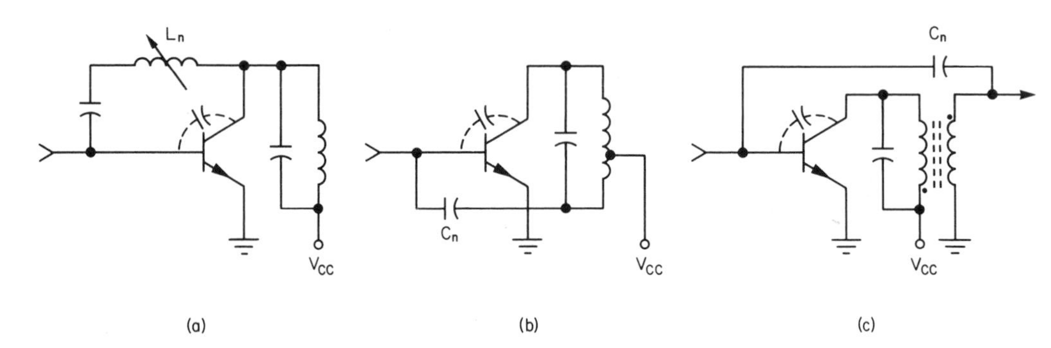

Figure 7-7 Neutralization techniques for RF amplifiers.

The second example in Figure 7-7(b) is more common. The tuned circuit at the collector is tapped to provide a signal out of phase with the collector. A small-value capacitor then forms a feedback path to the base, where cancellation occurs. If the tuned circuit were centertapped, the neutralization capacitor would have to be the same value as the internal capacitance $C_{b'c}$. Since this is in the neighborhood of 1 pF, the neutralizing capacitor value would be difficult to control. If the tuned circuit is tapped instead at about $1/4$ of its total turns, a neutralizing capacitor four times the value of $C_{b'c}$ could be used, a value that is a little easier to work with. This method will be effective over the full bandwidth of the collector tuned circuit and is also usable if the resonant frequency of the tuned circuit is varied.

A variation of this technique is shown in Figure 7-7(c). The secondary of the transformer is connected to provide the out-of-phase signal for the neutralizing capacitor. The turns ratio again will determine the ratio of C_n to $C_{b'c}$, a step-down transformer giving the most desirable ratio.

7-7.1 Neutralization Adjustment

For many applications, full neutralization is not necessary. A "ballpark" value of neutralization capacitor (or inductor) is inserted and adjusted so that the tendency to oscillate is reduced and the "alignability" is improved. For more critical applications, greater design and adjustment care is required. One useful technique is to turn the amplifier "backward" temporarily. A signal generator is applied to the output and an RF meter is connected to the input. Optimum neutralization occurs when the meter reading is lowest.

If adjustments are made that completely eliminate all reverse feedback, the amplifier has been unilateralized. This is rarely strived for, as both the capacitive and resistive feedback must be counteracted and the minor improvement in amplifier operation is rarely worth it.

7-8 MISMATCHING

Improvements in amplifier stability can also be obtained by mismatching the source and load. This means the use of source and load resistance values that do not provide maximum power transfer, and so power gain is reduced. However, if the gain can be spared, this is an economical solution to the stability problem and is useful over wide bandwidths. Transistors are inherently more stable at higher frequencies (approaching f_T), since less forward gain is available to encourage oscillations. Larger mismatches and greater gain reduction will therefore be needed at lower frequencies and, fortunately, this is where gain can be spared. The technique is therefore quite useful.

For best results, both the input and output should be mismatched by the same amount, preferably by using lower source and load resistances (higher admittances). The amount of mismatch can be found using Equation (7-2), Stern's stability factor. If K_s is set to some value between 2 and 3, good stability should be obtained for typical variations of transistors. Assuming, then, that both input and output are to be mismatched by the same amount, so that

$$G_S = m \cdot g_{ie} \tag{7-9}$$

$$G_l = m \cdot g_{oe} \tag{7-10}$$

The value of m will indicate how much larger the source and load conductances (G_s and G_l) must be than the short-circuit input and output conductances (g_{ie} and g_{oe}). The value of the mismatch factor (m) will be

$$m = \sqrt{\frac{|Y_{re} \cdot Y_{fe}| + \mathrm{Re}(Y_{re} \cdot Y_{fe})}{g_{ie} \cdot g_{oe}}} - 1 \tag{7-11}$$

assuming that we try for a stability factor of $K_s = 2.0$.

The flow of signal in the unneutralized amplifier is shown in Figure 7-6. If the source is perfectly matched to the transistor, one-half of the energy flowing back through $C_{b'c}$ will be absorbed by the source resistance and one-half by the input resistance of the transistor. If R_s is reduced to one-half the input resistance of the transistor ($m = 2.0$), 2/3 of the feedback will end up in the source and only 1/3 in the transistor. The feedback around the loop has therefore been reduced by 3.5 dB. The resulting input mismatch will also reduce the amplifier gain, but only by 0.5 dB. Similar results will be obtained by mismatching the output, but the improvement is harder to calculate, since the input impedance of the $L, C_{b'c}, L$ network would have to be known.

EXAMPLE 7-1

Our sample transistor (Figure 7-1) is potentially unstable at 20 MHz. Find the source and load resistances that will result in a stable amplifier.

Solution

At 20 MHz, from Figure 7-1:

$$Y_{ie} = 2.5 + j4.0 \text{ mmhos}$$
$$Y_{fe} = 120 - j90 \text{ mmhos}$$
$$Y_{re} = 0 - j0.2 \text{ mmhos}$$
$$Y_{oe} = 0.1 + j0.5 \text{ mmhos}$$

The mismatch factor will be

$$m = \sqrt{\frac{|Y_{re} \cdot Y_{fe}| + \mathrm{Re}(Y_{re} \cdot Y_{fe})}{g_{ie} \cdot g_{oe}}} - 1$$

$$= \sqrt{\frac{|(-j0.2)(120 - j90)| + \mathrm{Re}[(-j0.2)(120 - j90)]}{2.5 \times 0.1}} - 1$$

$$= 5.93$$

Therefore,

$$G_s = m \cdot g_{ie} = 5.93 \times 2.5 = 14.83 \text{ mmhos}$$
$$G_L = m \cdot g_{oe} = 5.93 \times 0.1 = 0.593 \text{ mmhos}$$

so that source and load resistances will be

$$R_s = \frac{1}{G_s} = \frac{1}{14.83 \text{ mmhos}} = 67.5\ \Omega$$

$$R_L = \frac{1}{G_L} = \frac{1}{0.593 \text{ mmhos}} = 1686\ \Omega$$

7-9 GAIN CONTROL

When used in receiver circuits, amplifiers will often encounter a very wide range of signal levels; typically 0.1 μV to 0.1 V, a 120-dB range. To prevent overloading, their gain must be reduced as the signal strength increases, and this is usually done automatically be changing the bias point of the transistor. However, in addition to the change in gain, the bias changes may also alter the input and output impedances and make the amplifier more nonlinear.

Amplifier gain involves two components. The one is the power gain of the transistor itself and the other is the loss of gain due to input and output mismatching. A poor design might then start out with a high-gain mismatched stage and change the bias so that the transistor gain dropped, but the input and output impedances changed to provide an improved transfer of power. A better design would start out with ideal matching when highest gain is required and then benefit from both the mismatch and the transistor gain loss as bias is changed. Such a design would likely require neutralization at the highest gain setting.

The gain of a transistor can be changed in two different ways. Figure 7-8 shows the typical variation of β with changes in collector current and collector-to-emitter voltage. The highest gain occurs with a collector current of 20 mA and a collector-to-emitter voltage of 5 V. Current changes either above or below the optimum value result in a loss of gain. At lower current levels, the voltage has little effect, while at the higher currents the collector voltage has a more noticeable effect. The two methods of automatic gain control will then depend on whether the collector current is either increased (forward control) or decreased (reverse control) from the optimum value.

7-9.1 Reverse AGC

The simplest method of gain control is obtained by reducing the collector current. This will simultaneously decrease the current gain and raise the input impedance. Both effects will decrease the power gain, and the increase in input impedance will result in a further mismatch loss. Most transistors can be used,

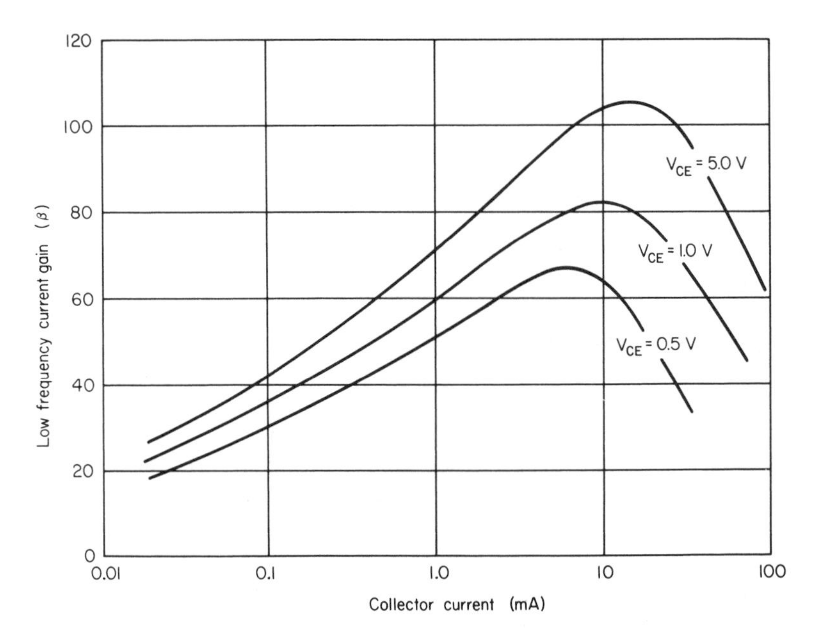

Figure 7-8 Typical variation of the current gain of a transistor with changes in bias current and voltage.

and the technique is good at all frequencies. However, the transistor is operating at very low current levels just at the time it is being asked to handle higher signal levels. Distortions of different types are therefore quite likely to occur as the gain is reduced. Reverse gain control should therefore be used only where signals are at a low level and where intermodulation distortion is not likely to be serious. Another problem associated with reverse gain control is the change in input and output impedances that accompany the change in collector current. These changes will result in shifts of the center frequency and bandwidth of any narrow-band filters associated with the transistor. These shifts can be minimized by loosely coupling the tuned circuits to the transistor and using separate resistances to determine the loaded Q's of the circuits. Power gain is lost as a result, but the overall frequency characteristics are more constant. Figure 7-9 describes the variations of the input and output admittances of a transistor at 50 MHz as the collector current is changed.

Reverse gain control is often used for the 455-kHz IF amplifiers of AM radio receivers, where the narrow-band tuned circuits provide some protection against intermodulation. A typical schematic of such an amplifier is shown in Figure 7-10. The detector stage is also included, as it provides the dc voltage that automatically reduces the bias on the IF amplifier as the signal strength increases.

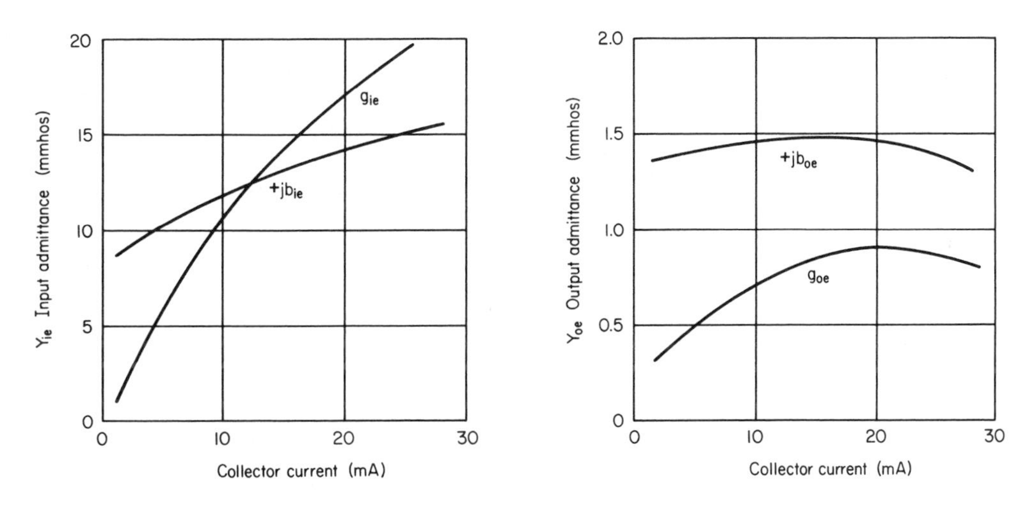

Figure 7-9 Changes in the 50-MHz input and output short-circuit admittances with collector current.

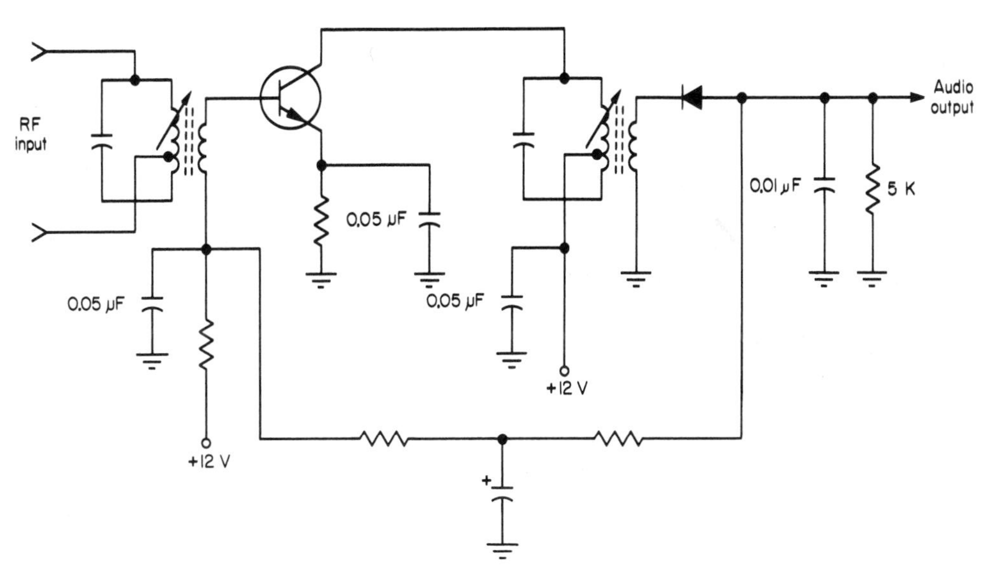

Figure 7-10 455 kHz IF amplifier and detector circuit using automatic reverse gain control.

7-9.2 Forward Gain Control

The main advantage of forward gain control is the increased ability of the transistor to handle larger signals without distortion even though the gain is decreasing. Not all transistors can be gain-controlled in this fashion, since some will require excessively high collector currents before the current gain starts to drop off. The operating frequency may also be limited to the higher frequencies near f_T.

At lower frequencies, the resistive part of the input impedance changes along with the collector current, thereby partially canceling any changes in power gain. At higher frequencies the input impedance is lower and more strongly dominated by the reasonably constant value of $r_{bb'}$; changes in collector current will therefore have little effect on R_{in}. This provides the additional benefit of a constant loading resistance for any tuned circuits.

The range of gain that is available through forward control of the collector current is more limited than for the reverse control method. Therefore, the collector-to-emitter voltage is reduced at the same time. This is easily done by including a series resistor in the collector supply line so that the higher current creates a larger voltage drop.

A schematic diagram of a forward gain-controlled amplifier is shown in Figure 7-11. Note the reversed polarity of the rectifier diode from the previous figure and the extra collector resistor used. Neutralization can be added to either circuit if required.

7-9.3 Comparison of AGC Methods

The graphs of Figure 7-12 compare the two methods of gain control, forward and reverse. The amount of gain reduction is about the same in each case, 35 dB. The bandwidth of the amplifier using reverse control will become narrower as the gain is reduced, owing to the increased input resistance of the transistor. This would be minimized if the loading on the tuned circuits was made more dependent on the output resistance of another noncontrolled stage and losses in the coil. Crystal, ceramic, and mechanical filters would show very little bandwidth change but could easily show a change in passband ripple if the input impedance were not swamped with a fixed-value resistor.

The forward controlled amplifier shows an increase in bandwidth as the gain is reduced, owing to the lowering of the input resistance with increased current. A second problem is that the center frequency may rise as the gain is reduced. This happens because the capacitance $C_{b'c}$ increases with reduced collector voltage, making the total input capacitance increase. The detuning can be minimized by keeping any external shunt capacitors large in comparison with the input capacitance.

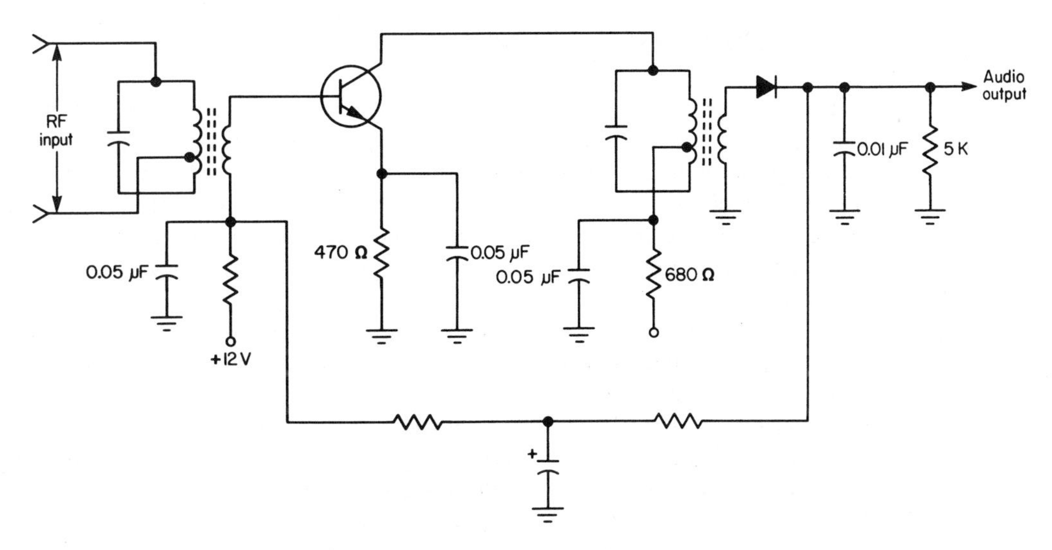

Figure 7-11 Amplifier and detector circuit to provide automatic forward gain control.

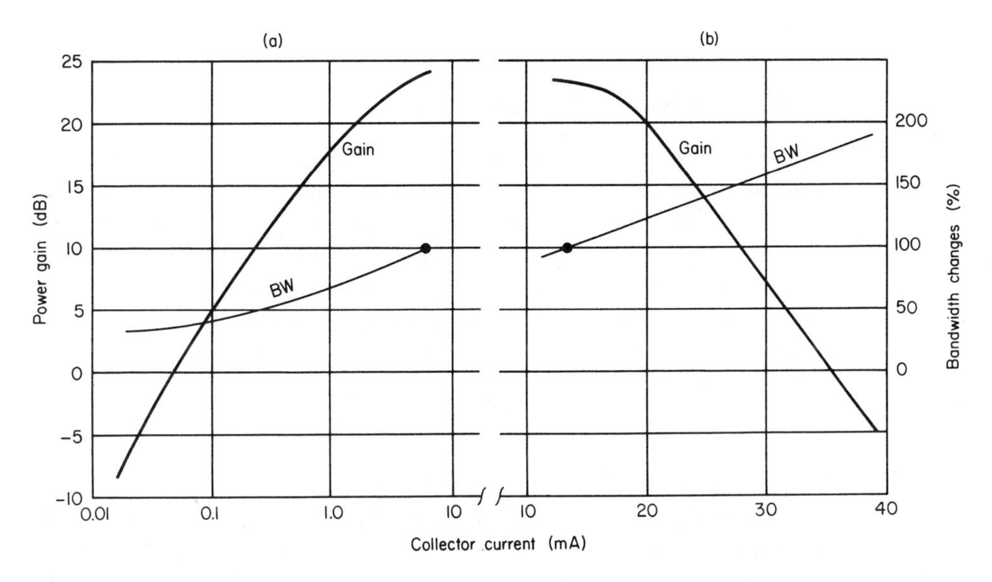

Figure 7-12 Gain reduction and bandwidth changes of a reverse gain-controlled amplifier (a), and a forward gain-controlled amplifier (b).

7-10 LOW-NOISE DESIGN

When used as the first stage of a receiver, the noise generated within an amplifier becomes more important than the power gain. The conflict between gain and noise figure arise because lowest noise figure depends on careful control of emitter current and source resistance. A typical variation of noise figure with bias current and source resistance is shown in Figure 7-13(a), and the variation of the optimum source resistance with frequency is shown in Figure 7-13(b).

These bias current and source resistance values are different from the corresponding values necessary for maximum gain, and so gain suffers. For most transistors, the loss is only a few dB and is rarely of any consequence, since low-level RF stages should operate with low gains anyway (see Chapter 10).

Any resistive losses in tuned circuits, transmission lines, and matching networks before the transistor will directly add to the overall system noise figure. For example, a transistor capable of a 1.5-dB noise figure that is used with an input-matching network having a 0.5-dB loss will result in an overall noise figure for the amplifier of 2.0 dB. These losses must therefore be kept low, and to do this a high ratio of loaded to unloaded Q must be used.

$$IL\,(\text{dB}) \approx 20\log\left(1+\frac{Q_L}{Q_U}\right) \qquad \text{for } Q_U > 5Q_L$$

(A 10 : 1 ratio will produce an insertion loss of 0.83 dB.)

7-10.1 Low-Noise Example

As an example, a low-noise amplifier was designed for use at 100 MHz. The transistor characteristics are those shown in Figures 7-1 and 7-13. The optimum emitter current is indicated as 1.5 mA and the optimum source resistance is 190 Ω. The transistor itself is capable of a 2.0-dB noise figure at this frequency.

By experimenting, it was found that an inductor could easily be made with an unloaded Q of 100. To keep the insertion loss below 0.3 dB (total noise figure would then be 2.3 dB), the loaded Q will have to be about 1/30 the unloaded Q.

$$IL = 20\log(1+1/30) = 0.285\,\text{dB}$$

The loaded Q should then be 100/30 = 3.33.

The parallel input resistance of the transistor was measured as 60 Ω when the transistor was biased at 1.5 mA and with a load resistance of 330 Ω. All the values necessary for the input circuit design are now known and are shown in Figure 7-14. Note the large mismatch between the transformed source resistance of 190 Ω and the transistor parallel input resistance of 60 Ω (which results in a 1.5-dB mismatch loss).

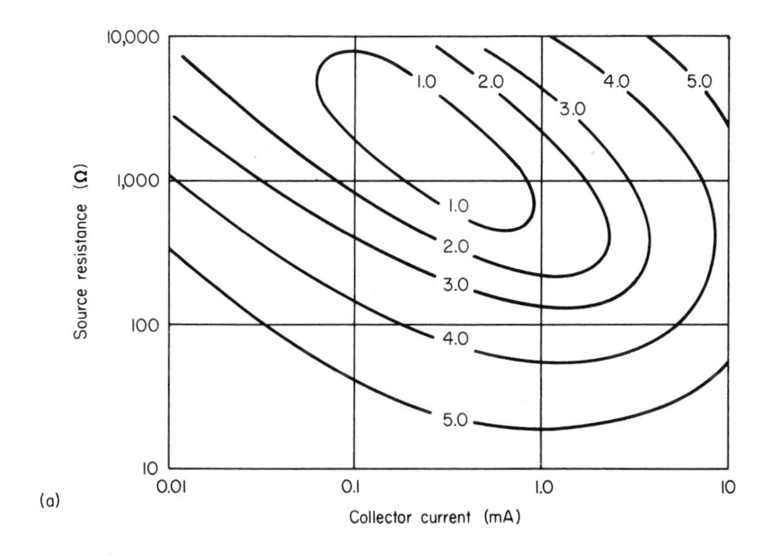

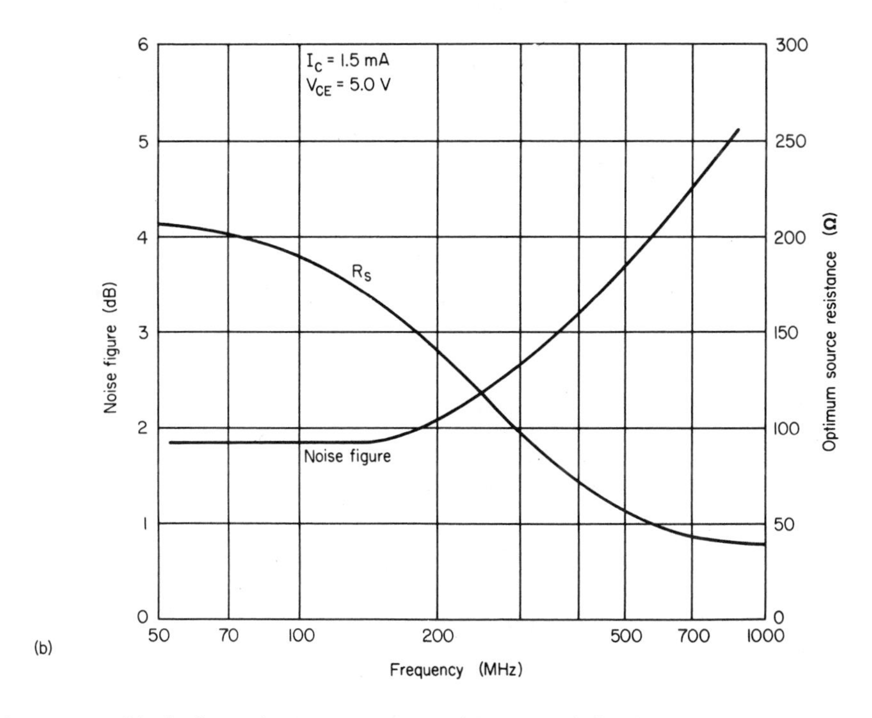

Figure 7-13 Variation of 1.0-MHz noise figure with both source resistance and emitter bias current (a), and the variation of optimum noise figure with frequency (b).

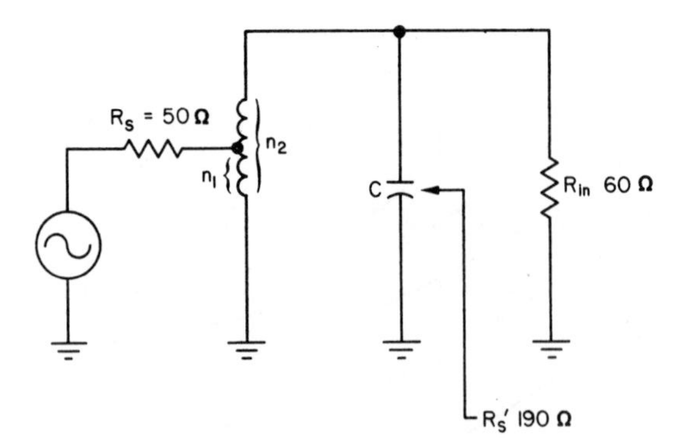

Figure 7-14 Input matching network for a low-noise design.

The turns ratio on the inductor will be

$$\frac{n_2}{n_1} = \sqrt{\frac{190}{50}} = 2:1$$

The total load on the inductor will be

$$60\,\Omega \| 190\,\Omega = 45.6\,\Omega$$

(the coil losses will have an insignificant effect on the total loading).

The reactance of the inductor and capacitor will be:

$$X_L = X_C = \frac{R_{\text{total}}}{Q_L}$$

$$= \frac{45.6}{3.33} = 13.7\,\Omega$$

The inductor value at 100 MHz is

$$L = \frac{X_l}{2\pi f} = \frac{13.7}{2\pi \times 100 \text{ MHz}}$$

$$= 0.022\,\mu\text{H} \quad \left(\text{tapped } \tfrac{1}{2} \text{ way}\right)$$

The capacitor value at 100 MHz is

$$C = \frac{1}{2\pi f X_C} = \frac{1}{2\pi \times 100 \text{ MHz} \times 13.7}$$

$$= 116 \text{ pF}$$

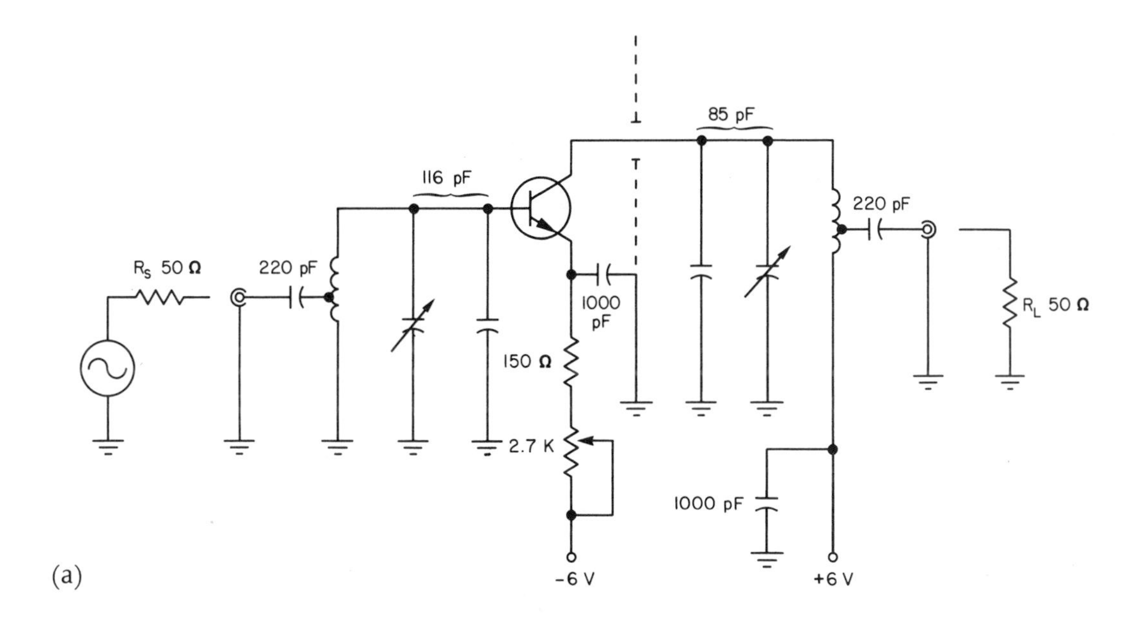

Figure 7-15 Circuit diagram (a) of the 100-MHz low-noise amplifiers and a photograph (b) showing the construction.

This should be made as a combination of a fixed and a variable capacitor so that the tolerance of the inductor and the input capacitance of the capacitor can be adjusted for. The final circuit diagram and photograph are shown in Figure 7-15. The bias arrangement used is one way of keeping any bias resistors from adding extra thermal noise.

Just for interest, the noise figure and the gain of the amplifier were measured both at the design current of 1.5 mA and also at higher and lower values. The results shown in Figure 7-16 illustrate two points. First, the value of collector current chosen does give the lowest noise figure. Second, reverse gain control, at least for this transistor, provides little increase in noise figure.

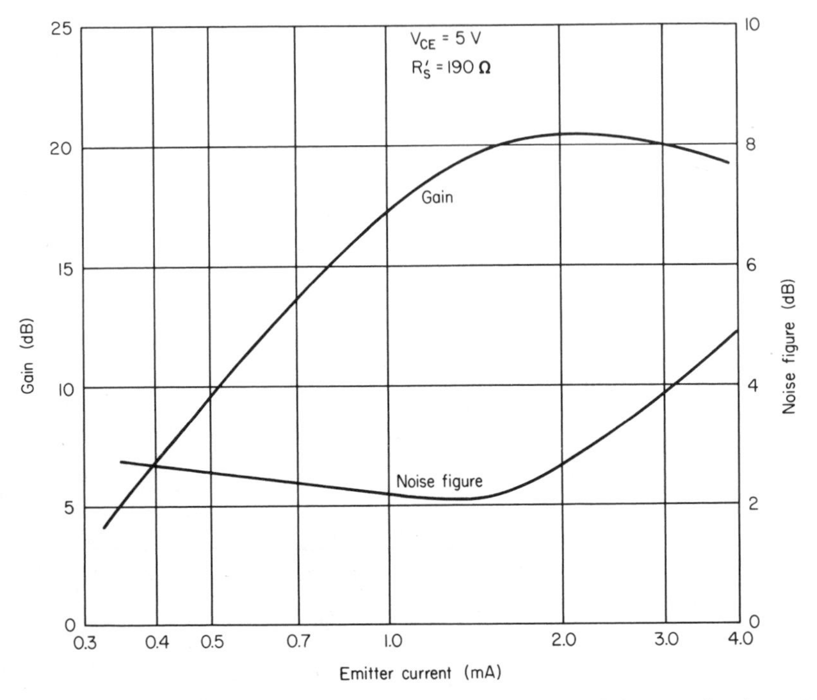

Figure 7-16 Noise figure and gain of a 100-MHz amplifier with changing collector bias current.

7-11 COMMON BASE/GATE

When operated near f_T, the power gain of a transistor can be increased slightly by using the common base configuration. The configuration also tends to be more stable than the common emitter circuit; partly due to the decreased internal feedback and partly due to the lower values of source impedance used. Linearity is also improved slightly in this configuration, so cross-modulation problems are reduced, particularly for the bipolar transistor. Figure 7-17 shows the bipolar and field-effect transistors operating in the common base/gate configuration.

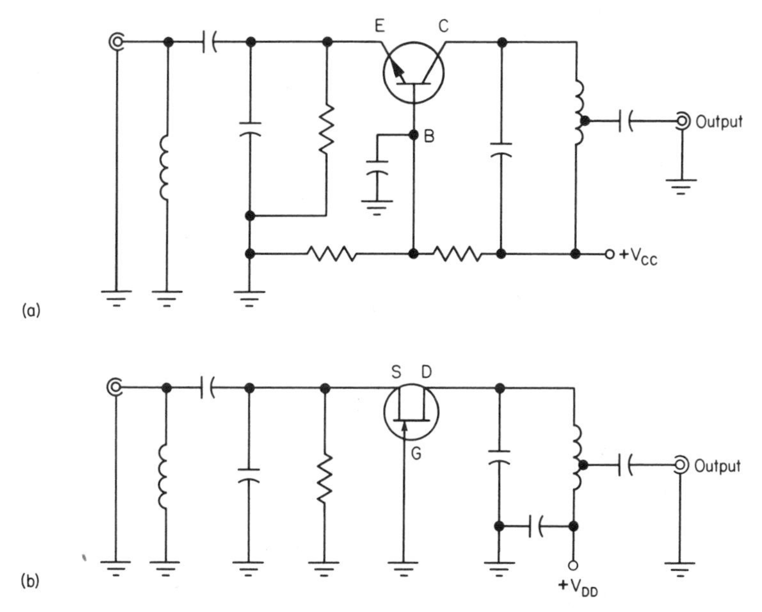

Figure 7-17 Common base bipolar amplifier (a) and common gate J-FET amplifiers (b) offer improved stability in a 200-MHz amplifier.

7-12 CASCODE CONFIGURATION

Another configuration that is inherently stable is the *cascode amplifier*, a name left over from the day of the vacuum tube. In the bipolar transistor version, the first transistor operates common emitter and sees as its load, the low input impedance of a common base stage. Neither stage has any great tendency to oscillate, since the first transistor is badly mismatched at its collector and the second stage has very little feedback because of its common base configuration. The cascode circuit will also have a low-noise figure and will be easy to design and align because of the very low output to input coupling. Figure 7-18 is a schematic of a wide-band amplifier built by the author. The transformers are 2 : 1 transmission-line types giving the amplifier a gain of 20 dB over a 5-MHz to 500-MHz bandwidth.

The same arrangement can be built with field-effect transistors. The dual-gate MOS-FET is already internally connected as a cascode arrangement and provides a very good combination of noise figure, cross-modulation rejection, and AGC operation. Most versions are unconditionally stable above about 200 MHz and can be used unneutralized over most of the VHF range. A schematic diagram of a 100-MHz amplifier using a dual-gate MOS-FET is shown in Figure 7-19.

Gain control can be provided for the amplifier of Figure 7-19 in a number of ways; either gate could be used for reverse or forward gain control. For best cross-modulation rejection, reverse AGC on gate 2 should be used and

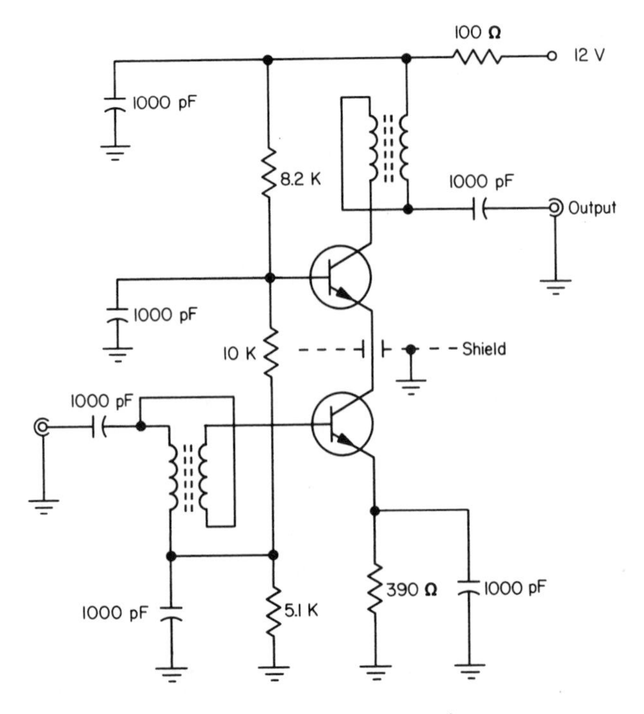

Figure 7-18 Wide-band amplifier using a cascode arrangement.

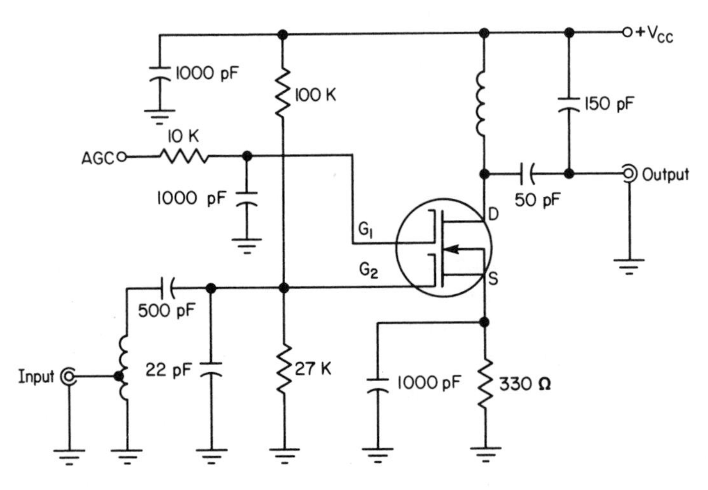

Figure 7-19 Dual-gate MOS-FET used in a 100-MHz tuned amplifier.

gate 1 held at a constant voltage: this will require a Zener diode at the source to maintain constant source voltage as the drain current changes. Gain reductions of 30–40 dB are easily obtained.

7-13 INTEGRATED-CIRCUIT RF AMPLIFIERS

When designing the IF stage of a receiver, best results are obtained if the filtering and the amplifying stages are separated into two individual blocks. The prepackaged crystal or ceramic filter is therefore very often used for the filtering and one or two linear integrated circuits supply the gain.

This packaging of the amplifiers provides several benefits for the designer:

1 High values of power gain are available in one package, 40–70 dB.
2 Internal feedback is minimized to the point where the circuits are usually unconditionally stable.
3 Input impedances are not sensitive to load impedances, so designing is easier: data-sheet input and output admittances are used directly.
4 AGC stages can be kept away from the input terminals so that preceding filters always see a constant input resistance, regardless of the amount of gain reduction.

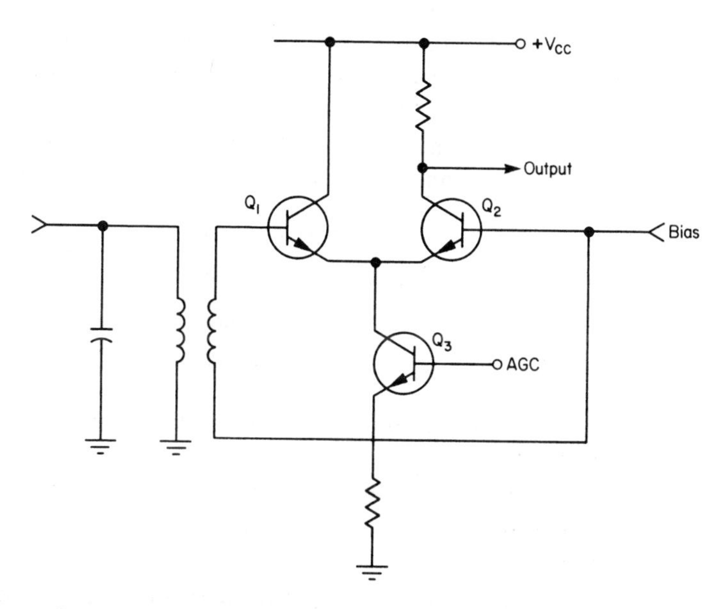

Figure 7-20 Emitter-coupled amplifier often used as the basic building block of high-frequency linear amplifiers.

The standard configuration for a linear IC amplifier is the emitter coupled pair shown in Figure 7-20. It is easily manufactured with closely matched transistors, its gain is easily controlled, and, for some configurations, the reverse feedback is much lower than for a single common emitter stage. Reverse gain control is obtained by reducing the base voltage on the current supply transistor Q_3.

For FM limiter applications, the emitter-coupled pairs are biased closer to cutoff. The incoming signal will then drive one transistor, then the other, alternately into cutoff. This provides a cleaner and faster limiting action than saturation does. It also avoids the lowering of the input impedance of a transistor when it saturates.

QUESTIONS

1 Calculate Stern's stability factor for a transistor with the characteristics shown in Figure 7-1. Use a frequency of 40 MHz, a source resistance of 180 Ω, and a load resistance of 1200 Ω. Is the amplifier stable?

2 Using the same transistor and operating frequency from Question 1, calculate the mismatch factor and optimum source and load admittances for a stable unneutralized amplifier.

3 A high-frequency transistor has an input resistance of 150 Ω, but for low-noise operation it is operated from a 250-Ω source. Calculate the loss in power gain (dB) due to this mismatch.

4 For a transistor with the following characteristics, calculate Y_{in} and Y_{out} for a source resistance of 75 Ω and a load resistance of 500 Ω,

$$Y_{ie} = 20 + j15 \text{ mmhos}$$
$$Y_{fe} = 30 - j100 \text{ mmhos}$$
$$Y_{re} = 0 - j1.2 \text{ mmhos}$$
$$Y_{oe} = 1.0 + j2.0 \text{ mmhos}$$

5 For the same transistor, source and load as in Question 4 and at a frequency of 25 MHz, what value components must be placed across the input and the output terminals to make each look purely resistive?

6 Explain why mismatching at the source and load ends of an amplifier can improve its stability.

7 Find the minimum Q needed to keep the added noise due to insertion loss below 1.0 dB for an amplifier at 40 MHz with an input tuned circuit bandwidth of 5.0 MHz.

8 Considering only the input side of an amplifier, when the source is matched to the transistor's input then only 1/2 of the signal fed back enters the transistor and the remainder is dissipated in the source resistance. By how many dB will the feedback to the transistor's input be reduced if the source resistance is 1/3 that of the transistor's input? (see Figure 7-6)

REFERENCES

Motorola Semiconductor Products Inc. *R.F. Small Signal Design-Using Admittance Parameters*, Motorola Application Note AN-215. Phoenix, Arizona.

Texas Instrument Inc., 1965. *Communications Handbook*. Dallas, Texas.

chapter 8

oscillators

Oscillators are common to both receiving and transmitting circuits, where they serve as carrier oscillators, local oscillators for mixers, and beat frequency oscillators for detection, to name a few. Their frequency may be fixed or tunable either continuously or in steps. Their power level may range from a few milliwatts up to the 1-W level.

8-1 OSCILLATOR SPECIFICATIONS

For any given frequency and power level, it is possible to design many different circuits, some very simple and others quite elaborate. The difference will be in the harmonic content and stability of the signal produced.

Harmonics of the oscillator frequency will always be present to some extent, and their amplitude can be reduced with properly designed low-pass or bandpass filters. Figure 8-1 shows a typical set of oscillator harmonics as they would appear on a spectrum analyzer. Whether the harmonics were too high or not would depend on the actual power level and the use the oscillator was intended for.

The stability of an oscillator is a more critical specification and will take the greater amount of a designer's time to meet. There are two types of

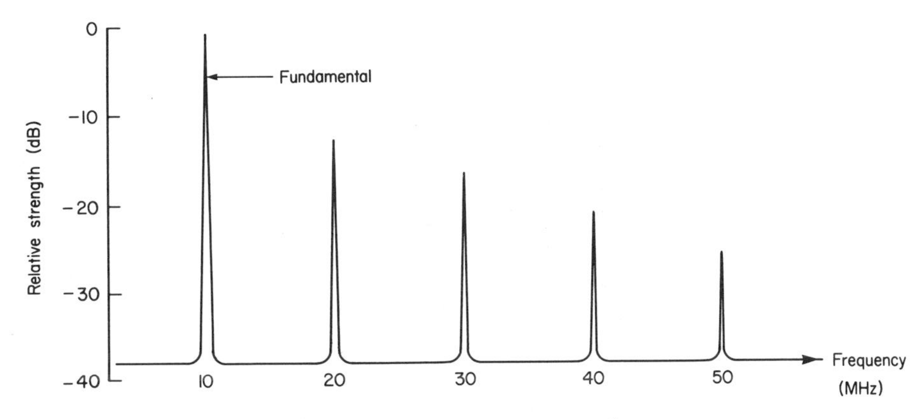

Figure 8-1 Oscillator harmonics as seen on a spectrum analyzer.

stability, long term and short term. *Long-term stability* describes the slow drift of the output frequency over a period of minutes or hours. A receiver operator can correct for this slow drift by turning the dial to retain the station he/she is listening to. Many television and FM receivers use automatic frequency control circuits to correct for the drift of the local oscillator in the tuner. For other applications, such as transmitters, drift cannot be tolerated, so much greater long-term stability is needed. Since drift is usually a problem of temperature changes, careful control of temperature and use of compensating capacitors and the like will reduce the problem.

Short-term stability is the hardest problem of all. If the oscillator rapidly changes its output level up and down, this represents amplitude modulation, so some sidebands must appear. Similarly, if the oscillating frequency shifts back and forth at a rapid rate, the resulting frequency modulation must also produce sidebands. Finally, any random noise generated within the transistor will cause both AM and FM and so create noise sidebands. The problem with these sidebands is that they are usually quite close to the main frequency and, once generated, are extremely hard to remove with any reasonable filters.

The spectrums of Figure 8-2 are commonly seen when designing oscillators.

8-2 CONDITIONS FOR OSCILLATION

An oscillator circuit consists of two parts, some type of active device, such as a transistor or vacuum tube that will produce power gain, and a feedback network that determines the operating frequency. Of the two, the feedback network is the more critical, as it determines:

1 The frequency of oscillation.

2 The thermal drift rate.

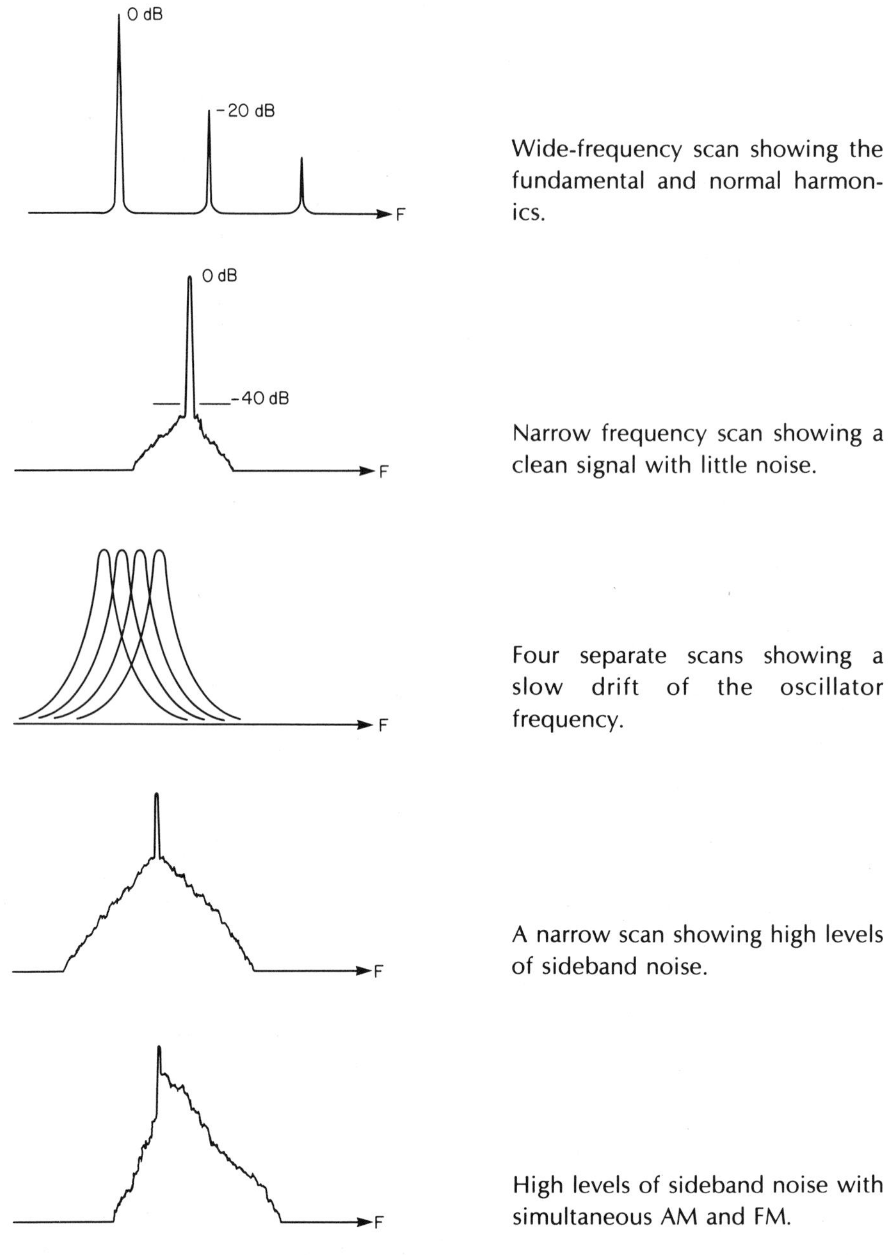

Figure 8-2 Common spectrum analyzer presentations seen when working with oscillator circuits.

3 The noise level of the output.

4 The harmonic content.

5 The sideband signal level.

The transistor itself will also affect each of the above, but to a much smaller extent if the circuit is properly designed.

To oscillate, the network must feed enough power back to the input of the transistor to sustain oscillations. This is satisfied by having the product of the transistor gain times the network losses, including any mismatches, exactly equal to 1. Normally, though, this product is set slightly higher than 1 so that

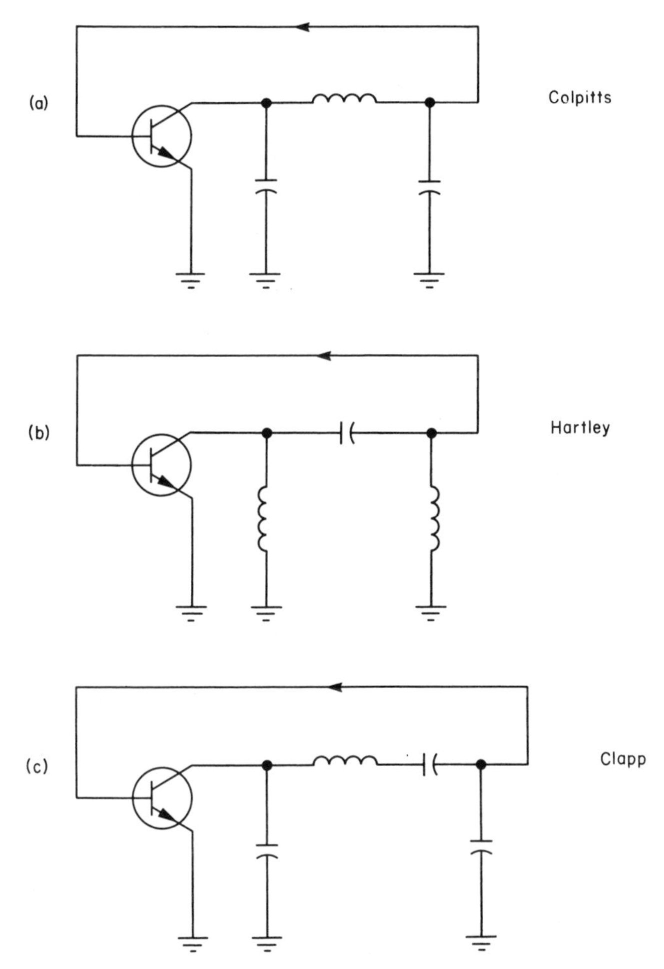

Figure 8-3 Three types of high-frequency oscillator circuits, named after the men who developed the feedback networks.

the oscillations can start and build up to a high level. Various nonlinearities within the transistor will eventually limit the amplitude at a constant value.

The network must also adjust the circuit phase shift, as oscillations can only occur if the energy fed back arrives exactly in phase with the signal already there. The total phase shift of the transistor and the network must, therefore, be exactly 360°.

Three common oscillator circuits are shown in Figure 8-3. They are named after the men who developed the feedback networks. Each network is essentially a Π configuration, with both shunt reactances the same type and an opposite type, series reactance.

8-3 COLPITTS OSCILLATOR

We will take a detailed look at the *Colpitts oscillator* now and see what component values are required to make it oscillate and what its frequency will be. First, let us look at a complete circuit for the oscillator and reduce it down to the essentials. A typical complete schematic is shown in Figure 8-4. The feedback network consists of L, C_1, and C_2. The load resistor (R_L) is connected to the collector by a large-value capacitor C_4 that is strictly for dc isolation. R_1, R_2, and R_E are biasing resistors; R_1 and R_2 are fully bypassed by the large-value C_3, so are not part of the active circuit. R_E is not bypassed, so remains a part of the active circuit.

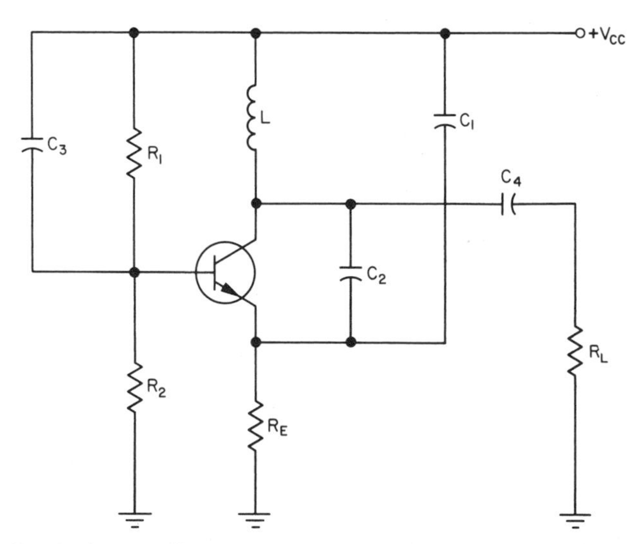

Figure 8-4 Typical schematic for a Colpitts oscillator.

Because the base is bypassed to the power supply, the transistor can be considered as operating in the common base configuration. The collector signal would then be fed back to the emitter through the tapped-capacitor C_1 and C_2 chain. Although this arrangement is perfectly valid, it means that common base characteristics must be available for the transistor, but they rarely are. Common emitter parameters from data sheets could, of course, be converted, but a certain amount of work would be involved. With one simple modification, the circuit can be simplified and common emitter parameters can be used directly. The problem is the load resistor (R_L). If the power supply and C_3 and C_4 are considered as short circuits at the operating frequency, the load resistor is actually connected from collector to base. If the voltage gain of the transistor is large, the collector-to-base signal would be about the same size as the collector-to-emitter signal, so we will consider the load resistor across the collector and emitter for simplicity. (We are ignoring any change in the transistor's input resistance by doing this.)

Our simplified equivalent circuit for the oscillator is shown in Figure 8-5. It includes the equivalent circuit of the transistor. R_{in} is the real part of the parallel input components when the load resistor (R_L) is connected to the output of the transistor. Any reactive input component will be lumped with C_1. A_i is the current gain of the transistor at the operating frequency. (For greater accuracy, R_L could be placed in parallel with L.)

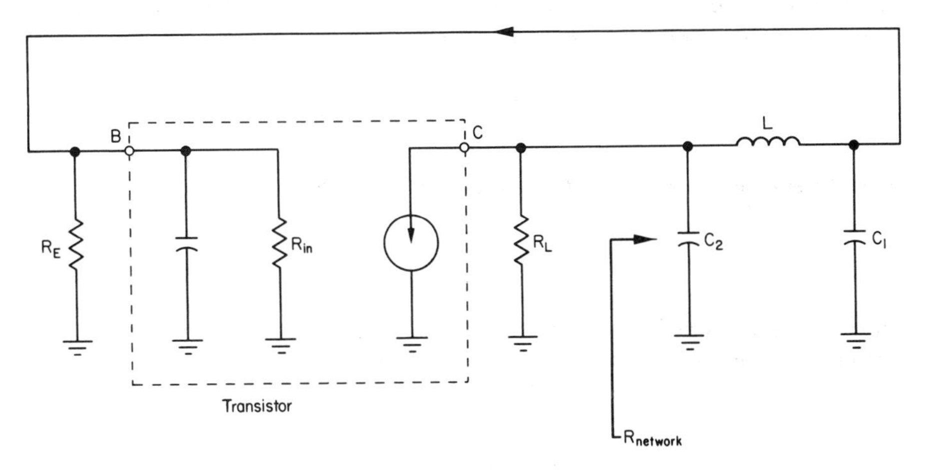

Figure 8-5 Equivalent circuit of the Colpitts transistor oscillator with the transistor operating common emitter configuration.

For best short-term stability and reduced drift caused by changes in the transistor characteristics, the feedback network consisting of C_2, L, and C_1 should have a high Q; 10 would be nice. This will mean that the phase shift of the network will change rapidly with frequency, and this will keep any frequency shifts small. This high Q also makes the design calculations simpler.

For the three-element Π network with high Q, the frequency of 180° phase shift will occur very close to the resonant frequency:

$$f_r \approx \frac{1}{2\pi\sqrt{LC}} \tag{8-1}$$

where: $$C = \frac{C_1 \times C_2}{C_1 + C_2} \tag{8-2}$$

At this frequency, the input impedance of the network when the other end is loaded by R_{in} of the transistor, will be

$$R_{network} \approx R_{in} \times \left(\frac{C_1}{C_2}\right)^2 \tag{8-3}$$

To sustain oscillations, a small amount of the transistor's output signal must be fed back to the base. If the transistor power gain, for example, was 10, $\frac{1}{10}$ of this power would have to be fed back, leaving $\frac{9}{10}$ of the total power for the load. Neglecting internal losses, all the power that enters the network will eventually reach the base, since only reactive components are involved. The feedback ratio must therefore be obtained as a mismatch ratio. If the network input resistance were set to 9 times the load resistance, the power entering the network would be $\frac{1}{9}$ the power entering the load. Using this ratio,

$$\frac{R_{network}}{R_{load}} \approx A_p \tag{8-4}$$

and the simplest formula for the power gain of a transistor,

$$A_p = (A_i)^2 \frac{R_L}{R_{in}} \tag{8-5}$$

plus the network resistance given in Equation 8-3, we can develop a simple formula for the minimum conditions for oscillation:

$$\frac{A_i \times R_L}{R_{in}} \geqslant \frac{C_1}{C_2} \tag{8-6}$$

C_1 will usually be the larger capacitor, and if the ratio of C_1 to C_2 does not exceed what really amounts to the voltage gain of the transistor (with R_L), the circuit will oscillate. The frequency of oscillation will be close to that given by Equations (8-1) and (8-2), how close depends on the total Q of the network and on the phase shift within the transistor. Since all oscillators will have some adjustable component, it is not worth calculating the operating frequency with any greater accuracy.

8-4 PRACTICAL CONSIDERATIONS

The previous section indicated the maximum capacitance ratio that could be used; it did not give an exact ratio, nor did it suggest actual component values.

If all our assumptions are correct and if we know the current gain and input resistance of the transistor accurately, the capacitor ratio suggested by Equation (8-6) will just barely sustain oscillations, if they get started in the first place. To account for variations in transistors and to make sure oscillations start, a lower capacitance ratio is used. This feeds back more power to the base than is required, but it does assure oscillator starting. But how much should the ratio be lowered? This usually depends on the application but let us look at the consequences.

If the ratio is lowered, the extra feedback must continue to build up the oscillations until something stops it. That something is usually the transistor going into saturation or cutoff or both. Extra feedback then means a higher output signal (which is good), but it also means distortion (which is not good), causing harmonics. A high Q in the feedback network will usually keep these harmonics down to manageable levels, and extra low-pass filters can be used if necessary. So far, then, the extra feedback does not look too objectionable.

The hardest design problem with an oscillator is the short-term stability mentioned previously. The transistor itself is the biggest problem. Any minor change in internal parameters will cause a change in power gain and overall phase shift. This results in frequency and amplitude modulation of the output waveform that creates hard-to-remove sidebands. Stability can be improved by increasing the feedback to the point where the transistor is cut off for a good part of the cycle; the oscillating conditions are then determined more by the network and less by the transistor. The final decision is best made while looking at a spectrum analyzer for harmonics and close sidebands. Normally, a ratio about one-half that given in Equation (8-6) provides good results.

Once the capacitor ratio has been more or less decided upon, the actual values must be selected. The prime consideration is to maintain a high Q in the feedback network to improve stability. The network itself is redrawn in Figure 8-6. The total Q of this network will depend on the loading at both ends and can be calculated from

$$Q_{\text{total}} = \frac{(X_1 + X_2)(Q_1 \cdot Q_2)}{X_1 Q_2 + X_2 Q_1} \tag{8-7}$$

where: X_1 = reactance of C_1

X_2 = reactance of C_2

$$Q_1 = \frac{R_{\text{in}}}{X_1} \qquad Q_2 = \frac{R_L}{X_2}$$

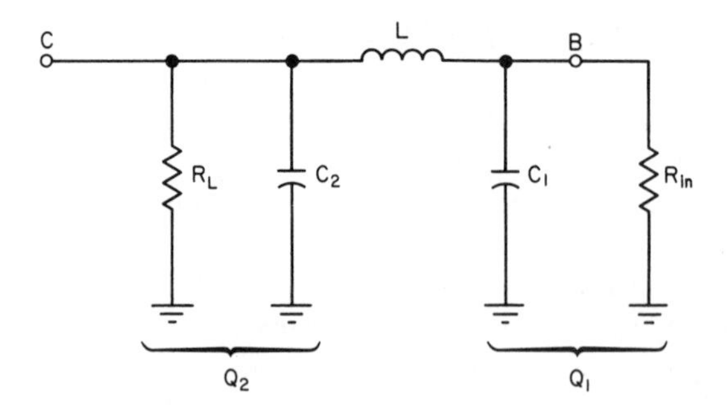

Figure 8-6 Loading of the feedback network of the Colpitts oscillator.

The total will always be someplace between Q_1 (which is usually the larger of the two) and Q_2 (usually the smaller), and for the typical oscillator application will usually be slightly higher than

$$Q_2 = \frac{R_L}{X_{C_2}}$$

We will therefore use the load-resistor end of the network to set the overall Q to about 10.

We should now have enough information to try our luck with an example.

EXAMPLE 8-1

Design a Colpitts oscillator circuit to operate at 25 MHz with a 10-V peak-to-peak signal across a 1000-Ω load resistance.

Solution

The first step is to select a transistor and bias it. Assume that we have obtained a transistor with an f_T of 250 MHz and that the manufacturer's literature recommends a bias current of 10 mA. The maximum output voltage from the oscillator could be as high as twice the supply voltage, but from this we would have to subtract any voltage drop across the emitter resistor plus a substantial RF saturation voltage for the transistor. A supply voltage of at least 9 V would then be about right. The emitter resistor usually drops about 20% of the supply voltage (or 1.8 V), and the bias resistors usually carry a current about one-fourth of the collector current (or 2.5 mA). The biasing is then shown in Figure 8-7.

Next come the large-value bypass capacitors. Values of 0.001 μF will have a reactance of 6 Ω and will be adequate for base bypassing and coupling the load to the collector.

Finally, we look at the feedback network. To obtain a Q of 10 at the collector end, we need a capacitor reactance of $R_L/10$ or 100 Ω. A 64-pF

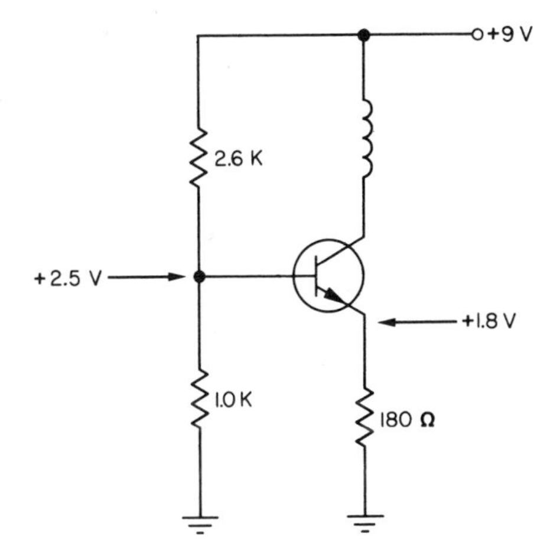

Figure 8-7

capacitor will do for C_2. The reactance at the other end can be found with the help of Equation (8-6). Since f_T is 250, the short-circuit current gain will be 10 at 25 MHz (the actual in-circuit gain will be lower). The data sheet indicates the parallel input resistance to be 120 Ω (the actual in-circuit resistance will also be lower). The maximum capacitor ratio will be

$$\frac{C_1}{C_2} = \frac{A_i \times R_L}{R_{\text{in}}} = \frac{10 \times 1000}{120} = 83.3$$

(The fact that A_i and R_{in} will both be lower tends to cancel.) We will use a capacitor ratio of one-half of this to ensure starting. Therefore,

$$\begin{aligned} C_1 &= \frac{83.3}{2} \times 64 \text{ pF} \\ &= 2666 \text{ pF} \end{aligned}$$

(2700 pF is a standard silvered mica value.)

This large value will override any variation in the input capacitance of the transistor and so improve the stability. The reactance of C_1 will be 2.4 Ω, bringing the total reactance of the two capacitors in series to 102.4 Ω. The coil reactance must approximately match this, so

$$\begin{aligned} L &= \frac{X_L}{2\pi f} \\ &= \frac{102.4}{2\pi \times 25 \times 10^6} \\ &= 0.652\ \mu\text{H} \end{aligned}$$

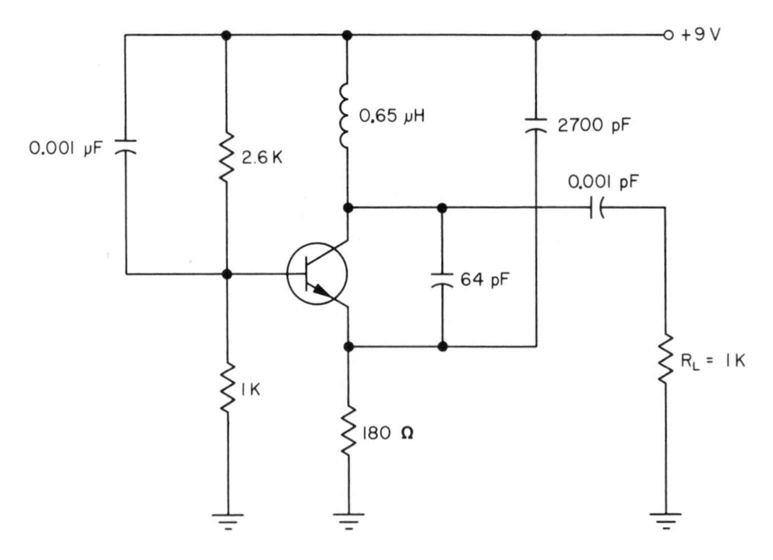

Figure 8-8

This is small enough to be wound on a nonmagnetic core such as a ceramic form. Either the coil or part of C_2 should be variable to allow fine tuning of the oscillator frequency. The coil can be made variable by including a small ferrite slug in the ceramic core. The temperature coefficient of the coil will now depend on the thermal expansion of the copper wire and the temperature characteristics of the ferrite. The end result is likely to be a positive coefficient, so a negative temperature coefficient capacitor could be used for C_2. (A coefficient around $N-100$ is usually necessary.)

The final oscillator diagram, as far as calculations are concerned, is shown in Figure 8-8. After construction, the final values may have to be altered somewhat while the output spectrum is checked for spectral purity and fundamental amplitude on a spectrum analyzer.

8-5 EMITTER RESISTOR

The emitter resistor (R_E) was included in the equivalent circuit of Figure 8-5, yet it never appeared in any of the discussion. This is because normally R_E is large compared with R_{in}, so can be ignored. However, another problem may exist. If excessive feedback is used, the transistor will be driven to cutoff. An *RC* time constant then exists between the feedback network capacitors and R_E, and this may control the time taken for the oscillator to turn back on. The result is called *squegging* and could appear on an oscilloscope as shown in Figure 8-9. The resulting amplitude modulation will cause a large number of sidebands to

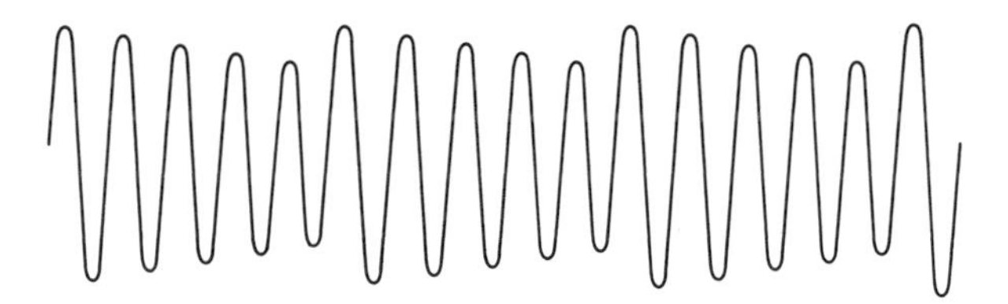

Figure 8-9 Squegging of an oscillator.

appear, some spaced a considerable distance away from the main carrier frequency. Adjustment of the feedback capacitor ratio and the addition of an emitter inductance will often solve the problem.

8-6 BUFFERING

If the load impedance that is connected to the oscillator changes in either magnitude or phase angle, the output of the oscillator will likely change in amplitude and frequency; in most cases this is an undesirable effect. A buffer amplifier can be placed between the oscillator and the load to minimize the problem and at the same time provide gain if necessary. A Colpitts oscillator circuit with a buffer amplifier added is shown in Figure 8-10. The circuit also shows the use of a regulated and filtered power supply for the oscillator section and the added emitter inductor.

The extra gain can be important when moderate output signal levels are required. It means that the oscillator level can be lowered, and this will improve the stability by reducing the heating in both the transistor and feedback network. The buffer amplifier does not provide complete isolation between the oscillator and load, for it must be remembered that transistors are not inherently unilateral; the input impedance does change with the load impedance. For best isolation, then, the buffer amplifier could be neutralized

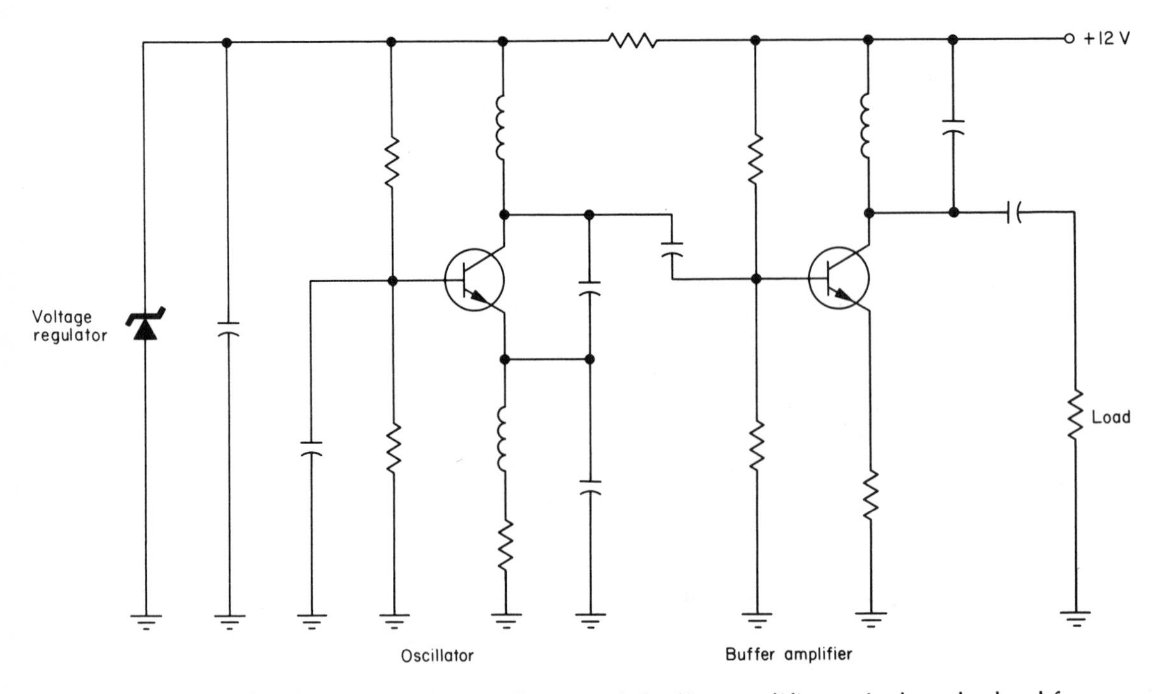

Figure 8-10 Colpitts oscillator with buffer amplifier to isolate the load from the oscillator.

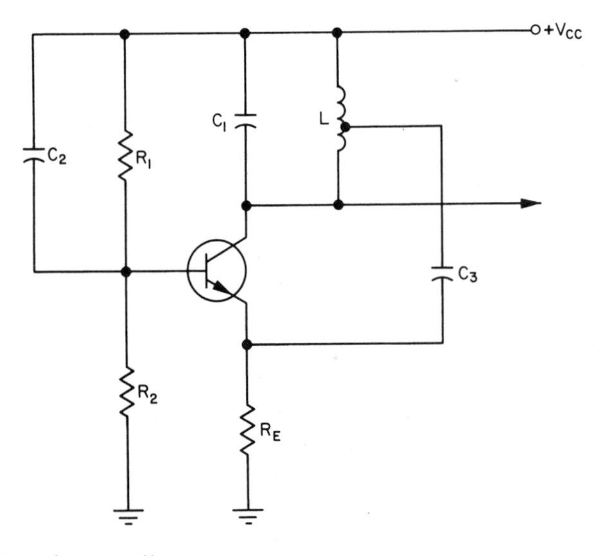

Figure 8-11 Hartley oscillator.

(rarely done) or mismatched by very loosely coupling it to the oscillator with a small-value capacitor and using low-value biasing resistors to swamp any input resistance variations.

Additional isolation is achieved with separate filtering and supply-voltage regulation for the oscillator. This would be especially necessary where changes in the load could significantly change the supply voltage.

8-7 HARTLEY OSCILLATOR

The *Hartley oscillator* is designed in exactly the same manner as the Colpitts circuit except that the inductor is tapped instead of the capacitor, but the ratio remains the same. A capacitor must now be included for dc isolation between collector and emitter. A typical Hartley oscillator is shown in Figure 8-11.

8-8 CLAPP OSCILLATOR

The *Clapp oscillator* is also very similar to the Colpitts in design and operation. The main difference is the extra capacitor in series with the inductor. At the operating frequency, the capacitor and inductor will still present the same equivalent inductive reactance as the Colpitts inductor, but, if the frequency were to drift, the equivalent inductive reactance would change very rapidly, much faster than a single inductor. This is shown in Figure 8-12.

In each case, the inductive reactance across the top of the network is $+j100\ \Omega$ at 1.00 MHz. But if the frequency were to increase 1%, to 1.01 MHz,

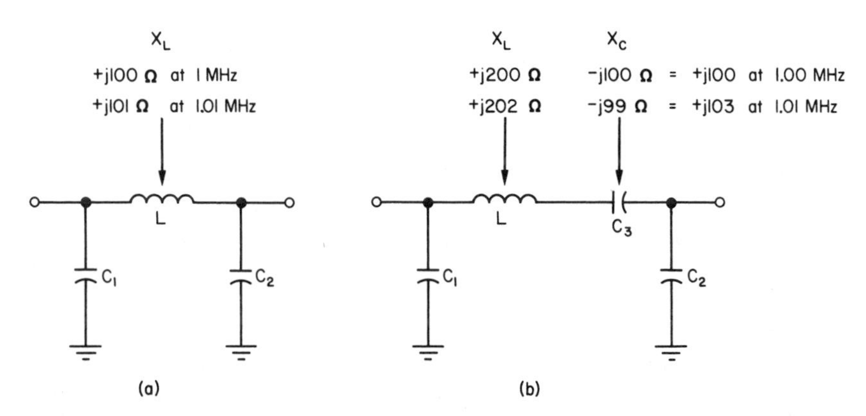

Figure 8-12 Comparison of the feedback networks for a Colpitts oscillator (a) and a Clapp oscillator (b).

the reactance of the Colpitt inductor would increase to $+j101\ \Omega$, and the equivalent position of the Clapp network would increase to $+j103\ \Omega$, three times the increase of the other circuit. The Clapp circuit *should* therefore be more stable, as much smaller frequency shifts will be needed to correct any phase errors that develop in the transistor. This stability, however, depends to a great extent on the inductor; if it drifts with temperature, the oscillator will drift at a faster rate than will the Colpitts circuit. Whether the Clapp circuit should be used or not will depend on the ability to construct high-quality coils and keep them at a constant temperature. A schematic diagram of a Clapp oscillator is shown in Figure 8-13.

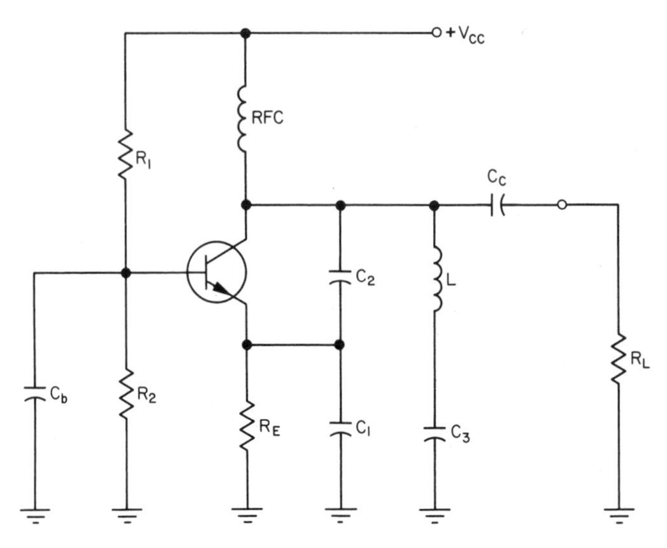

Figure 8-13 Clapp oscillator.

8-9 CRYSTAL OSCILLATORS

When extreme frequency stability is required, a quartz crystal can be substituted either for the tuned circuit or placed in the feedback loop. The typical tolerance of a crystal is about 0.002% over a wide temperature range, therefore providing a large increase in stability over any type of *LC* oscillator. The one disadvantage is lack of tunability; to change frequency a new crystal must be plugged in.

Recall from Chapter 3 that a quartz crystal is a very high *Q* tuned circuit with a series-resonant and a parallel-resonant frequency very close together. The equivalent circuit and the change of reactance with frequency are shown in Figure 8-14. The series frequency is the lower of the two resonances, and the equivalent reactance between resonant frequencies is inductive.

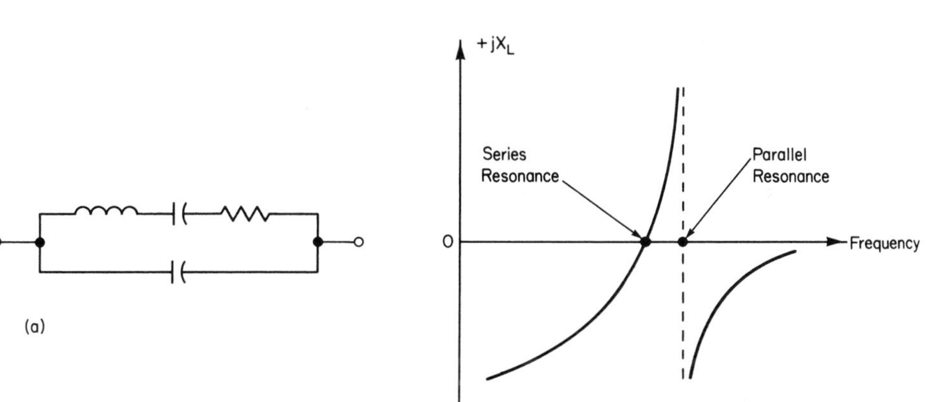

Figure 8-14 Equivalent circuit of a quartz crystal (a) and its impedance variation with frequency (b).

The crystal may be used in a number of different ways to stabilize the frequency of an oscillator, and each configuration could result in a slightly different frequency of oscillation. While a complete *LC* circuit does not have to be included in the circuit, it often is, since some crystal "cuts" may tend to oscillate on one of the crystal overtones. A relatively simple tuned circuit will encourage the oscillator to operate at the proper frequency.

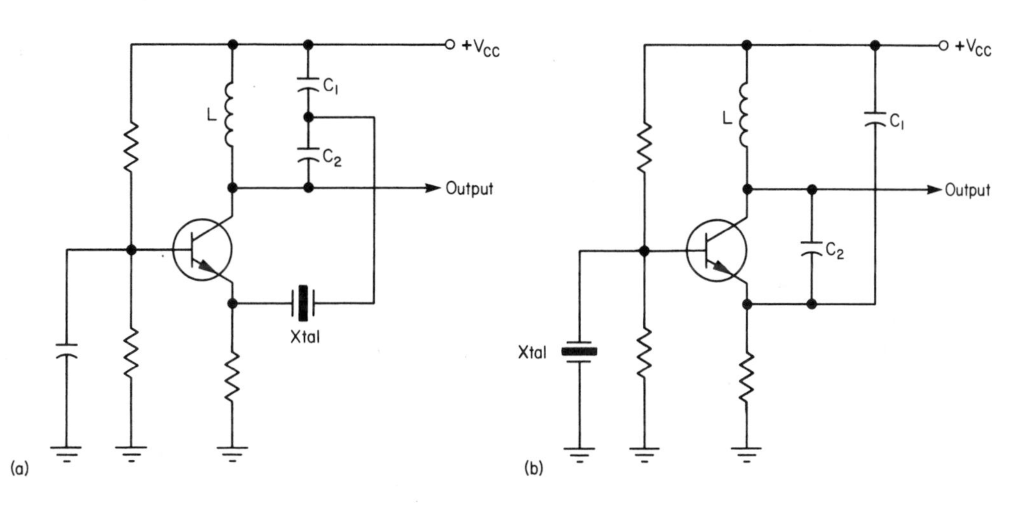

Figure 8-15 Two crystal oscillator circuits based on the Colpitts design. Operation will be at the series-resonant frequency of the crystal.

8-9.1 Series Mode

Several oscillator designs are available that operate at the series-resonant frequency of the crystal. Two of these are shown in Figure 8-15. The crystal in (a) is acting like a coupling capacitor and in (b) as a bypass capacitor. In either case oscillations can only occur when the crystal is a very low impedance, and this only occurs at the series-resonant frequency. Removal of the crystal should stop the oscillations, and replacement with a large-value bypass capacitor should cause oscillations at about the proper frequency.

At series resonance, the crystal is equivalent to a small-value resistor of about 20–200 Ω. If the RF current through the crystal is high (above 10 mA), the heating will cause the frequency to drift. A check of this current level is therefore worthwhile.

8-9.2 Parallel Mode

No circuit uses the crystal at exactly its parallel-resonant frequency, but some operate between the series and parallel points, where the reactance is inductive, and often this is *close* to the parallel frequency. Two *parallel-mode* oscillators are shown in Figure 8-16. Both oscillator circuits are shown with FETs in this case. While this is not absolutely necessary, it does emphasize the fact that better operation is obtained if the input impedance is high, particularly for (b). The crystal appears inductive in both cases while the drain tuned circuit of (a) appears capacitive and of (b) appears inductive. The circuit of (a) could be looked at as a Colpitts and (b) as a Hartley oscillator. In some cases, the internal capacitance of the transistor is sufficient to provide the feedback capacitance C_1 of circuit (b).

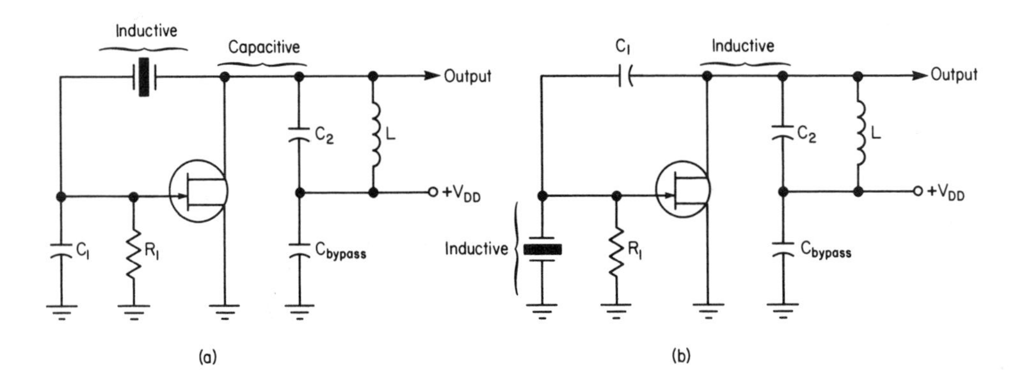

Figure 8-16 Two FET oscillator circuits operating close to the parallel-resonant frequency of the crystal. Circuit (a) is a Pierce oscillator and, (b) is a Miller oscillator.

8-10 FREQUENCY SYNTHESIS

The advantage of the crystal oscillator is its extreme accuracy and stability compared to any of the *LC* oscillators. Its disadvantage is that it cannot be tuned (series and shunt varactors can change the frequency of a crystal oscillator only 0.1%). For multichannel operation, a large number of crystals can be mounted and switched into the oscillator circuit one at a time. This can get expensive after a while and require a lot of space inside compact equipment. When VHF and UHF channels become closely spaced, even crystals with their low tolerances (0.002% or better) can produce uneven channel spacing.

The answer is the phase-locked loop synthesizer where only one or two crystals are required and the remainder of the work is done by digital circuits. As many as 4000 channels can easily be tuned, and the spacing between them can be kept constant. One adjustment is all that is required to accurately set all channel frequencies.

Figure 8-17 is a simple block diagram of a phase-locked loop (PLL) oscillator. The actual output signal comes from a voltage-controlled oscillator (VCO), a circuit whose output frequency is controlled by a dc input voltage.

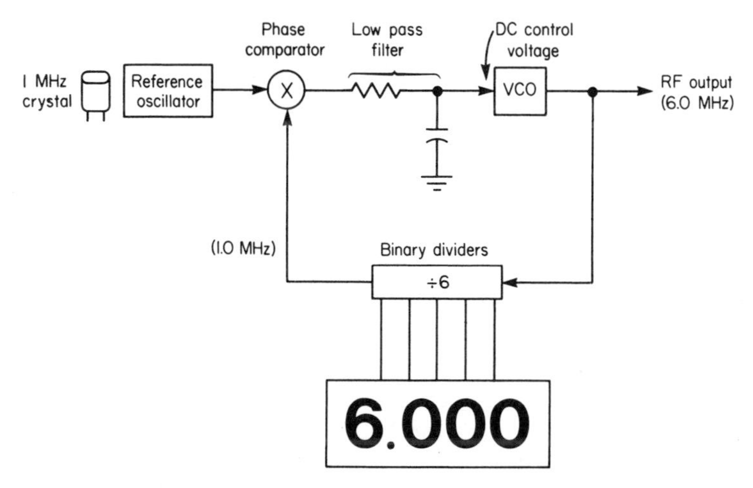

Figure 8-17 Block diagram of a phase-locked loop oscillator/multiplier.

A sample of the VCO output is fed back to a phase comparator (we will look at the divider shortly), which produces a dc output voltage proportional to the relative phase of its two input signals. If the reference and VCO inputs are at the same frequency, the phase comparator output will be a constant dc voltage. Any noise on this voltage is smoothed with a low-pass filter before it sets the frequency of the VCO. If the two comparator inputs are not identical in frequency, the control voltage will fluctuate and will attempt to drive the VCO to the correct frequency. Whether it succeeds or not will depend on gains within the loop, the comparator characteristics, and the cutoff frequency of the low-pass filter.

The reference frequency is derived from a crystal oscillator, as this will set the accuracy and frequency stability of the final RF output from the VCO. The PLL circuit has therefore given a variable-frequency oscillator (the VCO) the same *average* stability as the crystal-controlled oscillator. The word "average" is emphasized because a 1000-Hz low-pass filter, for example, will have a time constant of $1/2\pi \times 1000$ Hz $= 159$ μs. Any frequency changes over a shorter time period than this would go undetected.

To improve the usefulness of the circuit, a series of binary counters or frequency dividers can be placed between the VCO and the phase comparator. If the counters divide the VCO frequency by 6, the VCO could operate at 6.0 MHz and still supply a phase-related signal at 1.0 MHz for comparison with the reference signal. The phase comparator is happy, the dc control voltage is stable, and the VCO is operating at precisely 6 times the crystal frequency. Again, the VCO could wander around in frequency a bit, but the period of the wandering would be shorter than 159 μs.

As long as the VCO has the frequency capability, the dividers can be set to virtually any number, and the VCO output will always be that many times higher than the reference frequency. A typical CB application would involve a reference frequency of 10 kHz and dividers with 40 different settings. Forty frequencies could then be generated for the transmitter or receiver with crystal stability at 10-kHz intervals. Additional mixing would be required to get the frequencies up into the 27-MHz band. The low-pass filter, in this case, would likely have a cutoff frequency of 10 Hz.

The short-term stability of the voltage-controlled oscillator is very important, since the feedback loop will not have any control over frequency changes that occur faster than the low-pass filter bandwidth. Amplitude changes in the VCO, either short or long term, are not controlled at all. Long-term frequency stability, or drift, is the only parameter that is fully controlled. The designer has to be sure that, with a well-filtered dc voltage applied to the control input of the VCO itself, the RF output spectrum is relatively free of unwanted noise.

8-10.1 Initial Capture

When a PLL system is first turned on, it would be very unlikely to find the VCO at the correct frequency. Therefore, the loop has to generate a dc control voltage that will move the VCO toward the reference frequency so that phase lock occurs; the reference oscillator will then have "captured" the subharmonic of the VCO that is coming from the binary dividers. With proper design, capture will occur over a wide range of frequency differences between the reference and divided VCO frequencies. If, however, the desired frequency is outside this capture range, the VCO must be given a little help to find the right frequency.

To show how capture occurs, consider first that the reference oscillator and the divided VCO are very close in frequency. The output of the phase comparator could then contain a difference frequency that falls within the bandwidth of the low-pass filter. The VCO would immediately move to a new closer frequency as its control voltage changes and will quickly lock on to the reference signal. The output of the phase comparator will normally not even complete a full cycle before locking is complete. It should be kept in mind that this is a feedback loop and that the possibility of instability exists if the design is not carefully handled. Instability would mean that the control voltage would contain a sustained sine-wave component and that the VCO would keep swinging above and below the correct frequency.

If the reference and divided VCO frequencies are far apart initially, a very high difference frequency would appear at the comparator output, much higher than the cutoff frequency of the low-pass filter. The signal that finally appears on the control input of the VCO would be a very small sine wave, how small would depend on the initial frequency difference and the low-pass filter characteristics. The signal would be too small to immediately move the VCO to the right frequency, but it would cause it to move—alternately toward and away from the correct frequency. These frequency shifts, even though very small, are significant. When compared to the reference frequency, the relative phase will shift very rapidly when the VCO moves away from correct frequency and move slowly when the VCO moves in the right direction. The typical transient waveform that could be seen on the VCO control line during lockup is shown in Figure 8-18. The unequal size of the peaks produces an average component that is quickly amplified to speed the VCO toward the correct frequency.

8-10.2 Voltage-Controlled Oscillator

Each component of the PLL system will be examined briefly now to show some of the characteristics of the different circuits. All components are available in digital integrated circuit form so that final circuit details and characteristics are best obtained from data sheets.

The VCO must produce an output frequency over the full range needed by the designer and must be able to move under the full control of a dc control voltage. The oscillator must have very good short-term stability so that no significant sideband noise is created. In particular, random amplitude modulation must be prevented as the loop cannot correct this, no matter what the modulating frequency. Long-term drift will be controlled by the loop, so need not be considered by the designer.

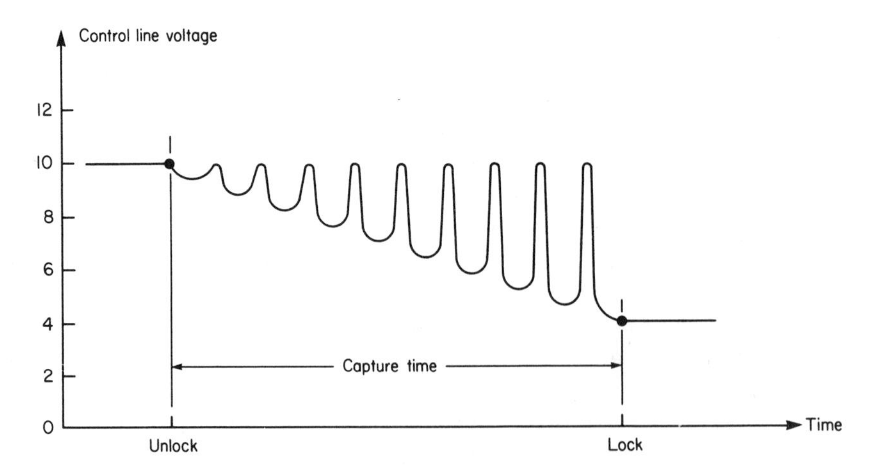

Figure 8-18 Transient waveform generated on the VCO control line as the loop pulls the VCO toward the correct frequency.

The relationship between dc control voltage and output frequency, oddly enough, should be linear. The reason is that the VCO is part of the feedback loop and while the output frequency and waveshape will not be affected, the time required to obtain lock-up and even the stability of the system could change from one end of the VCO range to the other.

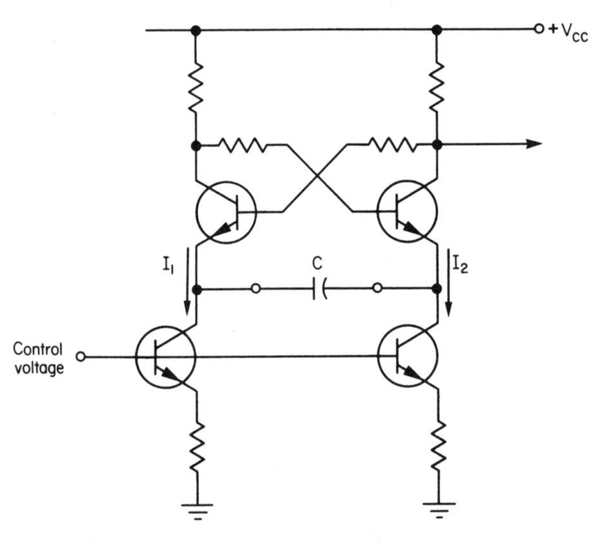

Figure 8-19 Three voltage-controlled oscillator circuits: (a) an emitter-coupled multivibrator, (b) a varactor-controlled Colpitts oscillator, and (c) a variable-frequency function generator.

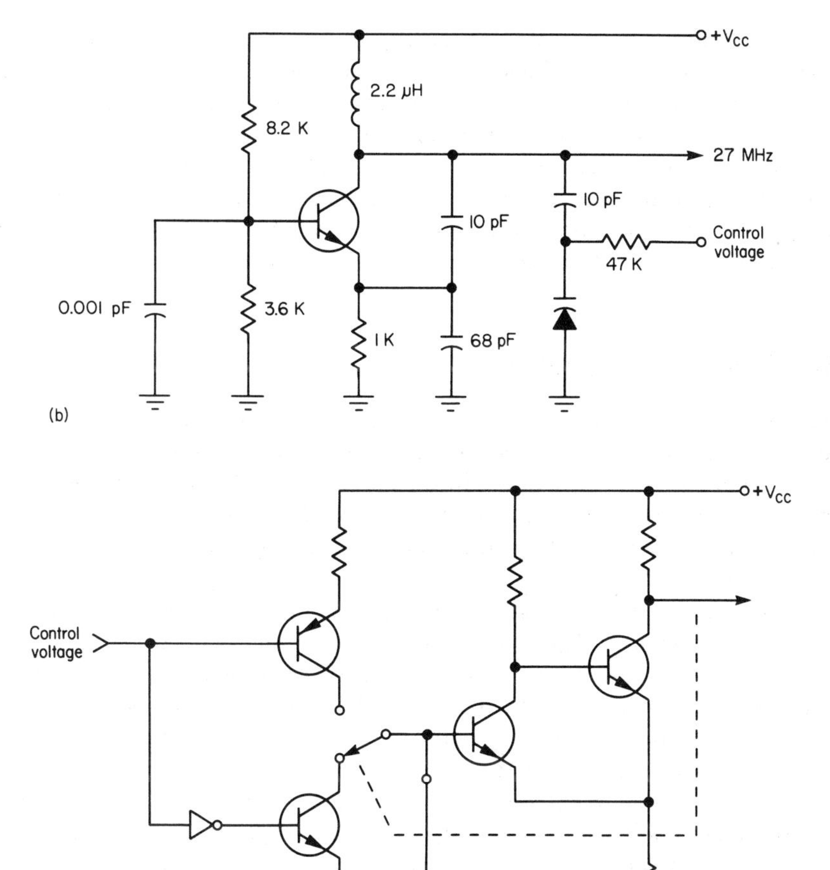

Figure 8-19 (cont'd)

Three samples of VCO circuits are shown in Figure 8-19. The first is an emitter-coupled multivibrator: the second is a normal Colpitts *LC* oscillator whose frequency is controlled by the bias voltage on a varactor diode. The third circuit is a variable-frequency *function generator* that simultaneously produces a square and triangular output. The emitter-coupled multivibrator produces a square-wave output, is available in integrated circuit form, and can operate up to about 20 MHz. It is tunable over at least 10:1 frequency range. Its frequency is adjusted by controlling the charge and discharge rate of the capacitor through a set of transistor current generators.

The output of the varactor-controlled oscillator is a sine wave and will operate only over a 2:1 frequency range. However, the operating frequency can extend right up to microwave frequencies. The function generator circuit consists of an integrator and Schmitt trigger circuit. The integrating capacitor is charged and discharged through constant-current generators, and the Schmitt trigger alternately switches on the charge and discharge circuits. Both a square and triangular output wave are available. The tuning range can exceed 10:1, but the top frequency is usually limited to about 2 MHz.

8-10.3 Phase Comparator

The comparator is available in several forms, each with its own characteristics and applications. Although discrete components could be used to build a phase comparator, digital comparators in integrated-circuit form are far more popular because of their low cost and ease of use. This does mean that the input signals must conform to the voltage levels for the logic family used, but this is rarely a problem.

The simplest comparator is made with an exclusive-OR gate, whose operation is shown in Figure 8-20. The output of this gate will be HI (+3.5 V for the TTL logic family) if one input or the other but not both inputs are high. Three separate sets of inputs are shown. For the first, input A leads B by a small amount, and the resulting output is a series of short pulses whose average voltage level will be close to zero. In the second case, A leads B by about 90°; the output is a square wave with a 50% duty cycle and will have an average voltage about one-half the peak output voltage. In the third case, A leads B by almost 180°; the average output is close to the voltage level corresponding to the HI. The phase to voltage conversion is summarized in (d). Note that the pattern reverses after the first 180°. One slope will drive the VCO away from the correct frequency, the other, toward the correct frequency. Lock will therefore always be obtained on the one slope and never on the other—which one depends on the inverting characteristics of any filter amplifiers and the VCO control input. The inputs to this comparator must always be a symmetrical square wave (rarely a problem), and this results in a fairly high immunity to noise on either input. The capture range of the PLL system with this comparator will depend to a great extent on the bandwidth of the low-pass filter that follows it. One disadvantage of this simple comparator is that it could cause the loop to lock onto harmonics of the reference input when the VCO is operating near the center of its range (average comparator output half of its maximum output). Another minor problem is that the divided VCO input to the comparator and the reference input will have relative phase angles some place between 0° and 180°, depending on the exact frequency.

A second phase detector is shown in Figure 8-21. This is a flip-flop that is SET (Q output terminal goes HI) by a short pulse from the reference oscillator

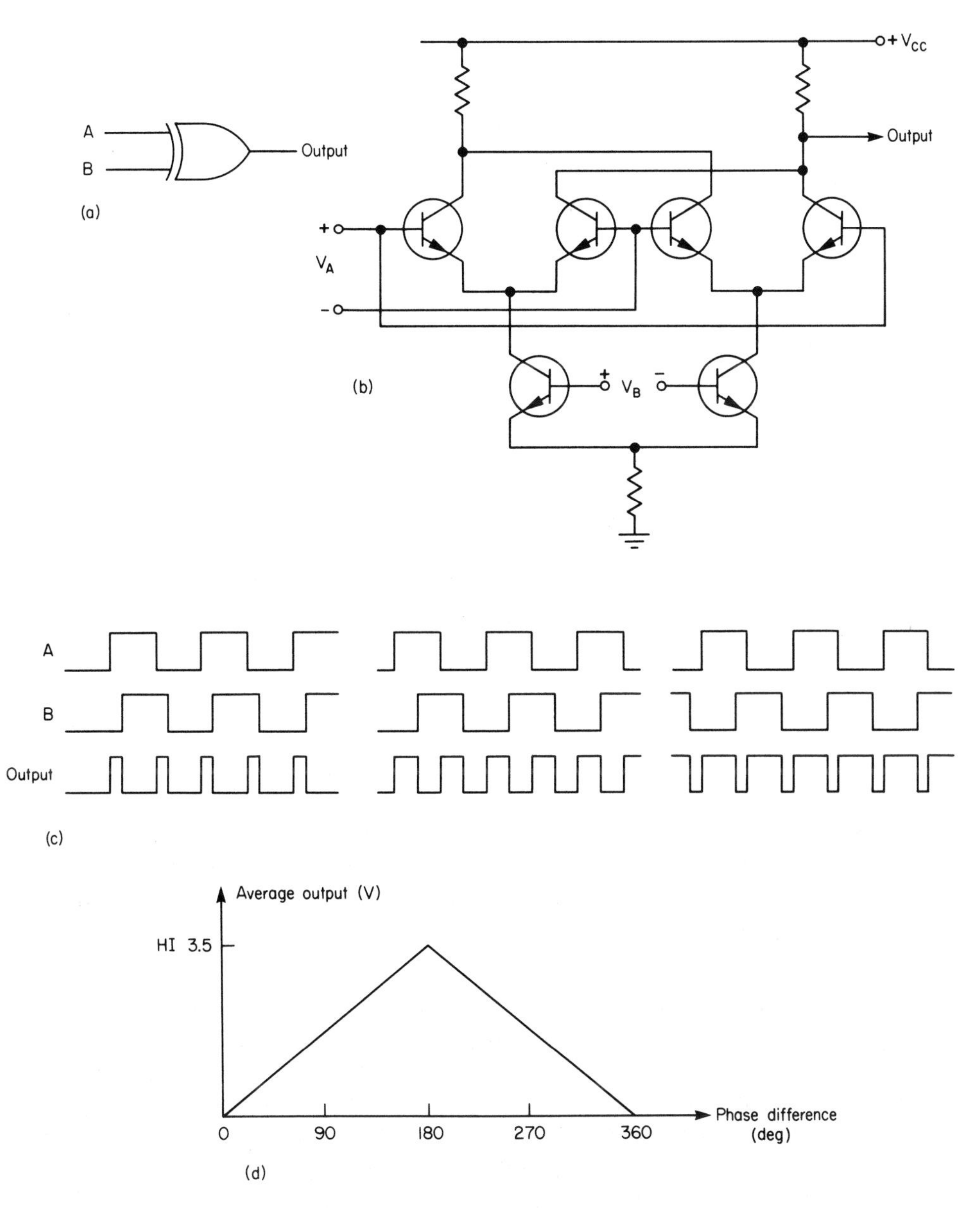

Figure 8-20 Symbol for an exclusive-OR phase comparator (a) and schematic (b) along with three samples of input signals and corresponding outputs (c). The average output voltage as a function of phase difference is summarized in (d).

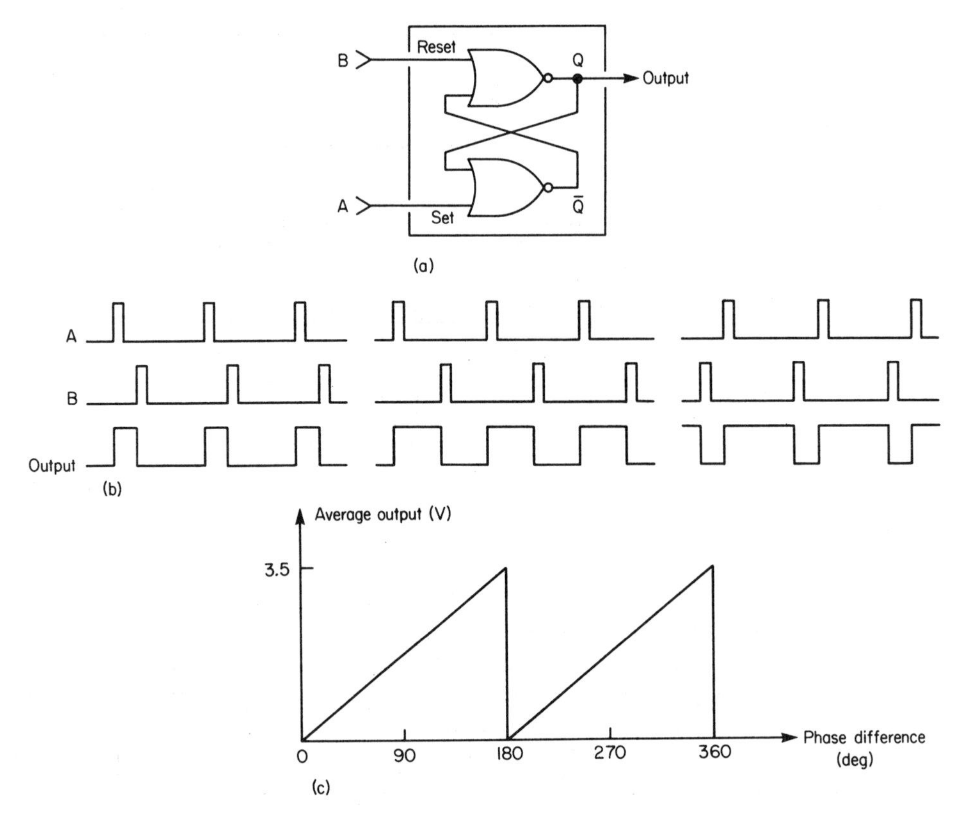

Figure 8-21 Digital-phase comparator consisting of a SET/RESET flip flop. The symbol and internal construction are shown in (a), sample input and output waveforms are given in (b), and the circuit operation is summarized in (c).

and then RESET (Q terminal returns to LO) by a short pulse from the divided VCO output. The operation is summarized by the three sets of input and output waveforms and the input phase versus output voltage chart of Figure 8-21(c). The biggest improvement is the extended range of input phase angles and the fact that this doubles the capture range of the loop. The disadvantage is the necessity of generating the narrow pulses (although some "edge-triggered" flip-flops avoid this) and the fact that the comparator is more noise-sensitive (because of the narrow pulses being indistinguishable from noise pulses) and is still apt to lock onto harmonics of the reference input.

The third comparator is the ultimate and is available at low cost in integrated-circuit form. It will cause the loop to lock only on the correct frequency and not on harmonics. The capture range of the loop will cover the full range of the VCO, and the two inputs will always be in phase when the loop is locked. As an added bonus, the comparator produces an output indicating whether it is in lock or not. The logic diagram is shown in Figure 8-22 for the TTL version. A slower C-MOS version is also available, usually with NOR gates instead of NAND gates.

The circuit consists of four SET/RESET flip-flops made up of gates 1 and 2, flip-flops 2 and 3, and gates 4 and 5. Gate 3 generates a RESET pulse when all its inputs are HI. The two output transistors are normally off so that no current can flow through either one to the filter capacitor. This would mean that the points labeled UP and DOWN are normally HI.

The two inputs, VCO and REFERENCE, are shown as symmetrical square waves, although, for this comparator, they could be pulses of any width. The phase comparison is made on the falling edge of the two inputs (for the C-MOS version, the comparison is made on the rising edge). The reference input is shown with a constant frequency as would normally occur in the PLL synthesizer. The VCO initially has a higher frequency, then has an indentical phase and frequency with the reference for a few cycles, and finally goes to a lower frequency, so its phase slowly lags the reference by an increasing amount. A point is reached where the VCO signal "appears" to be leading once again, but if a count of actual cycles is made, it can be seen that the lag now exceeds one full cycle, or 360°. Although a lag of more than 360° is, for simplicity, often referred to as a *leading angle*, the difference is important in this case, as some memory is built into the comparator.

The waveforms of Figure 8-22 were obtained with the VCO control input disconnected from the phase comparator. In this way the commands to correct the frequency go unheeded, just so we can see how the comparator operates.

For the short period that both input cycles match, the UP and DOWN points remain HI. Both output transistors stay off, so no charge is added to or taken away from the capacitor. When the VCO leads slightly, the DOWN output pulses LO for a time proportional to the phase difference. At each pulse, the *NPN* transistor conducts, causing the capacitor voltage to fall a small amount each time (read the waveforms from left to right as time increases in this direction). When the VCO lags, the UP output pulses the *PNP* transistor on and off so that the capacitor voltage charges up a small amount each time. The capacitor voltage will always stabilize at whatever value is needed to set both the frequency and phase of the VCO exactly the same as the reference. The capacitor voltage is then the integral with time of the phase difference.

If the VCO frequency is low and its phase keeps lagging more and more so that the angle eventually exceeds 360°, it is important to note that the UP output keeps pulsing, trying to raise the capacitor voltage and increase the VCO frequency. Simpler comparator circuits, such as those previously described, would have alternately tried to raise and lower the VCO frequency as the phase "appeared" to lead and then "lag." The capture range of the simpler comparators is therefore limited. The circuit of Figure 8-22 is providing a positive indication that the VCO is lower in frequency, and so the capture range of a PLL with this comparator would cover the full range of the VCO.

When the loop is out of lock, either the DOWN or UP point will be pulsing. If these two points were connected to an OR gate, its output would provide an "out-of-lock" signal. In a transmitter this could be used to disable one of the stages until lock is achieved, thereby preventing the transmitter from operating on the wrong frequency.

8-11 LOOP FILTERS

The main function of the loop filter is to remove most of the high-frequency components from the error voltage. It also plays a major role in determining the capture range, capture time, and stability of the loop. Rarely is any more than a single reactive element needed or even desirable. The two filters of Figure 8-23 are usually adequate. The cutoff frequency is chosen to provide high attenuation at the reference frequency so that it does not pass through and

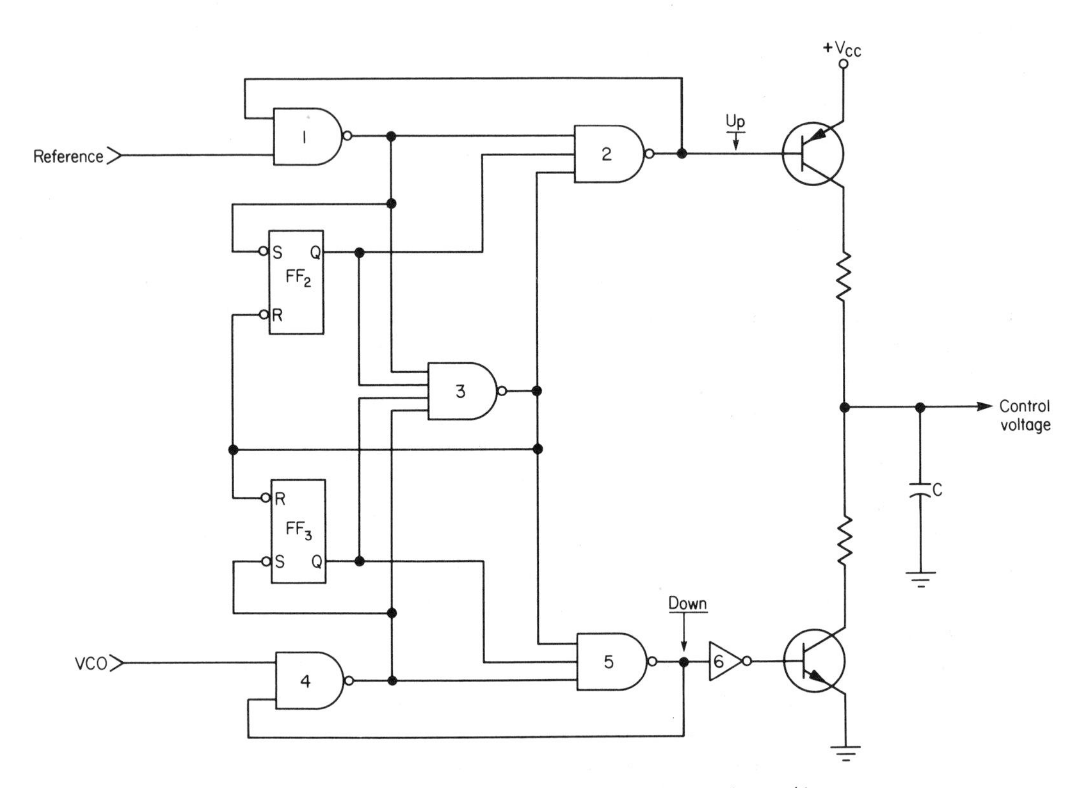

Figure 8-22 Logic diagram of a combination phase/frequency comparator (a) and the waveforms obtained at various points in the circuit as the relative phase angle of the two inputs shifts (b).

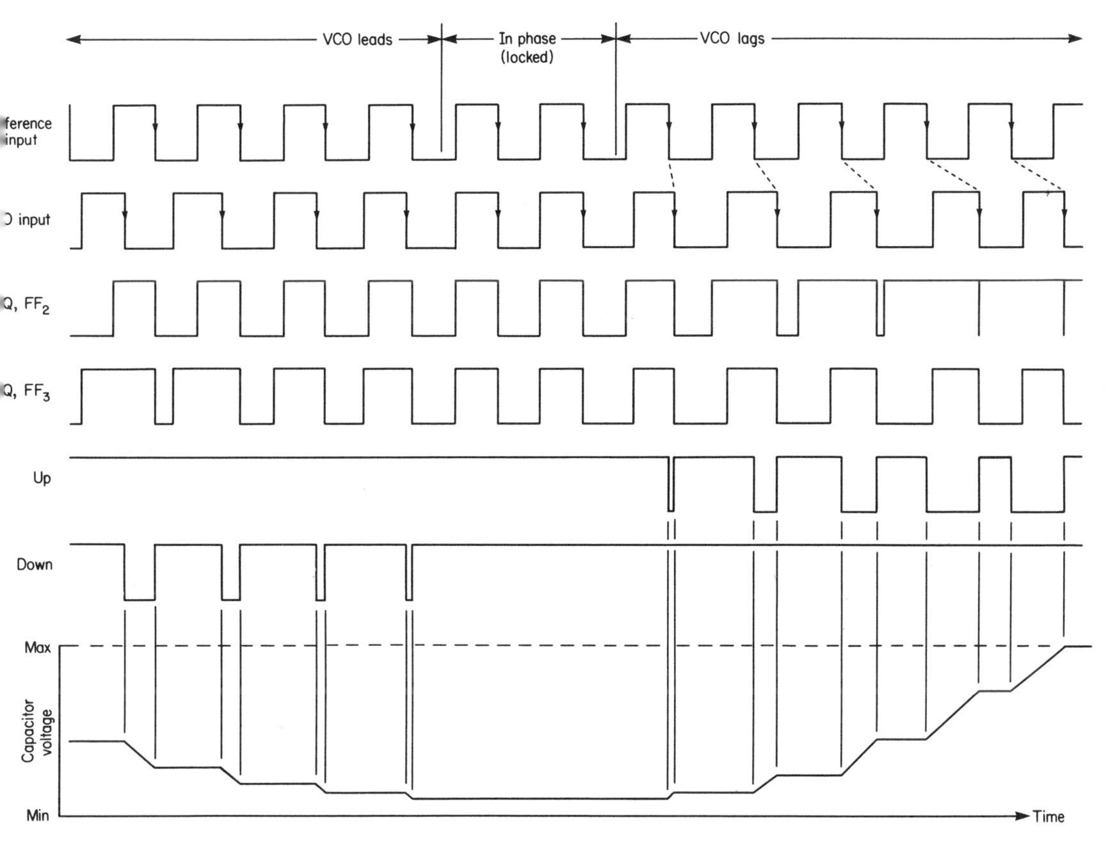

Figure 8-22 (cont'd)

modulate the VCO; typical cutoff frequencies will change with the application of the PLL and the reference frequency used but are usually in the range 5–1000 Hz. The lower cutoff frequency will make the loop slower to lock-up so that in the case of the 5-Hz cutoff frequency, lock-up time could approach 1 s. The lower cutoff frequency will also decrease the capture range of the loop (except when used with the phase comparator of Figure 8-22). Enough signal must pass through the filter when the VCO and reference frequencies are far out of lock to initiate the transient conditions of Figure 8-18 that moved the VCO toward lock. The lower the cutoff frequency, the lower the frequency difference that can be accommodated.

Once the phase comparator gains control of the signals, lock-up is still not a certainty. The entire loop could be unstable, and the VCO would then continually swing back and forth across the desired frequency. Even if the loop

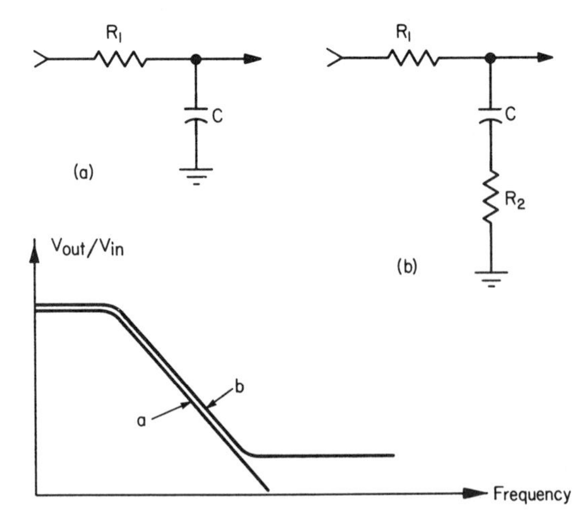

Figure 8-23 Two single-pole filters that can be used for the control line of the PLL. The filter of (b) will usually provide a more stable transient response.

is stable, the oscillations could still occur but would slowly decrease in amplitude until finally the VCO settles on the desired frequency. The slightest disturbance within the loop due to noise or power-supply transients could cause the oscillation to start up again. If such a PLL were used as the oscillator of a transmitter, a lot of people could be very unhappy. The filter of Figure 8-23(b) will usually provide a wider capture range and greater freedom from oscillation during the capture.

Once the PLL system is designed, a test of the loop's stability should be carefully made. With an oscilloscope monitoring the VCO control line, the channel selector switch (part of the binary dividers) should be abruptly moved and the transient on the control line noted. The waveforms of Figure 8-24 could be the result. These will be after any transients as shown in Figure 8-10 and are distinguished by the fact that these undesirable oscillations move above and below the final control voltage, whereas the other waveform slowly moved toward the final voltage without overshoot. This test should be made at a number of points across the full frequency range of the VCO. Any variations in the waveform will likely be caused by the nonlinear voltage to frequency characteristics of the VCO.

8-12 FREQUENCY DIVIDERS

The digital dividers must take the higher frequency from the VCO and divide it down to match the lower reference frequency, as shown in Figure 8-25. The division ratio is set by the positions of input switches used by the operator to dial in the desired frequency. The counters are called "programmable" since

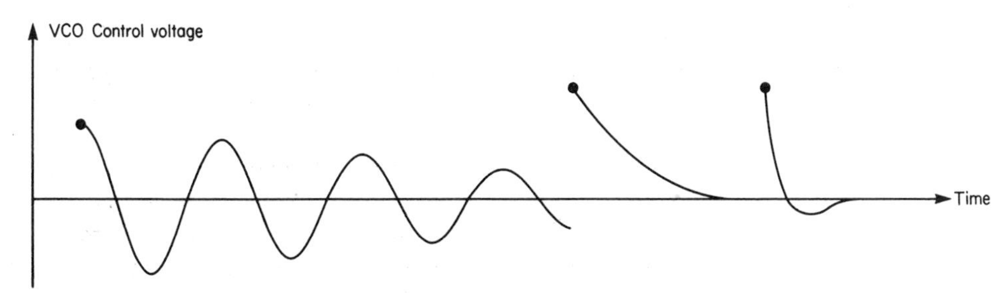

Figure 8-24 Control-line waveforms that could be noted as the PLL channel selector is changed: (a) shows too much oscillation and a long settling time; (b) shows no oscillation but the lock-up time is too long; (c) is a good combination of overshoot and settling time.

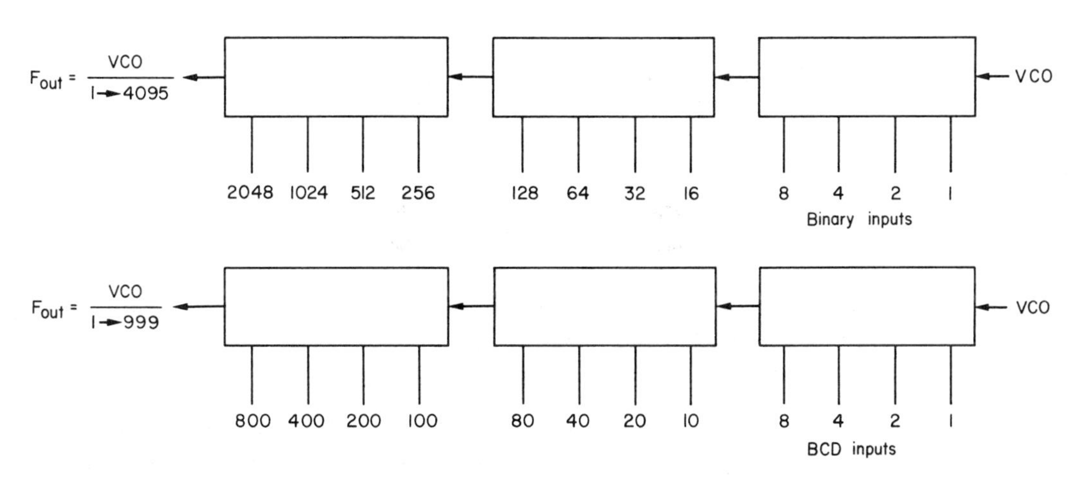

Figure 8-25 Programmable frequency dividers for (a) binary input, and (b) BCD input. Since division by zero has no meaning, the smallest division is 1, (i.e., the input and output frequencies are the same).

some type of binary information is fed to their control inputs, and this determines the whole number by which the counter will divide. Programmable counters are made to recognize two different binary inputs, BCD and natural binary. *Binary-coded decimal* (BCD) is a 4-bit code that represents only the numerals zero through nine. Natural binary can represent any number as long as enough control lines (bits) are used; four lines could represent 0 to 15, five lines 0 to 31, and six lines 0 to 63 (typical of 40-channel CB synthesizers). The higher numbers in BCD are accommodated by adding four more control lines for each digit needed.

The internal logic circuitry consists of a set of down counters that can be initially preset to any number. To understand the operation of the programmable counters and how they operate as part of the loop, we will consider the following example.

A synthesizer is required to produce 99 frequencies spaced 50 kHz apart, the first frequency 50 kHz. The highest frequency would be $99 \times 50 =$ 4.95MHz. (Designing the VCO for this 100:1 range would be interesting, but we are looking at the dividers at the moment.) The programming switches, and therefore the counter types, will be BCD. The circuit could look like Figure 8-26. For this example we will consider that the VCO is required to operate at 25 times the reference frequency.

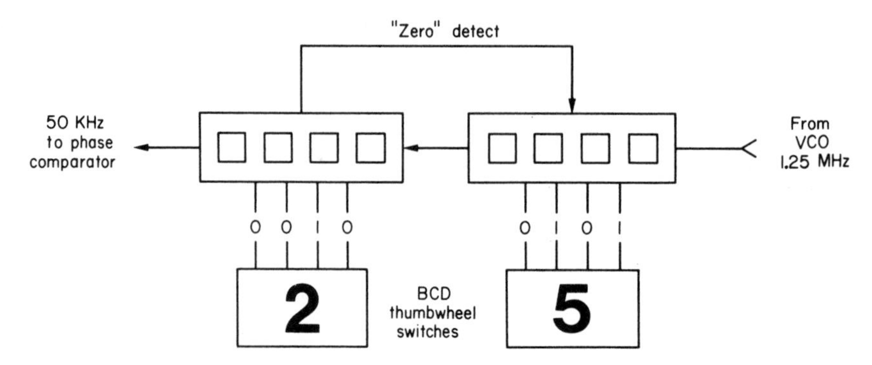

Figure 8-26 Two BCD counters are programmed by thumbwheel switches to divide the VCO frequency by 25.

The thumbwheel switches provide the BCD code for 2 (0010) and 5 (0101). When the cycle starts, the flip-flops in the counter are loaded (preset) with the 0's and 1's from the switches. The square-wave pulses from the VCO (at 1.25 MHz) start the least-significant counter downward. The first VCO pulse clocks the counter down to:

	4 (0100)	
then	3 (0011),	VCO pulse 2
then	2 (0010),	VCO pulse 3
then	1 (0001),	VCO pulse 4
then	0 (0000),	VCO pulse 5

The sixth VCO pulse sets this counter back to 9 (the switch inputs are ignored). Up until this time, the second counter, which had been preset to 2, does not change. As soon as the first counter goes back to 9, the second counter is provided with the appropriate clock pulse and counts down by one. The first counter now sits at 9 and the second at 1. Six pulses from the VCO have been used up. The count continues:

second counter (0001), first counter 9 (1001), VCO pulse 6
(0001), 8 (1000), VCO pulse 7
(0001), 7 (0111), VCO pulse 8
(0001), 6 (0110), VCO pulse 9
(0001), 5 (0101), VCO pulse 10
(0001), 4 (0101), VCO pulse 11
(0001), 3 (0011), VCO pulse 12
(0001), 2 (0010), VCO pulse 13
(0001), 1 (0001), VCO pulse 14
(0001), 0 (0000), VCO pulse 15

The 16th VCO pulse again sets the first counter back to 9 and again the second counter counts down by one. The total situation now sits at

second counter (0000), first counter 9 (1001), VCO pulse 16

The count continues:

(0000), 8 (1000), VCO pulse 17
7 (0111), VCO pulse 18
6 (0110), VCO pulse 19
5 (0101), VCO pulse 20
4 (0100), VCO pulse 21
3 (0011), VCO pulse 22
2 (0010), VCO pulse 23
1 (0001), VCO pulse 24
0 (0000), VCO pulse 25

On the 25th VCO pulse both counters now contain all zeros. This situation is sensed by a *zero detector circuit* and presets the counters immediately to 2 and 5.

Only now does an output pulse get passed on to the phase comparator. We have come back to our starting point, 25 VCO pulses have been used, and only one was fed to the phase comparator. The total circuit has therefore divided the VCO frequency by 25, the same number dialed on the thumbwheel switches.

QUESTIONS

1 Explain why the feedback network of any oscillator should have a high Q and why it should be constructed with temperature stable components.

2 Design the feedback network for a Colpitts oscillator at 10 MHz. The load resistance is 600 Ω and the loaded Q at the collector end will be 12. The input resistance of the transistor is 200 Ω and its f_T is 180 MHz.

3 Change the design of Question 2 to a Hartley oscillator.

4 Change the design of Question 2 to a Clapp network so that the equivalent reactance of L and C_3 (see Figure 8-12) changes three times as fast with frequency as the original L of the Colpitts' design.

5 A crystal oscillator operates at 27.5 MHz with a crystal tolerance of ±0.005%. Find the highest and lowest frequency that the circuit could operate at.

6 In block diagram form, design a phase locked loop oscillator that will produce 50 equally spaced signals from 15.0 to 15.98 MHz. Use BCD dividers, determine their highest and lowest binary inputs and also the frequency of the single crystal that would be required.

REFERENCES

Clarke, K. and Hess, D. 1971. *Communication circuits: analysis and design*. Addison-Wesley.

Zelinger, G. 1966. *Basic matrix analysis and synthesis*. Oxford: Pergamon Press.

chapter 9

transmitters

The purpose of a transmitter is to modulate a high-frequency carrier with some type of information: music, voice, data, video, etc., so that the information can be efficiently transmitted to remote receivers. The transmitter has to produce enough power to overcome any noise at the receiver location, it must hold its frequency accurately and confine its signals to the specified bandwidth. The modulated carrier fed to the antenna must be an accurate representation of the information fed in, meaning that amplitude and phase response and intermodulation distortion must be considered at all points in the transmitter.

9-1 CHOICE OF MODULATION

When designing a transmitter for a specific application, the designer does not normally have a choice of modulation type. This is set by the regulating government agency. Some exceptions occur in the radio amateur bands and, to a much smaller extent, in the 27-MHz citizens' band (AM or SSB).

If frequency modulation is used, the transmitter can be small and light, as the audio amplifiers need work at only a very low power level. The design of the final transmitting stages will be simplified by the constant-output power level and the inherently lower intermodulation problem. The complete FM transmission system offers a higher signal-to-noise ratio at the receiver end than the same-power AM signal; the improvement will depend on the frequency deviation used. FM does have one problem—at low signal levels, the received signal falls into the background noise level very rapidly, as shown in Figure 9-1. For critical situations, such as aircraft voice communications, AM is used because of its better signal-to-noise ratio over FM when weak signals are involved.

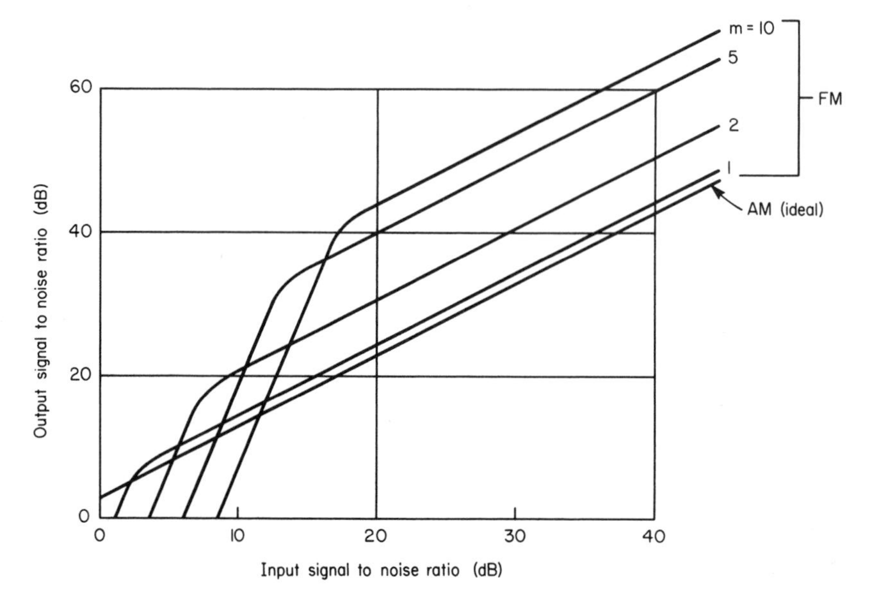

Figure 9-1 Comparison of the output signal-to-noise ratio as a function of received signal-to-noise ratio for AM and FM detectors.

The amplitude modulation transmitter requires a higher-power audio amplifier and closer attention to the modulated stages if maximum performance and freedom from distortion is to be obtained. The design is complicated by the fact that the output power must be capable of changing between 0 and 150% of the carrier power, in step with the modulating signal.

Single sideband makes more effective use of the frequency spectrum and also of the transmitter power level by getting rid of the one sideband and the carrier. The overall system improvements are the result of a 2:1 bandwidth reduction and corresponding decrease ($\sqrt{2}$:1) in noise level at the receiver and also the elimination of two-thirds of the required transmitter output power, owing to removal of the carrier. This high efficiency however, is partially offset by the higher costs involved, mainly in the low-level SSB modulator circuits and also in the greater attention to linearity required for the high-power amplifiers.

9-2 TRANSMITTER BLOCK DIAGRAM

Regardless of the type of modulation to be employed, the main parts of the transmitter remain the same, at least to first appearances. A typical block diagram is shown in Figure 9-2. The oscillator is the first stage of the RF chain of the transmitter. Its frequency stability and spectral purity will affect the overall performance of the transmitter. The oscillator will normally be crystal-controlled in some way, either with individual crystals or, if many channels must be tuned, with a crystal-referenced PLL frequency synthesizer. Power output will usually be low (less than 50 mW) to reduce the effects of heating on frequency drift. The oscillator frequency may or may not be the same as the transmitter output. Crystals operating on their fundamental frequency, for example, cannot operate above about 20 MHz; on overtones the approximate limit is about 150 MHz. Higher transmitter frequency therefore requires frequency multiplication. Frequency synthesis can operate at almost any frequency for which a VCO can be designed, since the reference oscillator and the phase comparator will operate at a much lower frequency. The first programmable dividers or prescalers must also be capable of operating at the VCO frequency, which places an upper operating limit at present of about 1500 MHz (emitter-coupled logic).

Instead of multiplying the oscillator frequency, it can be moved to a higher frequency by mixing with the output of a second mixer. This is commonly done in single-sideband transmitters, since the removal of the unwanted sideband is most effectively done with low-frequency filters and, once modulated, the carrier cannot be multiplied or the audio bandwidth would increase. FM transmitters often use frequency multiplication to increase the frequency deviation.

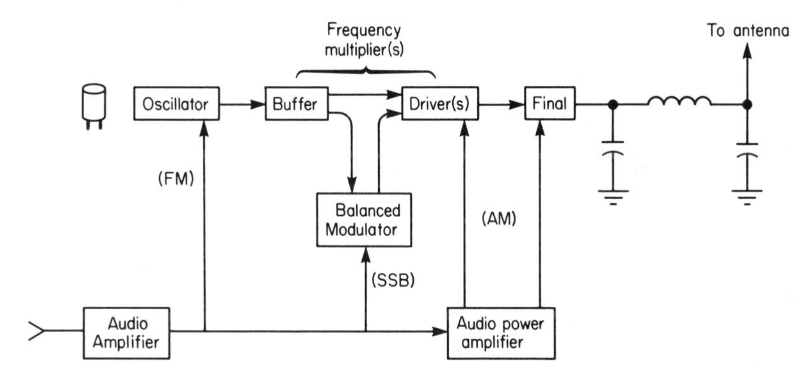

Figure 9-2 Block diagram of a general transmitter with the injection points for different modulation types.

The buffer amplifier has power gain, but its main purpose is to isolate the oscillator so that its operating frequency and amplitude are not affected by other changes due to tuning or modulation. For best buffering, this amplifier is often mismatched to the oscillator. The driver amplifiers are required to raise the oscillator power level as needed before the final power amplifiers are reached. For a low-power, low-frequency transmitter, where the final amplifiers have a higher gain, the driver amplifiers may not be needed. The drivers may also be frequency multipliers if required, but their power gain will then be reduced slightly.

The final amplifiers will take whatever form is needed to handle the required RF power while controlling harmonic levels and intermodulation distortion. The final amplifier might be a single transistor, a parallel combination, or a push–pull arrangement.

The point at which the modulating signal is applied will depend on the type of modulation. Frequency modulation is applied the earliest, since it must change the operating frequency of the oscillator. Single-sideband modulation is also performed at an early point where both the carrier frequency and its amplitude are low. This point should be after the buffer amplifier, to prevent simultaneous frequency modulation. Amplitude modulation is applied to the last stages, where the carrier power level is the highest. The audio system for AM therefore requires a power amplifier with an output capability at least one-half the total power input of the stages being modulated. Any RF amplifiers following either AM or SSB modulated stages must be very linear, to avoid intermodulation distortion products near the carrier frequency that could affect either in-band or adjacent channel operation. For "linear operation," distortion products should be 30 dB below the sideband level.

9-3 RF POWER TRANSISTORS

One of the more interesting features of a solid-state transmitter is the construction of the high-powered transistors used as driver and final amplifers. This will be discussed first, since many of the resulting characteristics control the design procedure to be used. Power amplifiers must handle large currents, and since current density in a transistor must be kept at reasonable levels, a large chip area inside the transistor is called for. Unfortunately, this area could result in very large junction capacitances that would very seriously limit the useful gain and bandwidth of the device. But all this capacitance is not necessary. Power transistors, particularly at higher frequencies, where gain is low, will have fairly large base currents. As shown in Figure 9-3, the base current flowing through the resistance of the base region will produce a lateral voltage drop. The highest base-to-emitter bias will be developed at the edge of the emitter region, and so the emitter current will be strongest at that point. The area directly under the emitter region will have a lower bias, so will contribute very little to the transistor action, only unwanted capacitance.

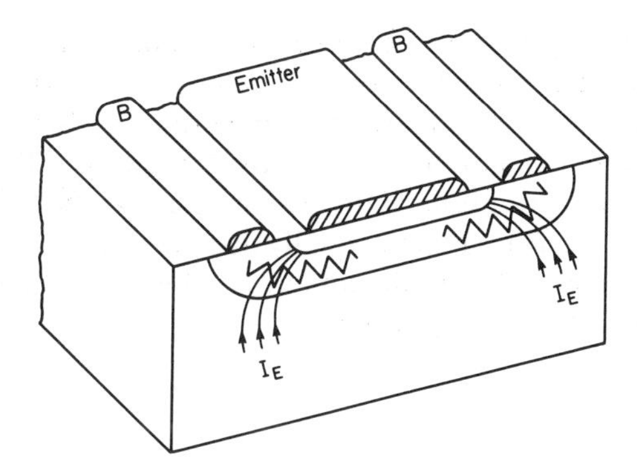

Figure 9-3 Cross section of an RF power transistor showing the crowding of emitter current near the emitter edges.

The practical solution is to make the horizontal dimension as small as possible; in fact, one figure of merit used in the design of power transistors is the periphery-to-area ratio, which should be kept high (8 : 1). As both operating frequency and power levels increase, the P/A ratio becomes increasingly critical, and the top view of the transistor shows a series of long narrow fingers for both the base and emitter metal contacts. Each emitter finger has two edges that concentrate the emitter current in the space down between the adjacent base fingers. The limit on the narrowness of these fingers is set by processing problems and by the minimum amount of metal required to handle the current levels without squirming or buckling. A view of a typical power transistor chip is shown in Figure 9-4.

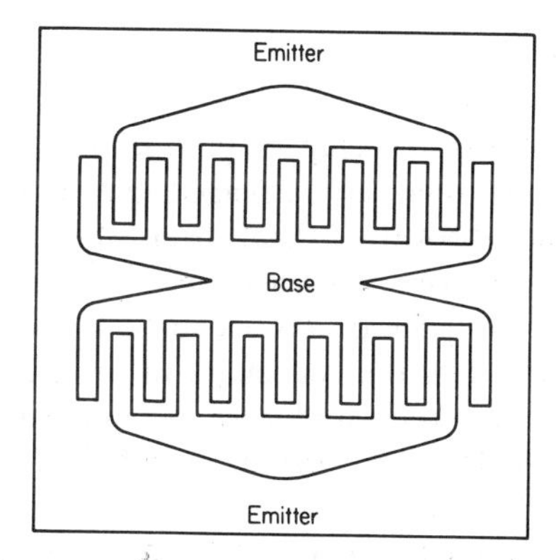

Figure 9-4 Top view of an RF power transistor with narrow emitter and base fingers to maximize the edges and minimize emitter area.

With all the long fingers, it becomes very important to ensure that each portion of the transistor shares the current equally. Any tendency for current to concentrate in one area will increase the temperature at that point, resulting in lower resistance and the possibility of thermal runaway. The resulting "hot spot" could damage the transistor if the temperature climbed high enough. The emitter current can be shared more equally and thermal runaway prevented if small resistors are included at each emitter. A schematic representation of this and a diagram are shown in Figure 9-5.

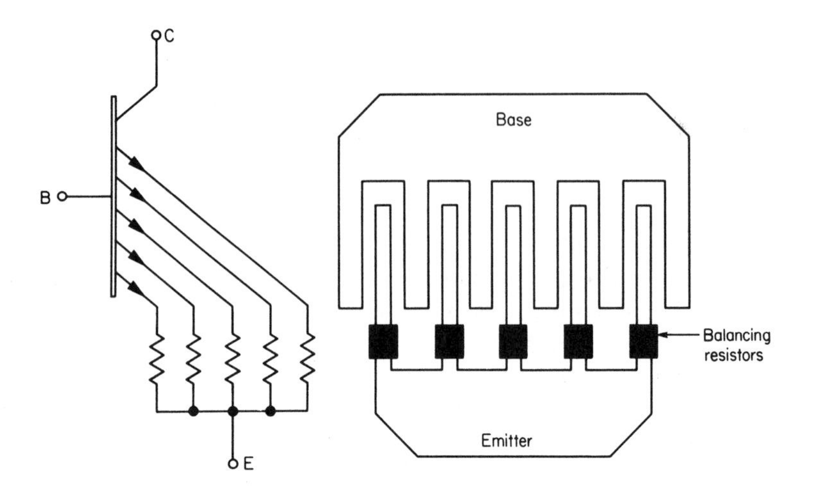

Figure 9-5 Small resistors can be added to the multiple emitters to encourage more-equal distribution of current and more reliable operation.

9-4 PACKAGING

When a good power transistor chip is produced, the job is only partially finished. The chip must then be packaged for final use. The two main considerations are the removal of heat from the chip and the electrical connections to the emitter, base, and collector.

For best heat removal, the collector can be connected electrically to the case, or if this is undesirable, the collector can be brought out as a separate lead. For minimum inductance and ease of connecting to printed circuit boards, the leads are usually wide metal tabs and the emitter is often a double tab.

The exterior and interior view of a popular package are shown in Figure 9-6. The interior view shows a matching network, consisting of a ceramic chip capacitor and controlled wire bonds, built between the internal base and the external base lead. This is done in some transistors for the designer's convenience, to raise the very low input resistance of the transistor (about 1 Ω or less) up to something a bit more manageable (several ohms) over a limited range of frequencies (less than one octave).

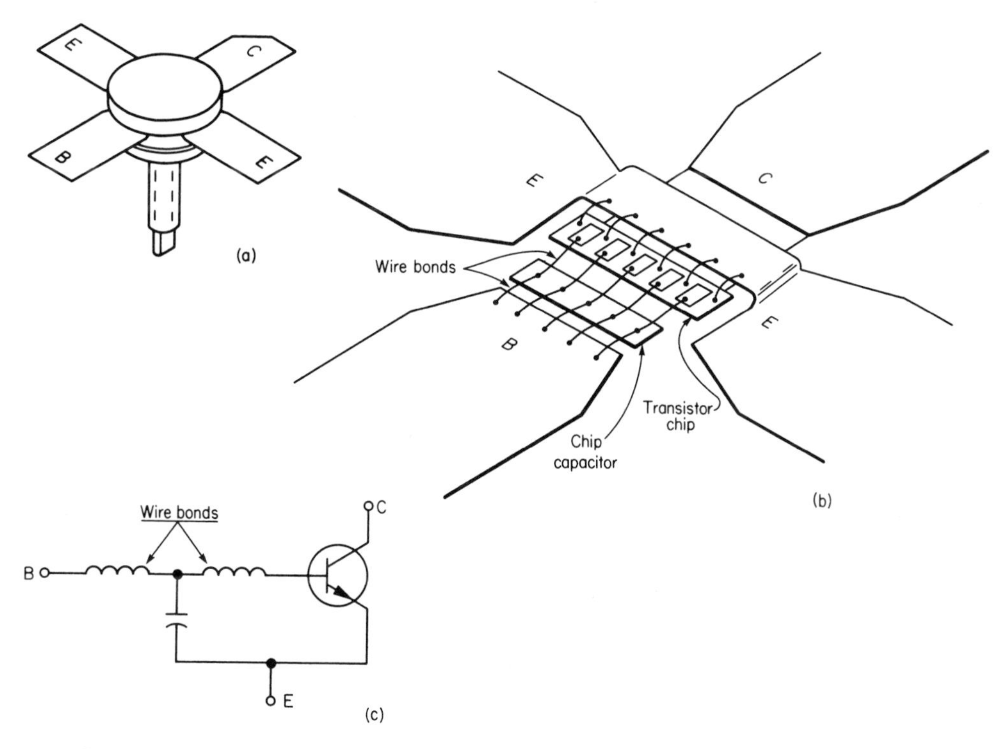

Figure 9-6 Exterior view of an RF power transistor (a), its internal view (b), and equivalent circuit (c).

9-5 BIPOLAR PARAMETERS

When used as RF power amplifiers, the important transistor parameters are slightly different from the small-signal characteristics. They tend to describe actual operating conditions rather than short-circuit parameters. One important graph is the relation between input power and output power at different frequencies, as shown in Figure 9-7. Such a graph can be used to determine maximum output power, power gain, and input power required at any frequency. For example, this graph shows that 10 W of output power can readily be obtained at 100 MHz with an input of about 500 mW. The resulting power gain would be 13 dB.

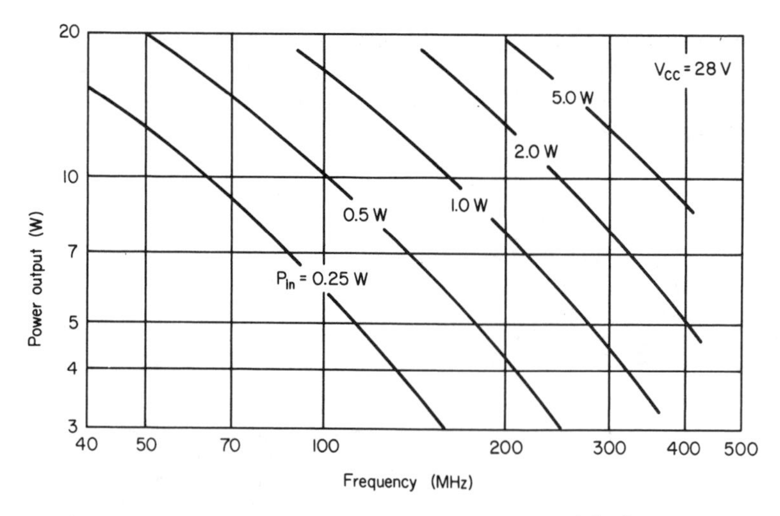

Figure 9-7 Input power versus output power over a wide frequency range.

The second parameter is the input resistance and reactance, either in series or parallel form. Since input resistance changes with average collector current, either this or the output power must also be specified. The two graphs of Figure 9-8 describe the parallel input resistance and reactance at three different power levels for the frequency range 100–500 MHz.

Several other parameters are important; one is the output capacitance of the device. This must be known when matching networks are designed. The output resistance is not of any great importance, since other considerations determine the load impedance.

The collector breakdown voltage (BV_{CES})[1] should also be known, as operation above this value could damage the transistor. Often the manufacturer will specify a transistor for 14- or 28-V operation, meaning that the breakdown voltage is safely more than twice this value.

9-6 V-MOS RF POWER TRANSISTORS

Recent developments on MOS-FETs have increased their frequency and power capabilities to the point where they can be considered for RF power amplifiers well up into the UHF band. The most significant improvement was in the ability to produce vertical MOS transistors with the help of a gate in the form of a deep V, as shown in Figure 9-9.

---

[1] BV_{CES} is the breakdown voltage between the collector and emitter with the base shorted to the emitter. The value will be much higher than BV_{CEO}, the same breakdown voltage except that the base is left open.

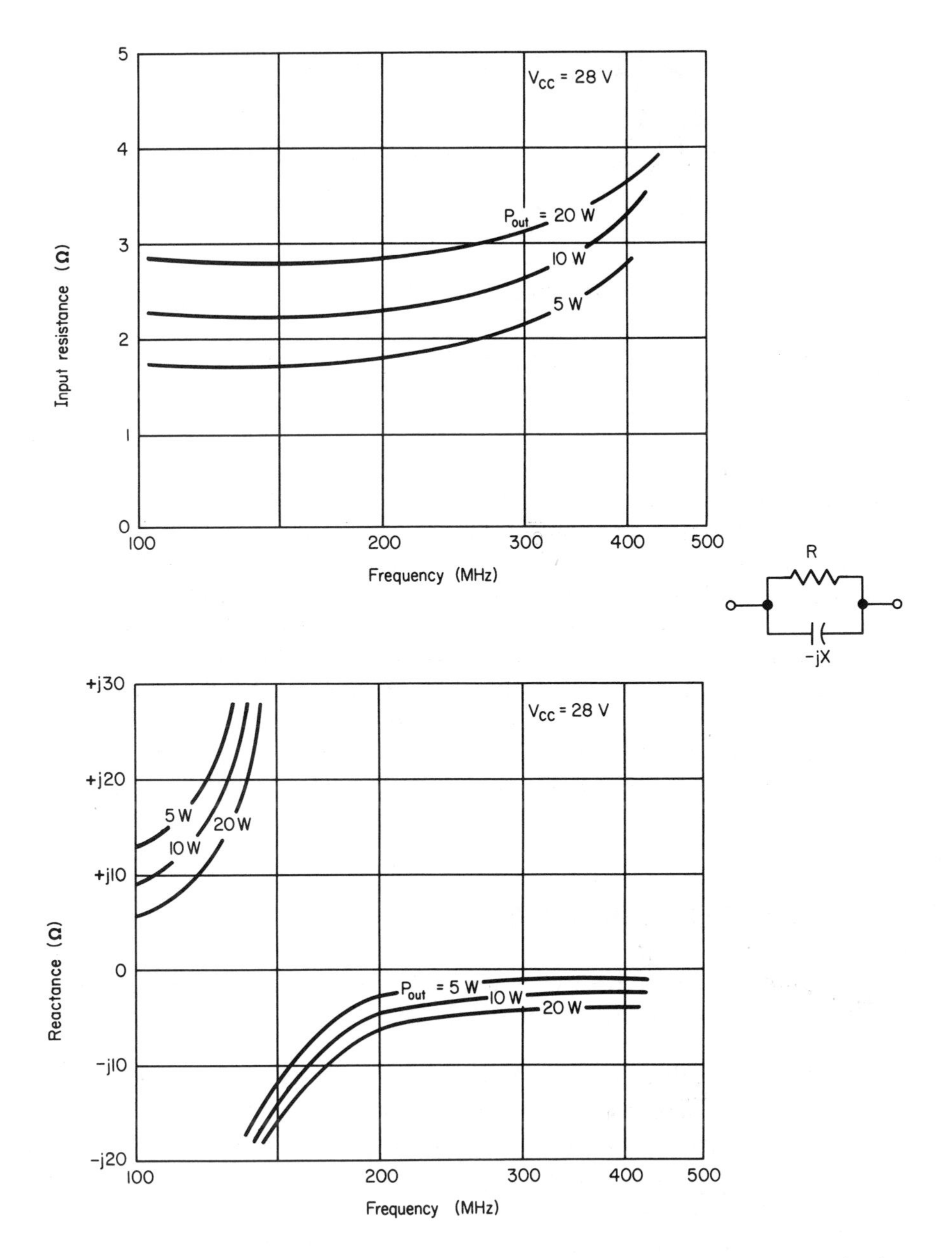

Figure 9-8 Parallel input resistance (a) and reactance (b) of a high-frequency power transistor.

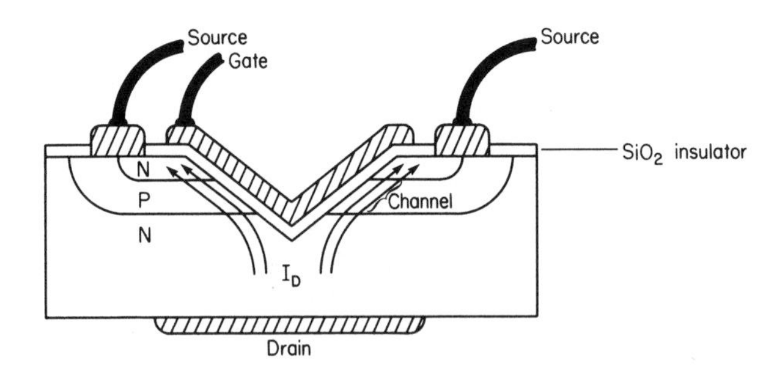

Figure 9-9 Cross section of a V-MOS RF power transistor.

With V-MOS, the vertical channel length between source and drain is very short and can be accurately controlled by diffusion depth. Current flow is in the same direction as a bipolar transistor—vertical. Two advantages of V-MOS transistors are the higher input impedances (about 10–100 times) compared to bipolar of the same power and frequency capability; and the freedom from secondary breakdown and thermal runaway since the temperature coefficient is positive—higher temperature increases channel resistance, decreasing current and reducing heating. One disadvantage is the slightly lower efficiency caused by the higher saturation voltage (about two or three times that of a similar bipolar device). Another is the higher input capacitance, which gives the input a higher Q, making broadband matching more difficult.

9-7 CLASSES OF AMPLIFIERS

Power amplifiers can be built with any of the four different classes of biasing. Each will have its own advantages and disadvantages.

1 *Class A.* The transistor is biased so that it never reaches cutoff. This is the most linear mode of operation, with low harmonic content and low intermodulation distortion. Efficiency is poor, less than 50%. Used only for low power levels.

2 *Class B.* The transistor is biased at cutoff so that current theoretically flows only for one-half of the cycle of the input signal. Harmonics are generated, but for amplifiers with less than a 1-octave bandwidth, they can be filtered out. Efficiency can approach 75%.

3 *Class AB.* Biased with a small collector current, the transistor is cut off for less than one-half of the input cycle. Other than the harmonics that are generated, the linearity can be very good, so in-band intermodulation products can be low (greater than −30 dB).

4 *Class C.* For a bipolar transistor, the base is kept at 0 V or below. Current flows for less than one-half of the input signal. Efficiency can approach 90%. Linearity is the poorest of all classes.

The amplifier class to be used would depend on the exact application. Class C is usually the first choice because of its high efficiency. For amplifying single frequency signals, it is the perfect choice. Power amplifiers for FM transmitters also use class C amplifiers, since they operate at a constant signal

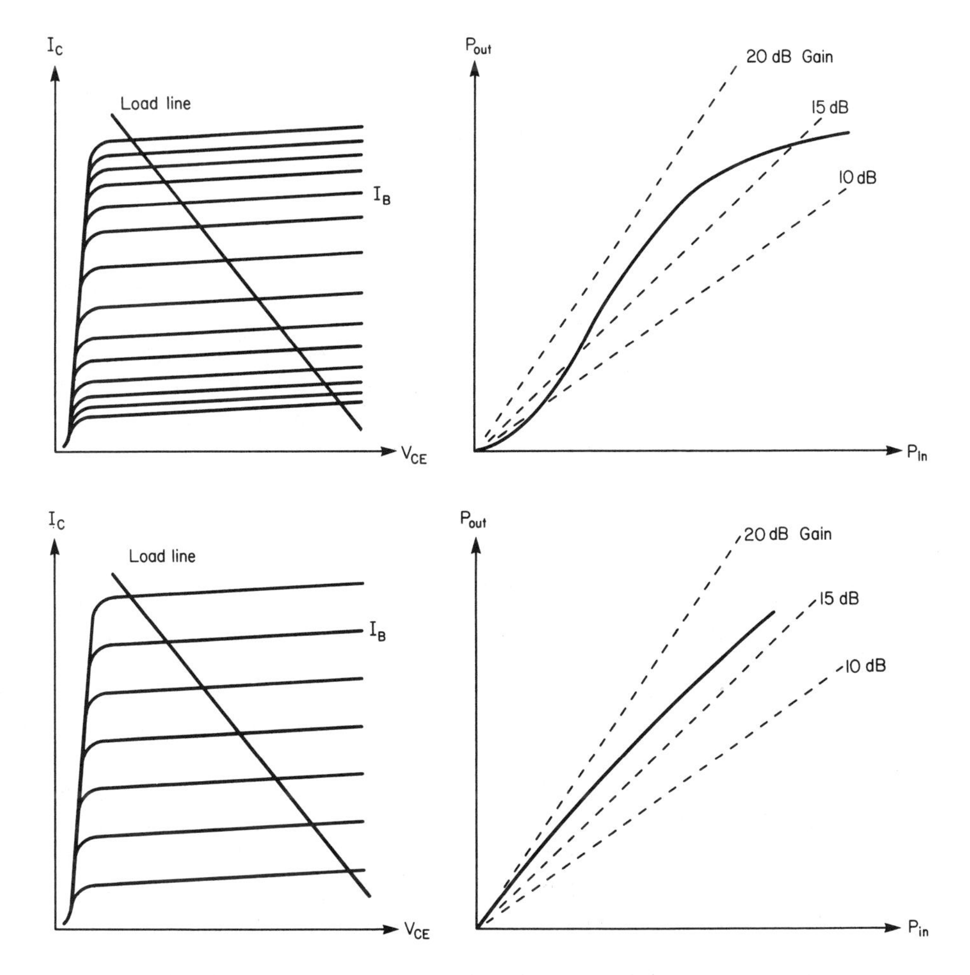

Figure 9-10 Amplifier linearity depends on the changes of the current gain β with collector current. The collector characteristics of (a) would result in the input–output power relationship of (b). The more even spacing of the lower set of collector characteristics produces a more constant power gain.

level. Stages to be amplitude modulated (collector) can also be operated class C; but amplifiers used to raise the power level of carriers that have already been amplitude- or single-single-band-modulated must be more linear, so class AB is a more common choice.

Linearity problems in power amplifiers are usually the result of current gain that changes with average collector current. As shown in Figure 9-10, current gain of a transistor normally increases with average collector current up to a point and then starts decreasing again. Careful manufacturing can produce a more constant current gain, which results in a more linear amplifier.

9-8 POWER AMPLIFIER DESIGN

A transmitter is normally designed with some particular power output in mind. The transistors are then chosen so that the low power level of the oscillator (perhaps 20 mW) can be amplified up to the required output power as efficiently as possible. High gain usually means matching the output resistance of the transistor to the load, so maximum power can be transferred. However, for RF power work, this is not desirable, for two reasons. First, the efficiency of such an amplifier will always be less than 50%, and second, the power-supply voltage will usually prevent the output voltage swing from reaching peak values high enough to create the power levels needed. If the collector voltage can swing between saturation and approximately twice the supply voltage, the maximum power fed to the load will be

$$P = \frac{V_{\text{rms}}^2}{R} = \frac{\left[(V_{CC} - V_{\text{sat}})/\sqrt{2}\,\right]^2}{R_L} = \frac{(V_{CC} - V_{\text{sat}})^2}{2R_L} \qquad (9\text{-}1)$$

The output signal from the collector of a transistor is shown in Figure 9-11. If the supply current is fed through an inductor, no dc voltage drop occurs and the collector voltage can then swing higher than the supply voltage, thanks to the stored energy in the magnetic field of the inductor. At the other end of the cycle, the transistor enters saturation and, if enough time is available, the saturation voltage will be low (about 0.1 V, depending on current). As the operating frequency increases, the average saturation voltage will rise and will reach typical values of 2–3 V for 12-V devices or 3–4 V for 28-V devices at the upper limit of these recommended frequency ranges. Efficiency therefore suffers at higher frequencies.

EXAMPLE 9-1

What value of load resistance would be required to obtain 1.0 W of RF output from a transistor that has a saturation voltage of 1.5 V at the operating frequency if the supply voltage were 10 V.

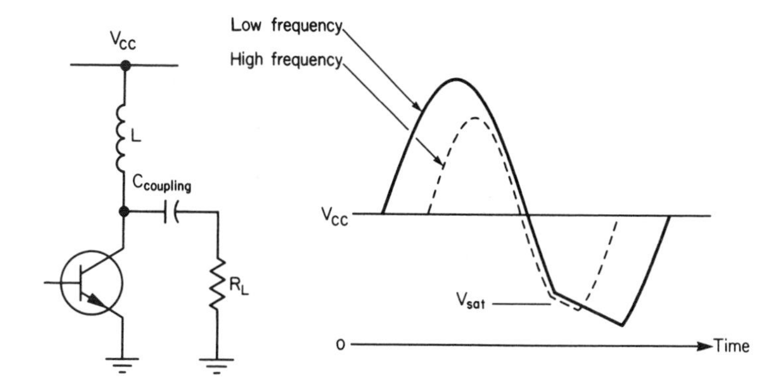

Figure 9-11 Maximum collector voltage swing; limited by the power supply and the collector saturation voltage.

Solution

$$R_L = \frac{(V_{CC} - V_{\text{sat}})^2}{2P}$$

$$= \frac{(10 - 1.5)^2}{2 \times 1.0}$$

$$= 36.1\ \Omega \text{ (resistive)}$$

For a specific application, the designer may be using the class C amplifiers of Figure 9-12. He/she has determined the amount of RF input power required to drive Q-2 to its full output power and knows the input resistance and reactance of the transistor of the frequency and power levels to be used (these should be available from the data sheets). The designer then simply designs a matching network to transform the input resistance of Q-2 up to the required load resistance of Q-1.

The matching network of Figure 9-12 very conveniently consists of only L and C. The inductor serves also to pass supply current to the collector of Q-1 and allows the collector voltage to swing up to almost two times the supply voltage. The series capacitor very conveniently provides dc isolation between collector and base in addition to its matching function.

The radio-frequency choke (RFC) plays only a minor part in the matching but is very important in the biasing of Q-2. There will be a dc current flow through RFC that will, in part, determine the operating point of the transistor. More important, it provides the path to ground for collector-to-base leakage current and, as such, sets the breakdown voltage of the transistor. If the dc resistance of this choke is high, the collector breakdown voltage will be lower

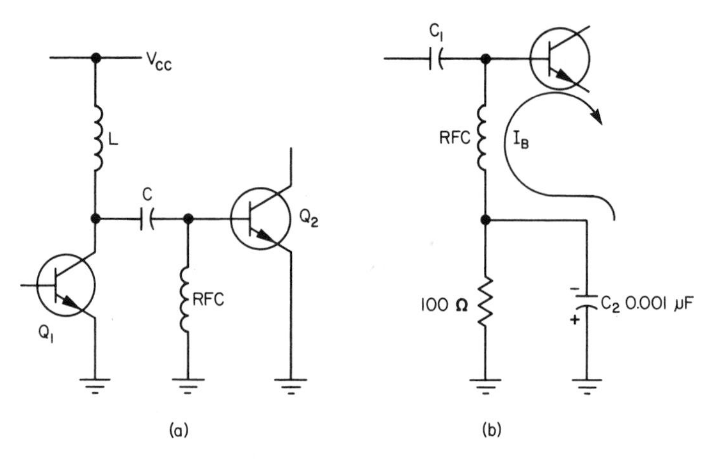

Figure 9-12 Interstage matching network for a power amplifier (a). To obtain higher reverse bias on the base, the network of (b) can be added.

(approaching BV_{CEO}) and transistor destruction would be likely to occur. The total impedance of this choke should be several times the input impedance of the transistor itself and, to reduce the possibility of low-frequency parasitic oscillations, the Q of the choke should be less than 5. Very often the choke is made by winding several turns on a lossy ferrite core. For increased reverse bias on the base, a small resistor (1 to 100 Ω) and filter capacitor can be added in series with the choke.

An example of a two-element interstage matching network design was provided on page 175. In that case the load was a low, complex impedance—very typical of an RF power transistor.

9-9 FREQUENCY MULTIPLIERS

As explained earlier, frequency multipliers are often needed to obtain power at VHF and UHF frequencies, particularly when crystal-controlled oscillators, with their limited upper frequency, are used. Since any nonlinear device will create harmonics of a sine wave, a resonant circuit tuned to one of these harmonics will be selecting a multiple of the original signal. All frequency multipliers work in this fashion, but some produce more efficient results than others.

The simplest frequency multiplier is a single diode that half-wave-rectifies the input signal and a tuned circuit that selects the harmonic. The efficiency will depend on the harmonic selected, the input signal level, and the switching speed of the diode. Operation should be limited to doubling and tripling and to input frequencies below 75 MHz. Step recovery diodes continue to conduct for a short time when reverse-biased and then abruptly turn off. They can provide good multiplication from VHF up to microwave frequencies.

A class C transistor amplifier can provide frequency multiplication with gain and so is very often used as part of a transmitter line up. A typical circuit is shown in Figure 9-13, together with some descriptive waveforms. One point to note is that the higher the order of multiplication, the shorter the conduction time of the transistor must be for best results. This is so the transistor can turn back off and allow the collector-tuned circuit to oscillate at its natural frequency. For example, a tripler should not conduct for longer than 100° of the input cycle.

Note that the oscillations die down between peaks of the input cycle, owing to the use by the load of the stored energy in the tuned circuit. This can be minimized by keeping the order of multiplication low and by using high-loaded Q's, ($Q_l = 15$ minimum for a tripler, higher for a quadrupler). Power gain of a tripler can be about 40% of that obtainable when the same transistor is operated as an amplifier at its fundamental frequency.

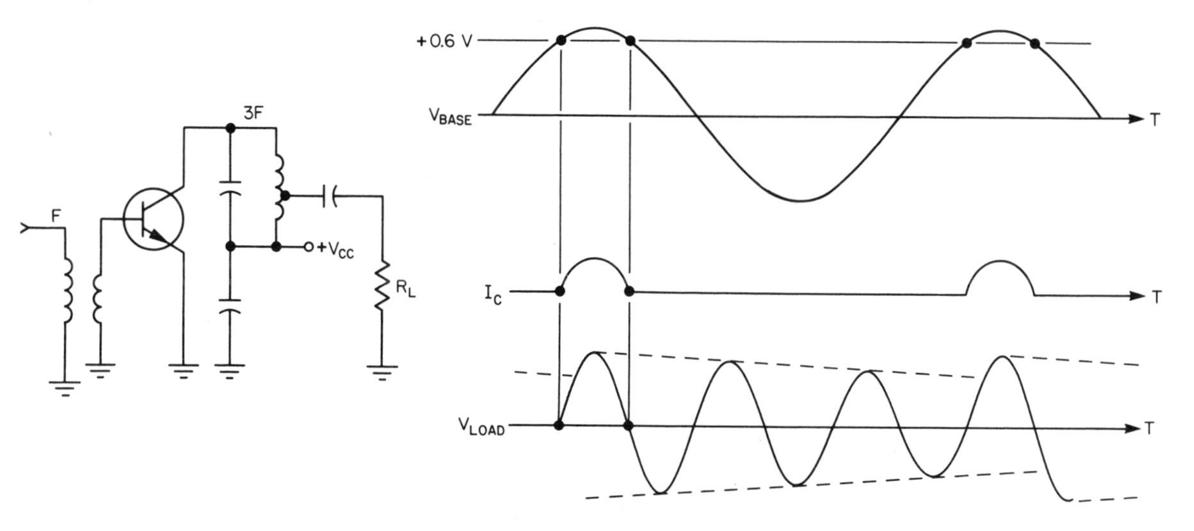

Figure 9-13 Class C frequency tripler and its input and output waveforms.

9-10 TRANSISTOR VARACTOR MULTIPLIERS

Many RF power transistors have a significant collector-to-base capacitance that is not directly underneath the emitter "fingers." Most of the series resistance into the base region ($r_{bb'}$) is therefore bypassed and a fairly high quality varactor diode then exists whose capacitance changes with collector-to-base voltage, as shown in Figure 9-14(a).

When used as a frequency multiplier, this transistor can provide noticeable improvements in power gain and efficiency, particularly when used near the upper-frequency limit if the transistor. The transistor portion is designed for

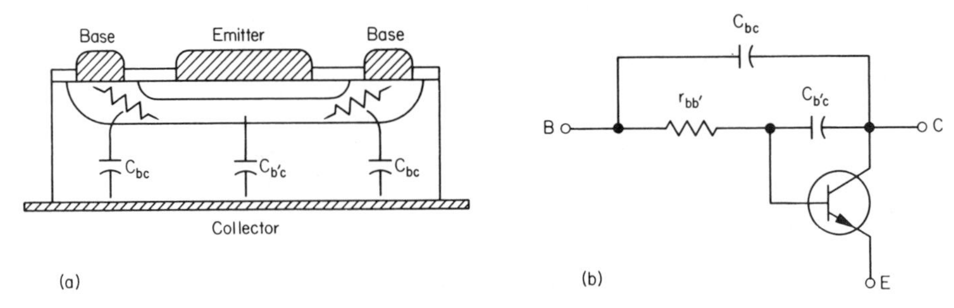

Figure 9-14 Fairly high quality varactor diode (C_{bc}) exists in many RF power transistors. Its origin is shown in (a) and position in the equivalent circuit in (b).

high gain at the fundamental frequency, and then the varactor portion provides efficient conversion of the fundamental energy into a harmonic. As shown in Figure 9-15, the circuit, even for a doubler, is slightly more complicated than for a basic class C multiplier.

The input components, L_1 and C_1, form a two-element matching network at the fundamental frequency (f). L_2 and C_2 are series resonant at the second harmonic ($2f$) of the input signal and place the base end of the varactor at ground potential. On the output L_3 and C_3 are series resonant at the fundamental frequency. This arrangement of tuned circuits ensures maximum signal flow through the varactor at the fundamental and second harmonic so that high conversion gain is obtained. The second harmonic alone passes through C_4 and L_4 to the load. Circuits built by the author showed an average 3-dB gain improvement of a varactor-enhanced 70- to 140-MHz frequency doubler over the same transistor operated as a class C doubler. At higher frequencies, where the transistor gain starts to decrease, the improvement would likely be higher.

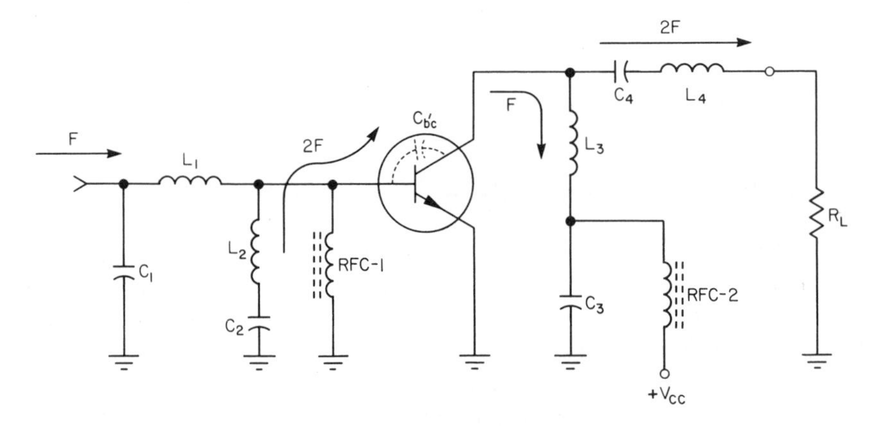

Figure 9-15 Class C frequency doubler using the internal transistor capacitance as a varactor to increase conversion gain.

9-11 AMPLITUDE MODULATION

The amplitude of a carrier can be modulated if the supply voltage of a power amplifier is increased and decreased. To be effective, this means that the amplifier must be overdriven so that the transistor is swinging between cutoff and saturation. The supply voltage changes are usually obtained by feeding the modulated amplifier through the secondary of a transformer that is connected in series with the supply voltage. The primary of the transformer is fed from an audio power amplifier.

A slightly simplified arrangement of a modulated carrier amplifier is shown in Figure 9-16. Capacitor C_1 is a bypass capacitor for the carrier frequency so that the modulated supply voltage appears to come from a very low impedance source and eliminates any transformer reactances from affecting the transistor's operation. The supply voltage at that point could swing between maximum limits of zero and twice the supply voltage. If enough drive power is available, the transistor will be driven between cutoff and saturation, and the collector waveform could appear as shown. Note that the collector voltage can hit a high of four times the power supply voltage, because of the collector

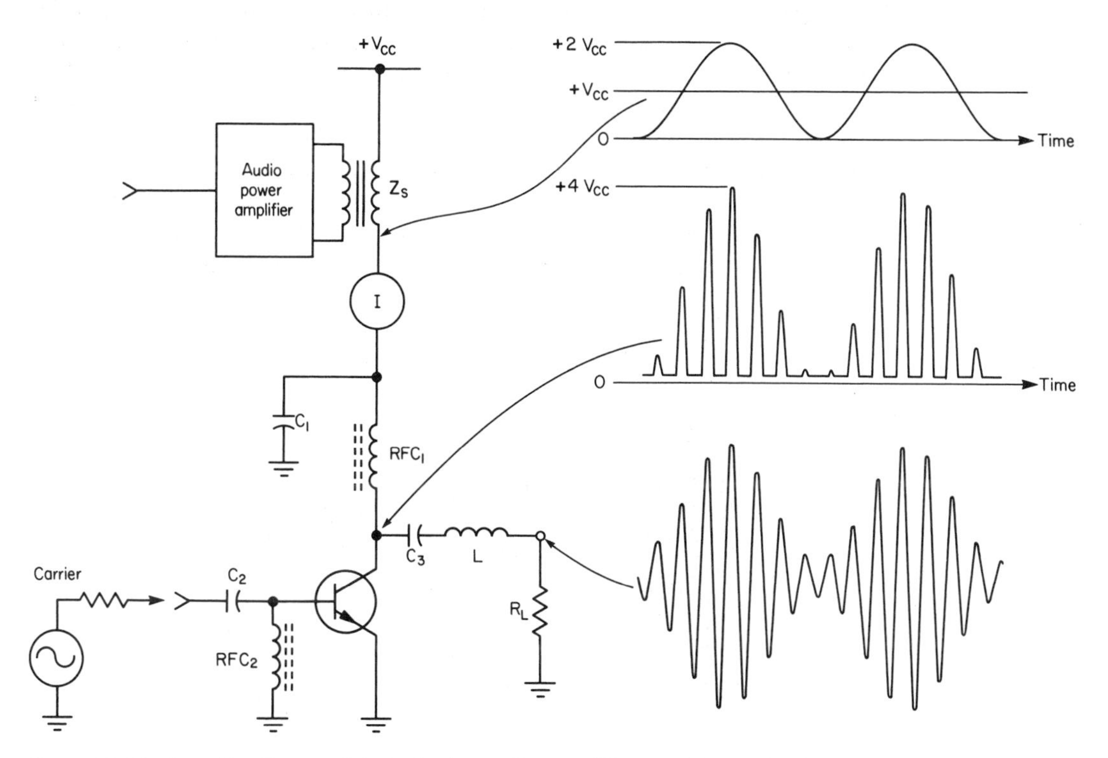

Figure 9-16 Amplitude-modulated class C amplifier with typical waveforms.

inductor. The transistor breakdown voltage (BV_{CES}) would have to be chosen with this in mind.

The collector waveform contains a number of unwanted components, one of which is the original modulating frequency, the others the harmonics of the modulated carrier frequency. Some form of bandpass filter will remove all the undesired signals, leaving only the carrier and its new sidebands; this is provided by the series-resonant filter consisting of L and C_3.

The audio power amplifier must be capable of supplying at least half the dc power input to the modulated amplifier. This dc input is the product of the current meter reading times the average supply voltage across the capacitor C_1.

$$P_{\text{dc}} = I_{\text{dc}} \times V_{CC} \tag{9-2}$$

The secondary of the transformer will "see" an impedance of

$$Z_S = \frac{V_{CC}}{I_{\text{dc}}} \tag{9-3}$$

As simple as this description seems, there are a number of problems that can arise; the drive level to the transistor is one of them. If enough drive is provided to maintain saturation and cutoff at the highest effective supply voltage ($2\,V_{CC}$), too much will be available when the supply voltage approaches zero. This excessive drive will not only be inefficient, but it will prevent the RF output of the transistor from going to zero, as it should if 100% modulation is to be obtained. The problem, at low collector voltages, is that the internal collector base capacitance increases and much of the drive signal sneaks through to the output. The waveform of Figure 9-17 shows a typical modulated output suffering from insufficient drive at the peaks and direct feed-through due to

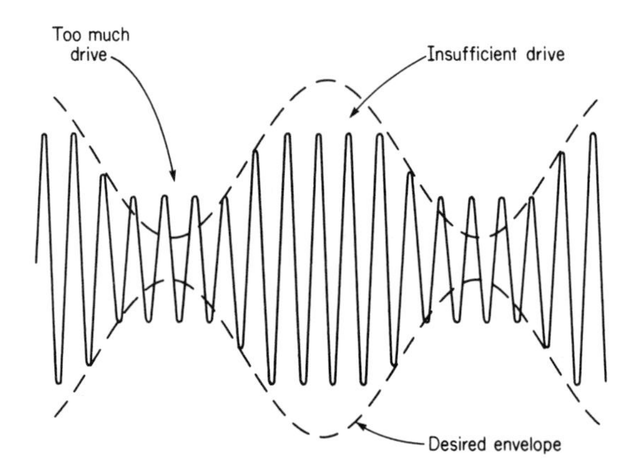

Figure 9-17 Desired modulated signal, shown as a dashed outline. The modulation actually obtained suffers from insufficient drive at the peaks and too much drive at the valleys.

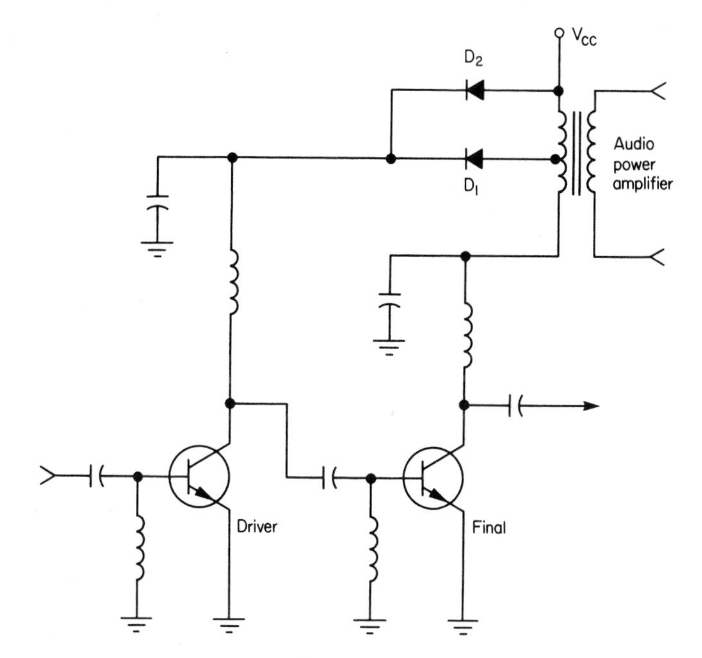

Figure 9-18 Driver stage partially modulated to improve the depth of modulation.

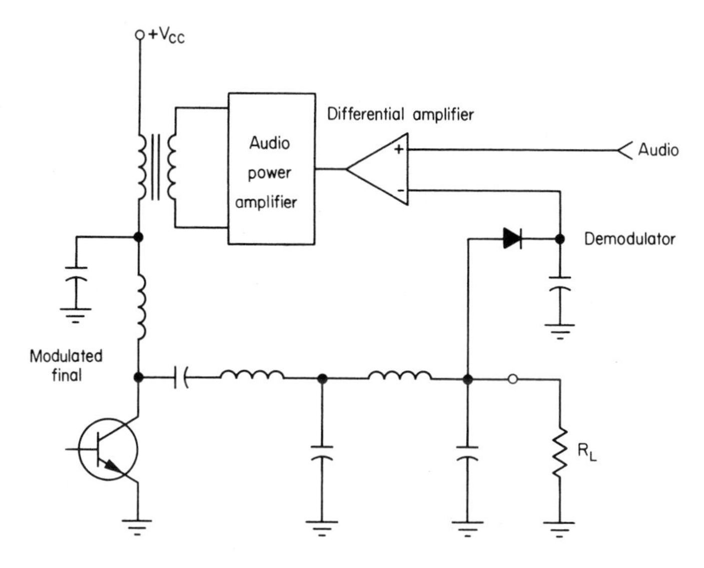

Figure 9-19 Improved modulation can be obtained by using a feedback circuit to compare the results with what is desired.

excessive drive at the valleys. The solution is to also partially modulate the stage used to drive the final output stage. A typical schematic is shown in Figure 9-18.

The secondary of the modulation transformer has been tapped so that the first stage is not fully modulated. The tap position usually has to be found experimentally. Occasionally, the driver is only "upward-modulated." The two diodes shown dashed form an OR gate, so the driver stage draws power from the transformer secondary through D_1 when that voltage is higher or from V_{CC} through D_2 when the modulated supply swings lower.

When very high audio quality is called for, a sample of the final output signal can be demodulated and compared with the desired audio signal, as shown in Figure 9-19. A differential amplifier can then provide the required distortion to the power amplifier so that the final modulation is much closer to what is desired.

9-12 OUTPUT NETWORKS

For any transmitter, regardless of the type of modulation used, an output network will be required between the collector of the final transistor and the antenna. Its purpose is twofold, first to transform the antenna impedance down to the value at the transistor required for the power output, second to remove both the audio-frequency component and the modulated carrier harmonics from the desired signal. Since the audio components are at a much lower frequency than the carrier, a single series capacitor is usually adequate; the remainder of the network is then essentially a low-pass filter.

The design of the output network must start with three pieces of information: the antenna impedance, the impedance to be presented to the transistor, and the amount by which the harmonics must be reduced. Typical values for a 4-W 13.5-V CB set might be: 50-Ω antenna impedance, 18-Ω load for the transistor, and a 60-dB reduction of all harmonics and other out-of-band signals. Without any filters, the output of any transmitter would likely contain harmonics of approximately the levels shown in Figure 9-20.

The second harmonic is seen to be the strongest, and to reduce this down to -60 dB below the carrier level, an extra 57 dB would be required. Since each filter component will produce a 6-dB/octave slope, it appears that 9 or 10 elements would be needed to provide the necessary harmonic attenuation. However, if the transmitter operates over a narrow bandwidth, higher-Q matching sections can be used that provide higher initial attenuations in the stopband before setting down to the 6-dB slope (see Chapters 4 and 5). This could result in a reduction of one or two components if a loaded Q of about 5 were used. Further improvements in number of components can be had if the second harmonic is ignored, temporarily, and an attenuation of 60 dB at the third harmonic is aimed for; four or five components would be adequate. A

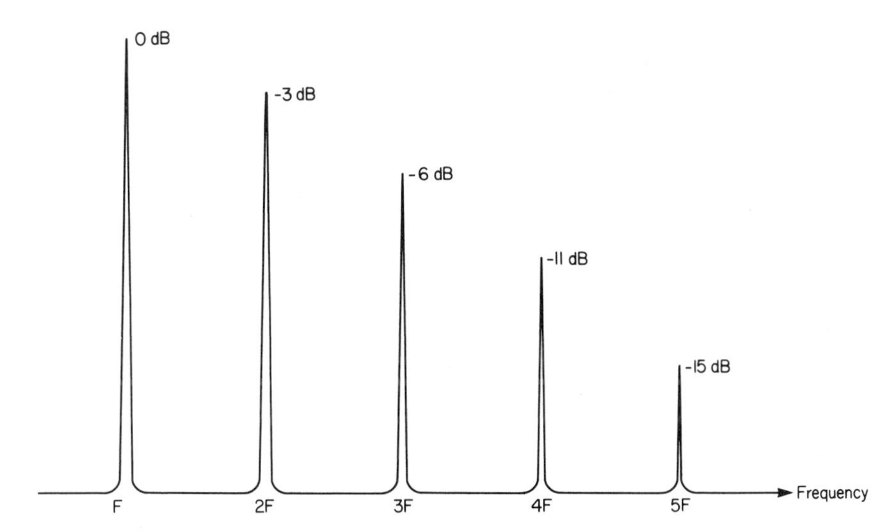

Figure 9-20 Approximate harmonic levels at the output of a class C power amplifier without any filtering.

series or parallel trap could then be added to provide a notch at the second harmonic, provided that the second harmonic of all possible channels of the transmitter are closely grouped.

One final problem: published curves for low-pass filters assume constant, resistive terminating impedances. When connected to an antenna, the load on the filter will be 50 Ω (or whatever) only over a very narrow range of frequencies; above this the antenna impedance is likely to rise rapidly and become quite reactive. The output impedance of the transistor itself is usually considerably higher than the design load value it is looking into. The true test of harmonic level is therefore best made as an on-the-air field-strength measurement.

9-13 EFFICIENCY

The input impedance of the filter at frequencies above the operating frequency should also be considered. Since the transistor is supposed to be switching between cutoff and saturation, it can do so more efficiently if it is not looking into a capacitive load that would extend its rise and fall times. Higher efficiency can be achieved with a series inductive input such as Figure 9-21(c). The disadvantage of this is that the harmonic levels then tend to be slightly higher. The efficiency improvement is most noticeable when the transistor is operating well below its maximum frequency, where its switching time is short compared to one cycle.

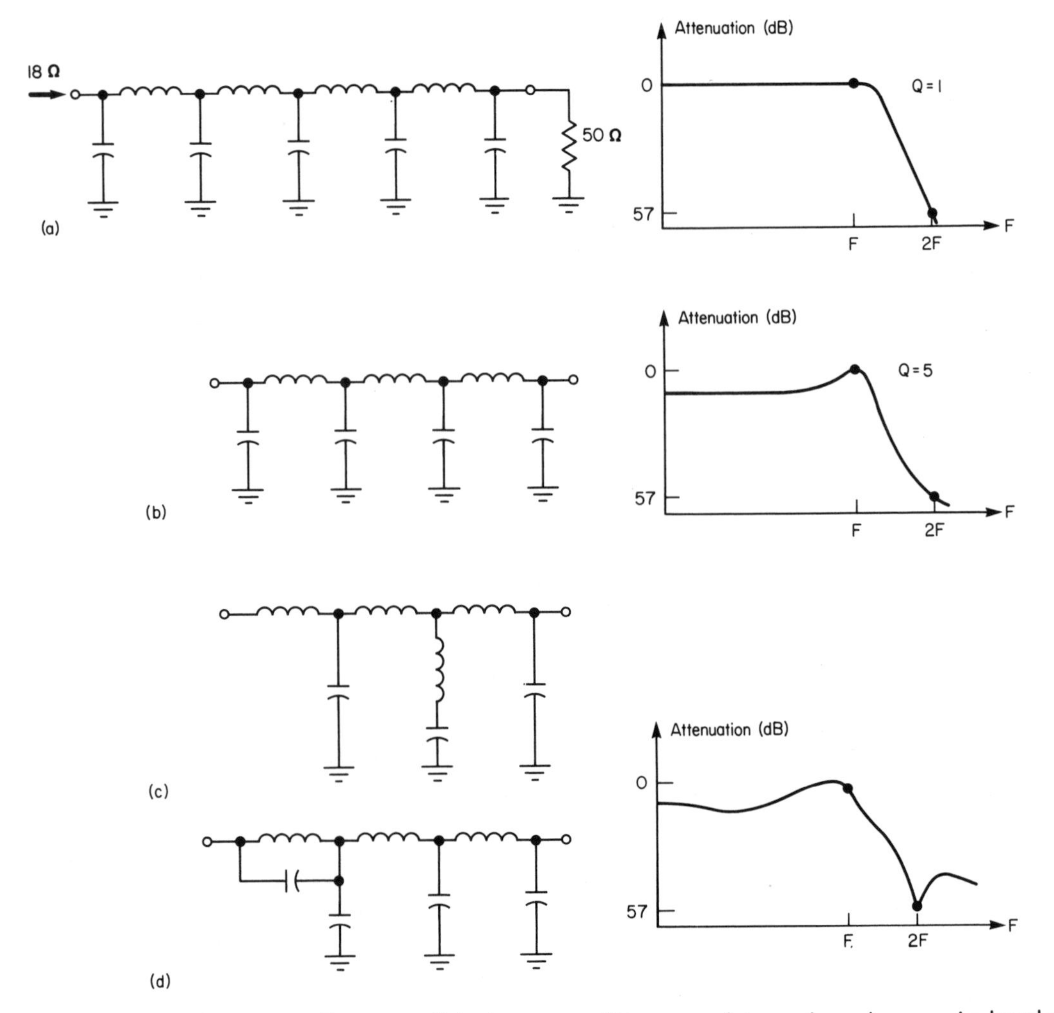

Figure 9-21 Four possible low-pass filters used to reduce harmonic levels below 60 dB.

9-14 STABILITY AND DAMAGE

When a new design is tested, it would be nice if the expensive final transistor does not instantly destroy itself. One possible reason for it doing so is the tendency to produce high-level oscillations at some other, usually lower, frequency. To prevent this tendency, the power-supply leads must be adequately bypassed not only at the carrier frequency but also at lower frequencies. This usually means a good-quality electrolytic capacitor in parallel with a good-quality ceramic. This reduces the possibility of the power-supply wiring accidentally creating a low-frequency high-impedance load.

The RF chokes at the base and collector should have an impedance no larger than two or three times that of the transistor input and the network input impedances, respectively. The Q of both should also be low (less than 5), to discourage unwanted resonances.

The load impedance itself is often a problem; rarely is it exactly 50 Ω (or whatever it should theoretically be). This means that the impedance presented to the transmitter will change with antenna cable length. The resulting impedance changes at the collector could encourage unwanted oscillations or even damage the transistor if the collector voltage were to rise high enough. Tests should therefore be made over the anticipated range of antenna mismatches and with a full half-length variation in cable length to ensure stable operation.

Newer power transistors are more immune to damage due to high VSWR than some earlier designs, but automatic shutdown circuits might be needed if the possibility of damage exists.

9-15 PUSH–PULL CONFIGURATION

For high output, power transistors can be grouped to increase their capabilities. One method is to parallel two or more devices, and while the technique works well, there is another possibility that offers some advantages. Operating in push–pull increases the total input impedance of the pair, which makes matching easier. More important, push–pull operation tends to cancel even harmonics of the output. For broad-band amplifiers, then, this greatly simplifies the output filters. The amount of even harmonic reduction will depend on matching of the two transistors used. Reductions of 15–20 dB are possible without any great care in matching, while reductions beyond 30 dB are possible using matched pairs in a common package.

The design of each half of the push–pull amplifier is straightforward; the main problem is the power and phase splitting at the input and the recombining of the signals at the output.

Two methods of splitting the input and recombining the output are shown in Figure 9-22. The first uses broad-band toroidal transformers as 1 : 1 Baluns and the second uses quarter-wavelength transmission-line sections. The transmission-line technique is limited to less than 1-octave bandwidth.

9-16 FM TRANSMITTER EXAMPLE

An example of a simple transmitter is shown in Figure 9-23. It is frequency-modulated with an approximate power output of 1.25 W at 140 MHz. The oscillator operates at 70 MHz with a third overtone crystal. It has a Hartley type of feedback circuit and produces an approximate output of 8 mW. The

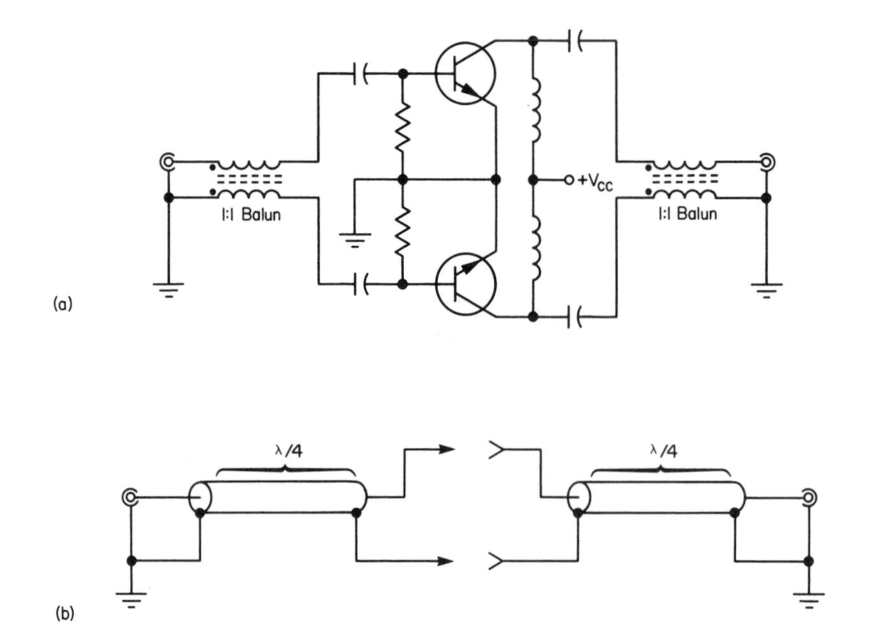

Figure 9-22 Even harmonics can be attenuated and power levels increased by using transistors in push–pull. The input and output circuits can be Balun transformers (a) or sections of transmission lines (b).

series combination of C_2 and C_3 determine the resonant frequency of the collector circuit along with L_1. C_2 and C_3 provide an impedance step down to the lower input impedance of transistor Q_2.

$$Z_{\text{in}} = 30 - j10\ \Omega\ (\text{at } 70\ \text{MHz})$$

L_5 and C_5 are series resonant at 140 MHz and L_4 is part of the input matching network at 70 MHz. The class C negative bias is established by C_4 and 1 kΩ. The output of the frequency doubler Q_2 will be a combination of 70- and 140-MHz signals. Both L_6, C_6 and L_8, C_8 are series resonant at 70 MHz to remove the fundamental and leave only the 140-MHz second harmonic. Power is fed to Q_2 through L_7. The dc current to Q_2 is about 9 mA, and the RF power output is about 60 mW, for an efficiency of 55%.

At 140 MHz the series combination of C_8, L_8 will be inductive, and this reactance, along with that of C_7 and C_9, form a T matching section down to the low input impedance of Q_3.

$$Z_{\text{in}} = 22 + j1.0\ \Omega \quad (\text{at } 140\ \text{MHz})$$

For an output power of 1.25 W, the load at the collector of Q_3 must be approximately 48 Ω. The antenna impedance is nominally 50 Ω so that the output filter can be symmetrical.

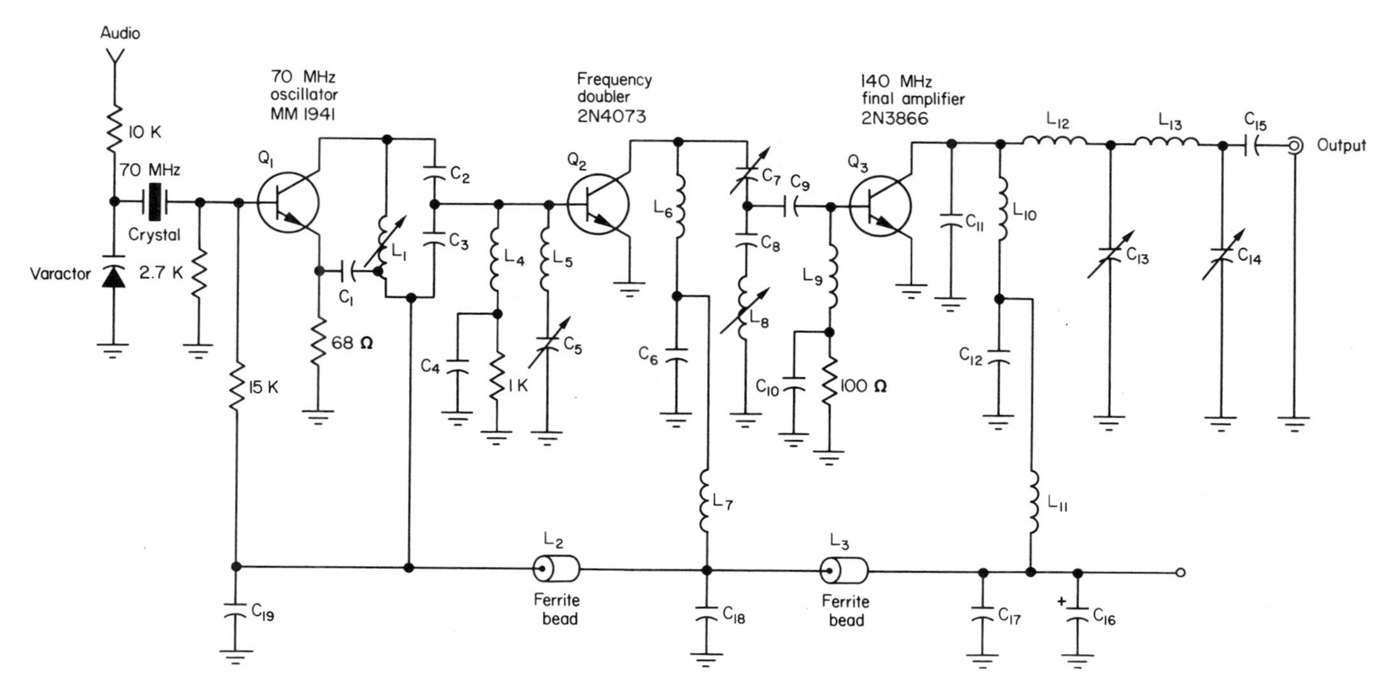

Figure 9-23 Schematic of a 140-MHz FM transmitter. The first stage is a 70-MHz oscillator using a fifth-overtone crystal. This frequency is doubled in the second stage and then further amplified by the final stage. Power output is 1.25 W.

TABLE 9-1
COMPONENT VALUES FOR FM TRANSMITTER.

C_1 ...18 pF	L_1 ... 0.13 μH adjustable
C_2 ...100 pF	L_2 ...Ferrite bead
C_3 ...68 pF	L_3 ...Ferrite bead
C_4 ...1000 pF	L_4 ...0.22 μH
C_5 ...5–35 pF variable	L_5 ...0.1 μH
C_6 ...52 pF	L_6 ...0.1 μH
C_7 ...2–8 pF variable	L_7 ...2.2 μH
C_8 ...100 pF	L_8 ...0.05 μH adjustable
C_9 ...33 pF	L_9 ...0.1 μH wound on 220-Ω resistor (low Q)
C_{10} ...1000 pF	L_{10} ...0.22 μH
C_{11} ...6.8 pF	L_{11} ...2.2 μH
C_{12} ...22 pF	L_{12} ...0.22 μH
C_{13} ...5–35 pF variable	L_{13} ...0.22 μH
C_{14} ...5–35 pF variable	
C_{15} ...560 pF	
C_{16} ...5 μF tantalum	
C_{17} ...1000 pF	
C_{18} ...1000 pF	
C_{19} ...1000 pF	

It is a five-element design using capacitors C_{15}, C_{17}, C_{18}, C_{19}, and C_{20} and inductors L_{11} and L_{14}. Capacitor C_{21} provides dc isolation and inductor L_{12} feeds power into the collector of Q_3. Average current into this stage is 160 mA for an efficiency of 65%.

At the output the following signal levels were measured relative to the 140-MHz carrier level:

70 MHz	−70 dB
140 MHz	0 dB
210 MHz	−45 dB
280 MHz	−55 dB
350 MHz	−60 dB
420 MHz	−65 dB

QUESTIONS

1 A 150-MHz transmitter with an RF power output of 15 watts is required. Using the information in Figures 9-7 and 9-8, find the drive power required, the amplifier gain (dB), and the parallel input resistance and reactance.

2 For a 15-watt RF output using a 28 volt power supply, find the load resistance that must appear at the collector. Assume a collector saturation voltage of 0.7 volts.

3 An FM transmitter at 150 MHz requires ±12-kHz deviation. The oscillator will operate at 16.667 MHz and its output will be multiplied up to the final frequency. How many kHz deviation will be required at the oscillator frequency and what % deviation is this?

4 Explain why the internal structure of an RF power transistor is so different from a low-frequency, small-signal transistor. Refer to Figures 9-4, 9-5, and 9-6 for the points to comment on.

5 A transmitter is required with 5.0 watts of RF power at the antenna terminals. The output-matching network has a 1.5-dB loss, the transistor efficiency is 60% and its saturation voltage is 0.4 volts. Find the collector load resistance and the dc input power for this stage if $V_{cc} = 13.75$ volts.

6 The transmitter of Question 5 is to be 100% amplitude-modulated. What audio power is required and what secondary impedance should the modulation transformer have if the power supply is 13.75 volts?

7 What minimum breakdown voltage (BV_{ces}) should the power transistor used in Question 6 have?

8 The third harmonic of a particular transmitter is only 8 dB below the fundamental level but should be 65 dB down to meet a specification. Using the information from Figure 4-21, how many elements would be required for a Chebychev filter to reduce the harmonic enough to meet the specification? (To minimize added insertion loss, use the filter at 0.9 times its cutoff frequency.)

REFERENCES

Motorola Semiconductor Products Inc. 1978. *Motorola RF Data Manual*. Phoenix, Arizona.

Motorola Semiconductor Products Inc. 1972. *Linear Amplifiers for Mobile Operations*. Motorola Application Note AN-762.

Motorola Semiconductor Products Inc. 1972. *A Broadband 4-Watt Aircraft Transmitter*. Motorola Application Note AN-481.

Motorola Semiconductor Products Inc. 1972. *Using Balanced Emitter Transistors in RF Application*. Motorola Application Note AN-521.

RCA Solid-State Division. 1971. *RCA Designer's Handbook SP-52, Solid-State Power Circuits*. Somerville, N.J.

Tatum, J. G. *VHF/UHF power transistor amplifier design, ITT semiconductors*. West Palm Beach, Florida.

chapter 10

receiver circuits

The purpose of a receiver is to select a desired group of frequencies from one transmitter, get rid of all unwanted signals and noise, and then demodulate the signal to obtain the modulating information. The better the receiver does its job, the closer the demodulated signal will resemble the original signal from the transmitter. Regardless of the type of demodulation required, the main functions performed by a receiver are filtering and amplifying. The superheterodyne receiver is the logical choice for the job.

10-1 SUPERHETERODYNE RECEIVER

Since it is easier to design narrow-band, steep-skirt filters and obtain high gains at lower frequencies, the "superhet" receiver is an efficient design. All incoming signals are mixed with the output of a local oscillator and the difference frequency is selected and amplified by the intermediate frequency amplifiers. The big benefit is that these amplifiers remain at a fixed frequency and only the RF amplifier and local oscillator need be tunable. Figure 10-1 is a block diagram of a typical superhet receiver. One further benefit is the fact that the gain is concentrated at two or sometimes three different frequencies. This

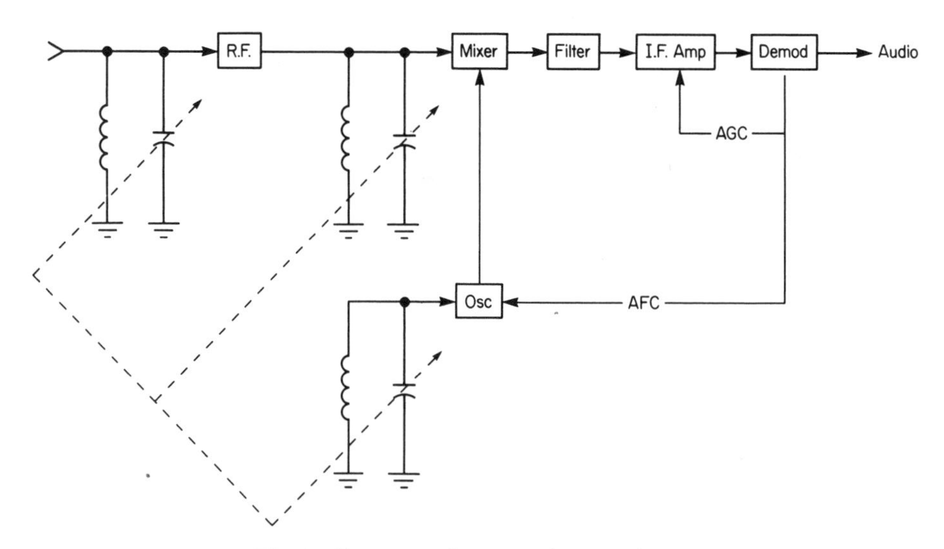

Figure 10-1 Block diagram of a superheterodyne receiver.

reduces the gain required at any one frequency and leads to more stable amplifiers. When over 120 dB of RF gain is involved, every little bit helps.

The function of each item in Figure 10-1 can be explained as follows:

1 *RF amplifier.* Should have just enough gain, usually about 10 dB, to establish the overall noise figure of the receiver. The tuned circuits at the input and output need only be selective enough to reject image signals and other spurious signals that could intermodulate and appear at the intermediate frequency. Some AGC may be needed to prevent overloading on strong signals. The RF amplifier may also be called on to suppress any tendency for the local oscillator to radiate out to the antenna and interfere with other listeners.

2 *Mixer and local oscillator.* The mixer has two inputs, one from the RF amplifier and one from the local oscillator. The nonlinearities of the mixer will create numerous intermodulation products, and one of these, the sum or difference frequency, will occur at the IF frequency. Usually, there will be a second frequency, the image, that can also mix with the oscillator frequency and produce an output at the IF. Depending on the type of mixer used, conversion gains from −10 dB to +30 dB are common. The local oscillator must be tunable, yet have a low drift rate and relatively low sideband noise, since this could increase the noise level of the receiver.

3 *IF filters and amplifiers.* This section establishes the overall bandwidth and adjacent channel selectivity of the receiver. The bulk of the receiver's gain will be concentrated here and some type of automatic

gain control will be included to adjust for variations in received signal strength. The IF is usually at a lower frequency than the RF, but, in some special cases, the IF may be higher to reduce spurious intermodulation and image problems.

4 *Demodulators.* For each type of modulation used (i.e., AM, FM, SSB, PM), a number of different circuits exist. Some will have gain, others a loss. Some will require a reference input (i.e., SSB and phase modulation), others won't. The demodulation may also be required to produce outputs to AGC or AFC circuits. The recovered audio level (or video, etc.) will determine the amount of gain required in the following audio or video amplifiers.

10-2 SPECIFICATIONS

Before beginning the design of a receiver, it is necessary to consider the specifications required of the final result. In most cases this ends up as a compromise between what the designer would like and what is possible. The determining factor will usually be financial limitations. The following should then be considered before proceeding:

1 *Tuning range.* What range of frequencies must be tuned and will it be tuned continuously or in discrete channels? A short-wave receiver, for example, must continuously tune from 3 to 30 MHz and will usually require some band switching. The local oscillator will be a continuously tunable type. Demodulators will be needed for AM, SSB, and CW, and IF bandwidths should correspond. For CB, a narrow range of frequencies from 26.965 to 27.405 are needed and will be tuned as 40 discrete channels. The local oscillator will therefore likely be a phase-locked loop synthesizer. Demodulation could be either AM or SSB.

2 *Sensitivity.* Often, too much emphasis is put on sensitivity without attention to other details. For example, a 100-kHz navigation receiver will pick up so much atmospheric noise that a 100-μV desired signal from the antenna could be obscured at times. On the other hand, a 0.1-μV signal at 150 MHz will often be readily distinguishable from background noise. Typical atmospheric noise levels at different frequencies are shown in Figure 10-2.

3 *Bandwidth.* When the modulation type and channel spacing are known, it is possible to determine the IF bandwidth and its skirt characteristics. For FM-stereo broadcasting, a bandwidth of 350 kHz is required. For AM aircraft communications, a bandwidth of 30 kHz is common—not to provide wide bandwidth for high audio-frequency response but to

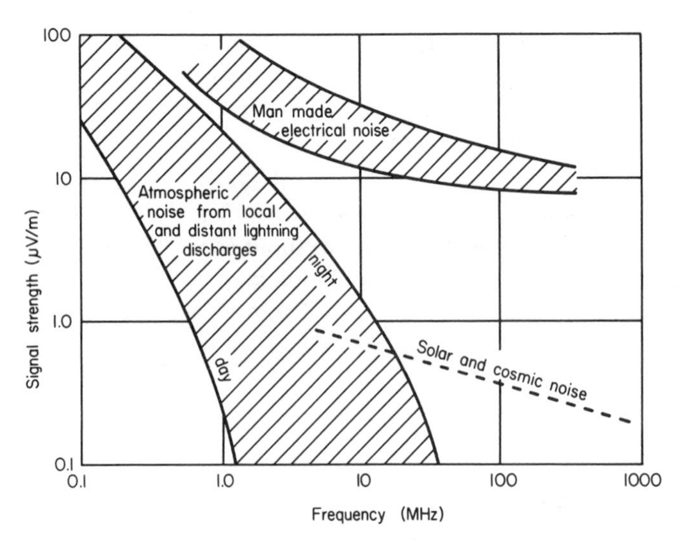

Figure 10-2 Typical noise levels at different frequencies.

accommodate frequency tolerances in the transmitters and receivers. The filter-skirt characteristics will be set to reject adjacent channel signals as required.

4 *Spurious signals.* An otherwise good design can be useless if unwanted signals can sneak into the receiver at the IF frequency(s), the image frequency, at various spurious frequencies related to intermodulation products, and through cross-modulation problems.

Typical specifications for several good receivers are as follows:

1 *FM stereo tuner:* frequency range 88–108 MHz

Sensitivity:	1.8 μV across 300-Ω input for 20 dB of quieting
Selectivity:	100 dB for channels 400 kHz either side of center frequency
Bandwidth:	350 kHz at −6-dB points
Image rejection:	90 dB
Spurious rejection:	90 dB
IF rejection:	90 dB
AM suppression:	65 dB
Capture ratio[1]:	1.5 dB

[1]If one FM signal is larger than another by at least 1.5 dB, only the stronger one will be heard.

2 *Shortwave receiver:* frequency range 3.0–30 MHz

Sensitivity:	0.5 μV for 10-dB S+N/N ratio
Bandwidth:	2.3 kHz at −6 dB, 5.5 kHz at −60 dB (SSB mode)
Image rejection:	60 dB
IF rejection:	75 dB

3 *CB receiver:* frequency range 26.965–27.405 MHz

Sensitivity:	0.5 μV for 10-dB S+N/N ratio
Bandwidth:	6 kHz at −6 dB 20 kHz at −60 dB
Image rejection:	60 dB

Once the specifications are carefully determined, it is time to start the design. But what is the best starting point? Generally, the most sensitive points will be the two nonlinear circuits, the mixer and the detector. The IF amplifier takes up the slack between the two, and the RF amplifier picks up the deficiencies of the mixer.

10-3 MIXERS

The mixer section of the receiver should ideally produce an IF output only at the difference (or sum, for up-conversion) of the two input frequencies. One of these inputs will be the local oscillator signal and the other will be the desired RF signal. Again, ideally, no other combination of input signals should produce an IF output. If such frequencies do exist, filters must be provided to remove them before they reach the mixer.

The closest thing to an ideal mixer is any circuit with a perfect square-law transfer characteristic, as discussed in Section 1-5. In addition to the input signals and their second harmonics appearing at the output, the sum and the difference will also appear. The difference is usually the one signal desired and so is selected by IF filtering. The amplitude of the difference signals will be proportional to the product of the original RF signal level and the local oscillator level. Any other two signals at the input could also produce an output at the IF if they are separated by an amount equal to the difference frequency. However, the output level they produce will be proportional to their signal levels.

Some discrimination against unwanted mixing products can therefore be had if all RF input levels to the mixer are kept as low as possible and the local oscillator signal kept as high as possible. The one desired signal will therefore be stronger than all the undesired ones. This is described in Figure 10-3. The

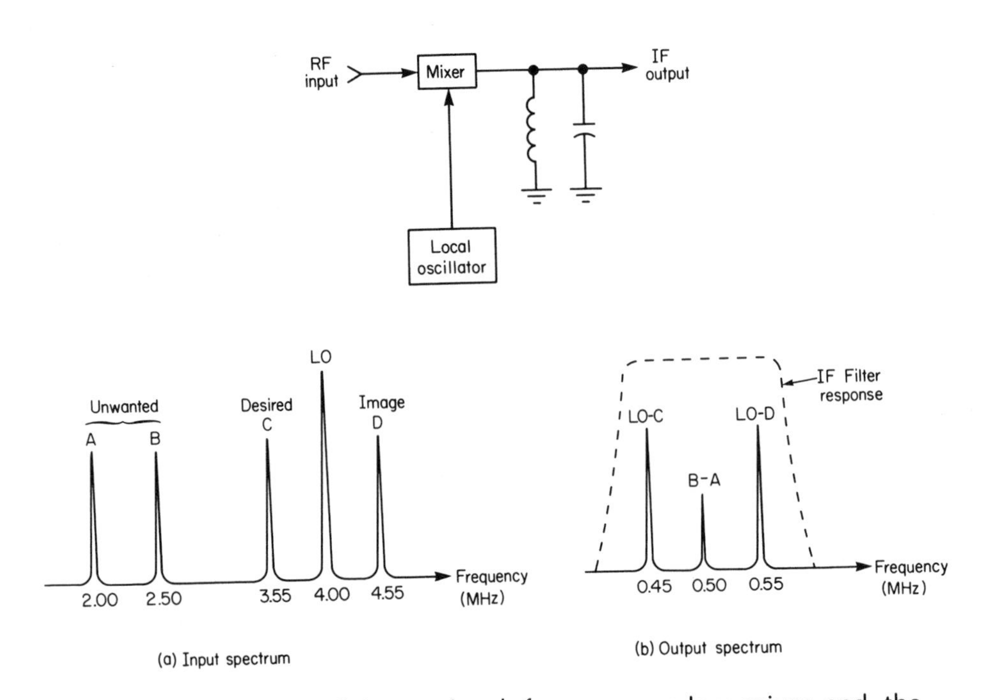

Figure 10-3 Spectrum of input signal for a square-law mixer and the outputs within the IF passband.

mixer circuit has four signals at its input, all of the same level. The local oscillator signal level is included for reference and is much higher than the other four. The IF filters only pass signals between 0.4 and 0.6 MHz. RF frequencies C and D can mix with the oscillator and produce outputs at 0.45 and 0.55 MHz, respectively, well within the IF passband. One will be the desired signal and the other is the image, which should be removed by filtering before reaching the mixer. Two other signals, A and B, happen to be separated by 0.5 MHz, so they will also produce a mixing product (which contains the combined modulation of each) within the IF passband. However, the amplitude of this signal will be much lower than the desired IF signal. Therefore, best results are obtained by:

1 Selecting a square-law mixer.
2 Using high local oscillator levels.
3 Maintaining low RF signal levels.
4 Providing proper filtering ahead of the mixer.

10-4 FILTERING REQUIREMENTS

If the ideal square-law mixer can be built, what is the minimum filter that is required ahead of it? We have already seen that the image has to be removed and also any group of frequencies that could themselves mix and produce an IF output. The limiting case is shown in Figure 10-4. The IF filters are placed at 5.0 MHz and have nearly vertical skirts. The RF filters ahead of the mixer also have nearly vertical skirts and cover the range 5.5–10.0 MHz, nearly a 1-octave range. The RF bandwidth is just narrow enough (4.5 MHz) that no two signals can exist within the passband to cause mixing. The image frequencies would lie in the range 15.5–20.0 MHz and are also outside the filter range. In a practical design, the filters would have wider skirts, so the useful range of the mixer would be an even smaller portion of the theoretical 1-octave bandwidth.

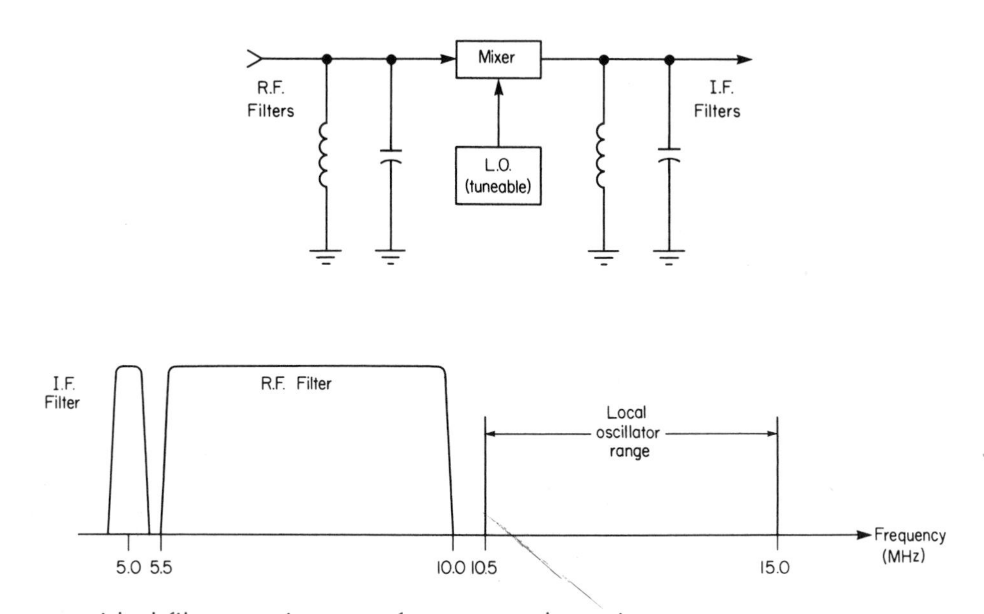

Figure 10-4 Ideal filter requirements for a square-law mixer.

10-5 CONVERSION GAIN AND NOISE

Since any practical circuit will generate excess noise, mixers are no exception. Each type of mixer circuit will therefore have its own noise figure. If the combination of this noise figure and any losses in the RF filters ahead of the mixer are low enough, no amplification is required nor even desired before mixing. Amplifiers would inevitably have some nonlinearities and would increase the signal input level to the mixer so that other mixing products could appear. When considering a mixer noise figure, it should be remembered that two frequencies could contribute to noise output at the IF frequency, the desired frequency and its image. Removal of the image by placing a filter between the antenna and the mixer is not always sufficient. If the mixer "sees" a resistive impedance at the image frequency, thermal noise will be added. The filter should therefore appear as a short circuit at the image frequency.

If the noise figure is too high, an RF amplifier will be necessary. Its gain must be just adequate to set the overall noise figure to the desired level and no more. In fact, a little negative feedback in this stage will improve the linearity and the resulting loss of gain will actually be welcome.

The mixer may produce an IF signal that is either higher or lower in amplitude than the RF input signal. The relative size would be indicated by the conversion gain or loss. Conversion gain should be a secondary consideration, as any deficiency can always be made up for with other amplifiers. The only real problem involves setting the overall noise figure of the receiver. With no RF stage and a mixer with a conversion loss, the first stage of the IF must have a very low noise figure.

10-6 MIXER SPECIFICATIONS

We have already identified the points to be considered when selecting a mixer; they are repeated here, along with a few extras.

1 How close is the mixer to a perfect square-law device?
2 What conversion gain does it have?
3 What will its noise figure be?
4 How much local oscillator drive is required?
5 How much of the RF and local oscillator signal will there be at the output, not considering any IF filtering? (This point refers to balanced mixers that can suppress the RF and/or local oscillator signals at the output side.)

To describe the nonlinearities of the mixer, a graph relating RF input level to the output level of the various mixing products can be drawn similar to those of Section 1-5. A typical input–output relationship is shown in Figure 10-5 for a constant level of local oscillator input. The highest line represents a linear conversion gain of 0 dB, so a −60-dBm RF input would produce a −60-dBm IF output, for example. The next line represents the conversion gain for the actual mixer. In this case it is a loss of 6 dB. For high input levels, the

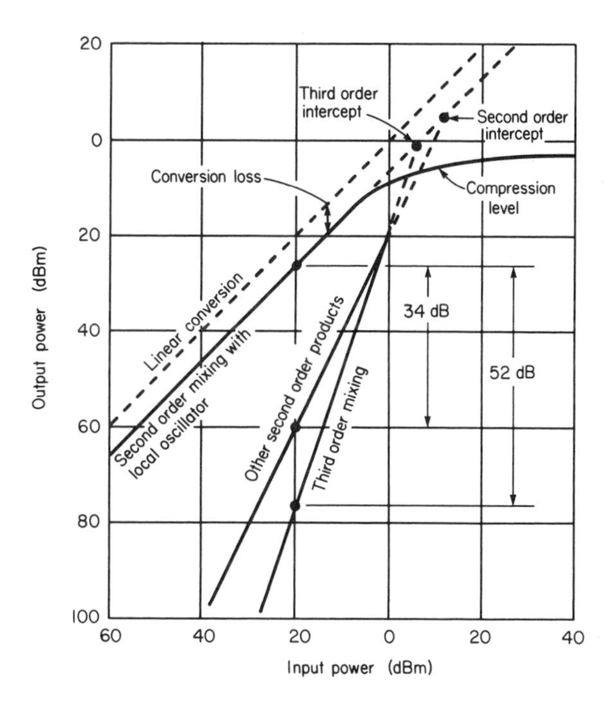

Figure 10-5 Input and output signal levels for a practical mixer. The local oscillator level is kept at a constant level.

output will not increase beyond the compression point, so inputs must be kept well below this.

The other second-order mixing products that result when any two inputs, other than the local oscillator signal, combine produce much lower-level outputs for most inputs (the "RF input power" marked on the horizontal axis would be the sum of the two inputs in this case). The strength of these outputs, however, climbs rapidly and they equal the desired mixing products (sum and difference) at the second-order intercept. Similarly, the third (and higher)-order products are low levels normally but climb very fast, producing their own intercept points.

Such a graph is very useful, as it shows what range of RF signal levels to use. The lowest input level will be determined by noise considerations, and the highest input level will be set by the need to avoid the output compression level and by the desired reduction in other second- and third-order mixing products. The vertical line represents an input level of -20 dBm. The second-order products will be 34 dB below the desired output, and the third-order products 52 dB below.

10-7 MIXER TYPES

A large number of mixer circuits are available; six are shown in Figure 10-6. Each circuit will require different levels of local oscillator signal and will have different conversion gains and losses. Some are balanced circuits, so that one or both inputs do not appear at the output.

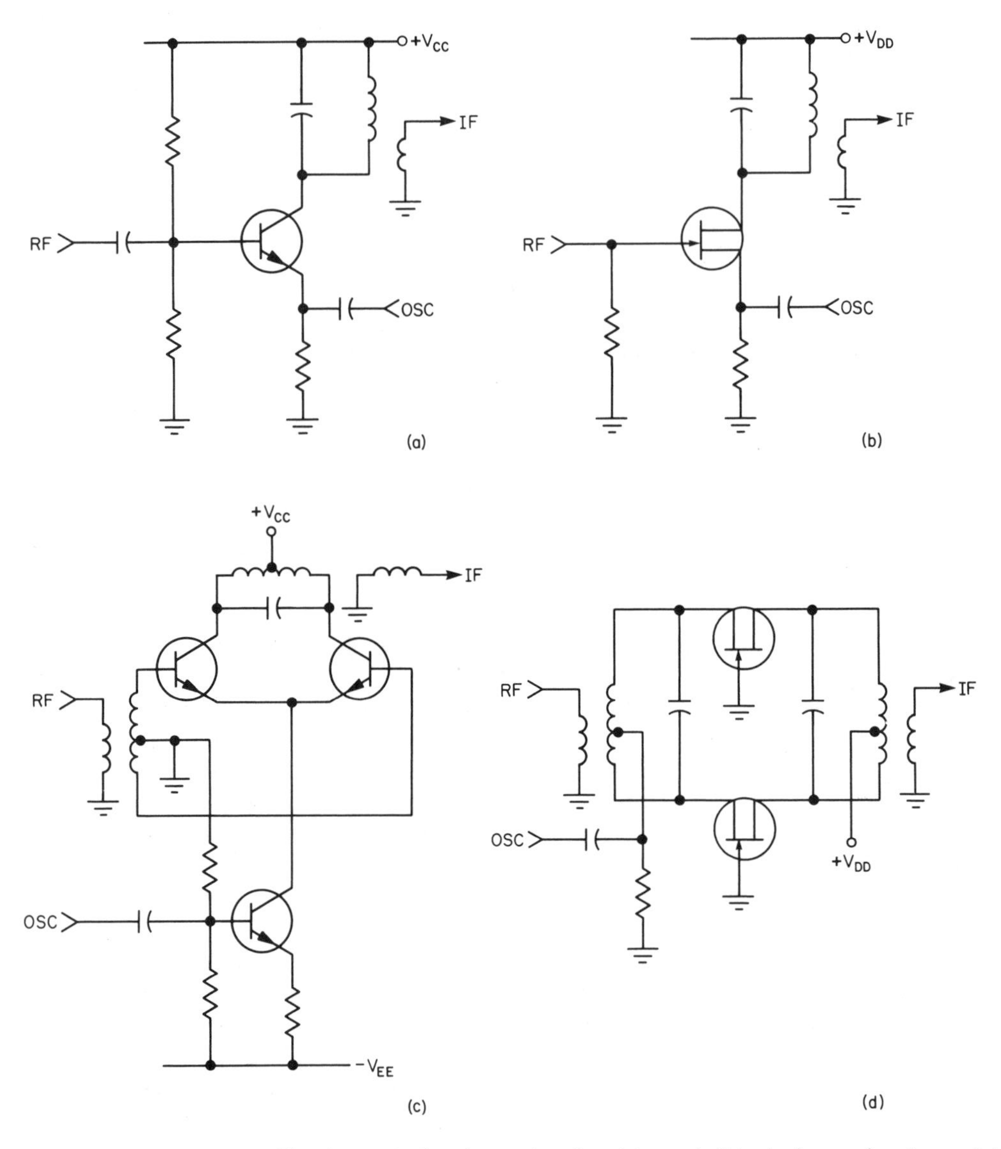

Figure 10-6 Single-ended mixer circuits (a) and (b); balanced mixers (c), (d), and (e); and a double-balanced mixer (f).

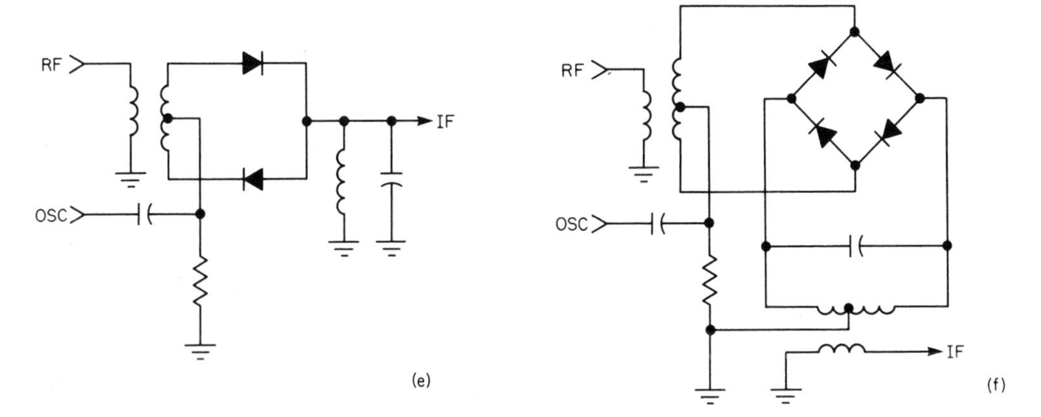

Figure 10-6 (cont'd)

The bipolar mixer of Figure 10-6(a) is probably the circuit most often used, but is probably the poorest. Its popularity is due to a high conversion gain, which makes it attractive economically, since lower IF gain is then required. Mixing takes place because the current gain of the transistor changes with collector current. The local oscillator signal injected into the emitter changes the bias and therefore the collector current. The relationship is not "linear," unfortunately, and so higher than second-order mixing also occurs and spurious interference results. The oscillator level (usually less than 100 mV rms) and the dc bias point must be chosen carefully to minimize higher-order mixing. Highest conversion gain is obtained by ensuring that the transistor stays nonlinear. Both the local oscillator and the RF signal should therefore come from low-impedance sources, thus reducing series resistances that tend to linearize the circuit.

The junction FET of Figure 10-6(b) is a closer approximation to the ideal square-law mixer but, unfortunately, is not used as often because its conversion gain is lower than that of the bipolar mixer. Oscillator level can be much higher, so better discrimination against unwanted second-order mixing products can be had. Conversion transconductance is about one-half that of its equivalent amplifier transconductance.

Both (a) and (b) circuits will have a strong oscillator signal at their output. Unless this is properly filtered out, the first stage of the IF amplifiers could also act as a mixer or, at the very least, as a nonlinear amplifier. For this and other reasons, such as suppressing oscillator radiation out to the antenna, balanced mixers are desirable.

The balanced bipolar mixer of (c), FET mixer of (d), and diode mixer of (e) will all suppress the local oscillator signal at the output by an amount depending on the matching of the diodes, transistors, and transformers. This makes them all suitable for use as balanced modulators for generating SSB signals. The balanced arrangement will also help suppress any tendency for mixing products to appear that contain even harmonics of the RF signal (i.e., $f_{osc} \pm 2f_{RF}$ and $f_{osc} \pm 4f_{RF}$. The FET and bipolar versions of the balanced mixers will have some positive conversion gain of 6–20 dB. The diode version will typically have a conversion loss of 6 dB. Because of difficulties in precisely matching transistors, the second-order mixing product rejection of the versions in (c) and (d) are not as good as the diode version of (e). The use of Schottky (hot-carrier) diodes allows much more careful matching in that case.

The balanced idea can be carried one step further, as shown in Figure 10-6(f). This double-balanced mixer keeps the oscillator signal from appearing at both the RF and the IF terminals. It is therefore useful as the front end of a receiver without an RF stage and with minimum filtering, as little oscillator signal would pass to the antenna.

Another benefit of any balanced mixer is associated with the elimination of the oscillator signal at the IF output. Since the oscillator input to the mixer is very strong compared to the RF input, any oscillator-noise sidebands could add very significantly to the overall receiver noise. Balanced mixers greatly reduce the problem. Oscillator noise can still be a problem, though, as strong, adjacent signals could still mix with a sideband of the oscillator. The problem, called *blocking*, can only be controlled by reducing the oscillator noise as much as possible. Filtering will have little effect.

10-8 RF AMPLIFIERS

Once the desired range of input signal levels to the mixer has been chosen, the RF amplifier can be designed (or eliminated) as required. Its gain should be just sufficient to bring the weakest signal from the antenna up to whatever level is needed to override noise generated in the RF amplifier and mixer. The total noise factor of the receiver will be given by

$$F = F_1 + \frac{F_2 - 1}{G_1} \qquad (10\text{-}1)$$

where: F_1 = noise factor of the first stage
F_2 = noise factor of the second stage
G_1 = power gain from input to second stage

For the arrangement shown in Figure 10-7, the RF amplifier has a gain of 12 dB (8 : 1) and the mixer has a loss of 4 dB (0.398 : 1). The noise figure of the RF stage is 2.0 dB (1.585) and of the IF amplifier is 2.5 dB (1.778). The

Figure 10-7 Noise figures, gains, and losses of the first three stages of a receiver.

overall noise figure would then be

$$\begin{aligned} F &= F_1 + \frac{F_2 - 1}{G_1} \\ &= 1.585 + \frac{1.778 - 1}{8 \times 0.398} \\ &= 1.829 \quad \text{(or 2.62 dB)} \end{aligned}$$

The RF amplifier has therefore provided enough gain so that the overall noise figure of 2.62 dB is only 0.62 dB higher than that of the RF amplifier itself. Higher gain would provide little overall improvement and would simply cause more problems with the mixer.

After the gain and noise figure are set, the next requirement is the filtering associated with the RF amplifiers. Part of this will depend on the mixer and part on the amplifier itself. The minimum filter requirements needed to complement the perfect square-law mixer were discussed in Section 10-4. If the mixer also has a significant third-order component, several new frequencies could end up in the IF passband. These would be:

(a) IF = RF ± 2OSC, RF = 16 or 26 MHz
(b) IF = 2RF ± OSC, RF = 2.75 or 7.75 MHz
(c) IF = 3RF, RF = 1.667 MHz

The examples shown alongside assume an IF of 5.0 MHz and the local oscillator at 10.5 MHz, a situation taken from Figure 10-4. The first new frequencies (a) at 16 MHz and 26 MHz would be outside the passband of the minimum 1-octave filters (5.5–10 MHz) and, also, if balanced mixers are used, the mixer would not function at even harmonics of the oscillator. This frequency then does not present any problem. The second pair (b) represents signals at 2.75 or 7.75 MHz. The latter lies right in the middle of the 1-octave filter range, so that if the mixer has significant third-order distortion, added filtering would be needed. The one choice is half-octave filters, the other is narrow-band, continuously tunable filters (with their tracking problems). Other spurious signals can be created by harmonics produced within the RF amplifier itself; such is the case with frequency (c) where a third harmonic created by amplifier nonlinearities could pass straight through the mixer. The 1.667-MHz input can easily be eliminated with filters ahead of the RF amplifier.

The total spurious frequency problem therefore depends to a great extent on the linearity of the RF stage, on filters before and after this stage, and on the mixer itself. The big problem involves gain control. To maintain low-level signals at the mixer input, the gain of the RF stage may need to be reduced at some time. For automatic gain control, the amplifier must have a nonlinear transfer characteristic so that a change of bias produces a change in gain. The resulting second-order nonlinearities could then produce spurious signals, which would cause mixing products to appear within the IF passband, particularly since the filters ahead of the RF stage are usually minimal. If AGC is used, good RF filtering is required.

A better approach is to make the amplifiers very linear, even by going to the extremes of balanced amplifiers with negative feedback as shown in Figure 10-8. Gain control can then be manual—either turning a potentiometer or switching in resistive pads, or automatic if linear devices are used.

The idea of a linear, two-terminal device that will not distort a signal yet can change its resistance with a voltage change may seem strange. A small incandescent lamp is one example. If a dc voltage is applied across the lamp and slowly changed, the current flowing into the bulb will not change linearly with the applied voltage. As the filament heats up, its resistance will increase. Any rapid voltage changes, however, will cause linear current changes, since the thermal time constant of the filament will be long enough to hold the

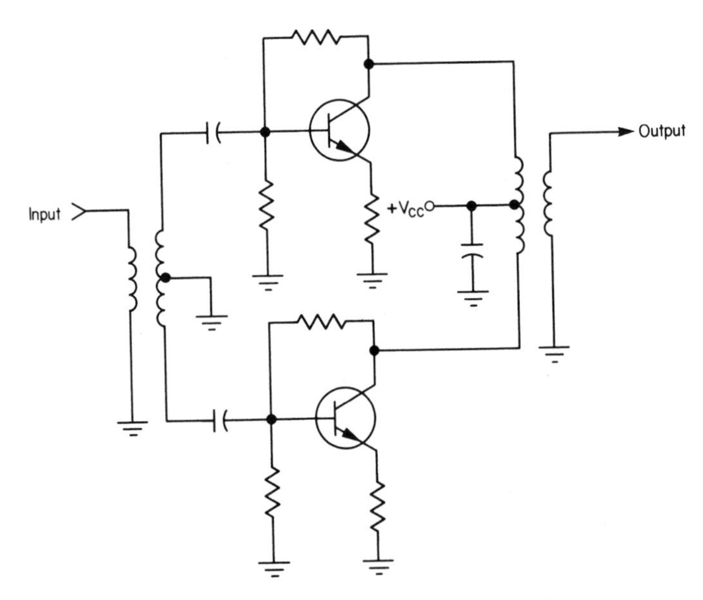

Figure 10-8 Linear RF amplifier without AGC capabilities.

resistance constant for a while. Such an idea has been used for automatic level control of good-quality audio oscillators for years. The modern equivalent of the lamp is the PIN diode. At low frequencies the device acts like a diode, but at higher frequencies it acts like a variable resistor, since the lifetime of its charge carriers is quite long (up to 500 ns). Above about 10 MHz (depending on the particular diode), a linear attenuator can be made that can be varied with a dc control voltage. A diagram of the lamp attenuator and a PIN diode attenuator are shown in Figure 10-9.

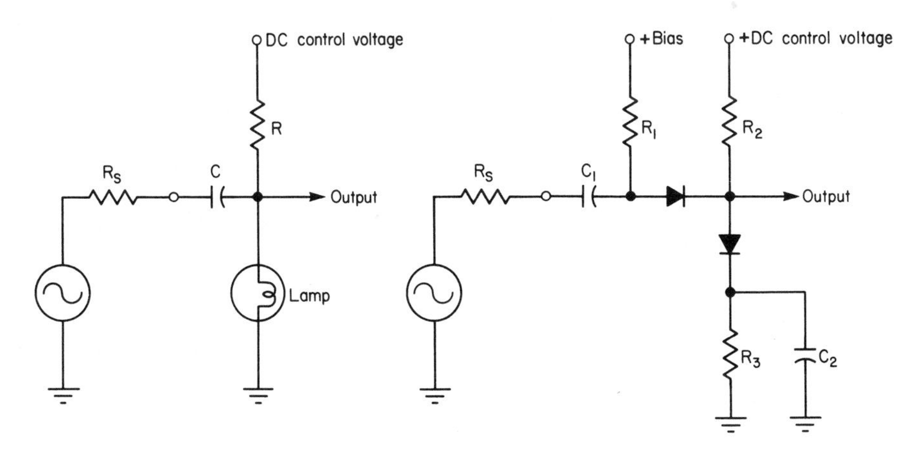

Figure 10-9 Linear attenuator circuit using an incandescent lamp (a) and PIN diodes (b).

10-9 IF AMPLIFIERS

The intermediate frequency section of a receiver is placed between the first mixer and the final detector circuits. It must:

1 Provide a high amount of gain, 60–100 dB, and reduce this when strong signals are present.
2 Filter out all unwanted signals outside the passband.
3 Limit amplitude variations in the case of FM signals, thereby determining the FM capture ratio.
4 Limit the amplitude of noise pulses in the case of AM and SSB signals.

These tasks must be performed without destroying the noise figure set by the receiver's front end and without introducing distortion products within the desired passband.

For the majority of receivers, a total gain of at least 20 dB will exist in the RF and mixer stages, so the IF noise figure is usually not significant. For the few cases where no RF stage is used and the mixer operates with a conversion loss, the IF noise figure will be very important. Any losses in the IF filters ahead of the amplifying stages must be considered; for if an RF gain of 15 dB, a mixer loss of 6 dB, and an IF filter loss of 9 dB occur, the IF signal level will be right back to where it was at the antenna terminals. The IF noise figure would then be very important.

Attention to noise figure itself is not sufficient, as the total noise bandwidth must also be considered. One part of this has already been pointed out; the image frequency from the mixer will add thermal noise in addition to the possibility of interfering signals. The total noise bandwidth could then be twice as wide as the IF filter bandwidth. The other noise problem can occur whenever separate filters and amplifiers are used.

As shown in Figure 10-10, there is no filtering after the integrated circuit used for the gain. The total noise output to the detector would then be the narrow-band noise through the IF filter from the RF amplifier and mixer plus the broadband 10 or 20 MHz generated within the IC. Some additional noise bandwidth filtering should therefore be provided between the IC and the detector.

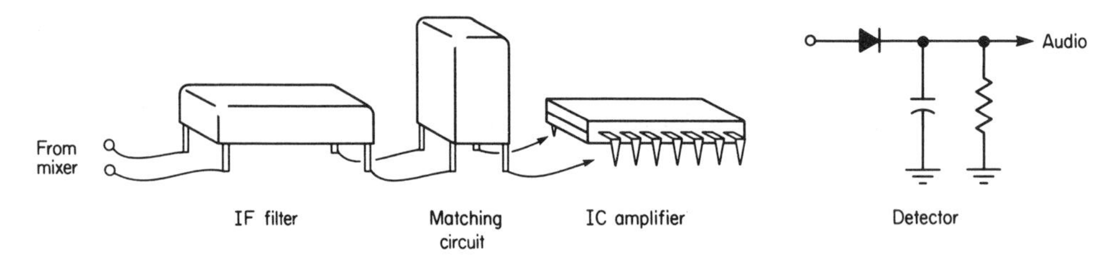

Figure 10-10 IF amplifier with a packaged filter and integrated-circuit amplifier. There is no filtering between the IC and the final detector, which could lead to a very wide noise bandwidth.

Distortion must also be considered. Since the IF amplifiers must be capable of automatic gain control, they must have a square-law or second-order transfer characteristic (V_{out}/V_{in}). Gain can then be changed by varying the amplifier's bias. As long as the second-order characteristic is maintained, no distortion will occur, assuming that the IF filters are relatively narrow band, less than 1 octave. All harmonics and intermodulation products will then fall outside the passband, so the amplifier will appear linear. But if any odd-order distortion is present, undesirable in-band mixing products and compression will appear. This type of distortion can obviously occur in the bipolar transistors normally used for the IF amplifiers, and the amount can be controlled through careful biasing and selection of the transistors used. It can also occur in some not-so-obvious components. Any quartz, ceramic, or mechanical filters involve physical movement of their internal elements and there will be symmetrical limits to this linear motion. The filters themselves can therefore be a source of distortion, particularly if the applied signal level is too high. Ferrite materials commonly used in filters can also be a source of nonlinearities. For very demanding applications, then, each component of the receiver must be carefully analyzed for its contribution to the final performance of the receiver.

10-10 AM IF AMPLIFIERS

The amplifiers of an AM or SSB receiver should appear linear to the signal and should supply a constant signal level to the final detector. As the signal strength to the antenna increases, the IF gain should be reduced proportionally so that the signal-to-noise ratio at the detector will improve, as shown in Figure 10-11.

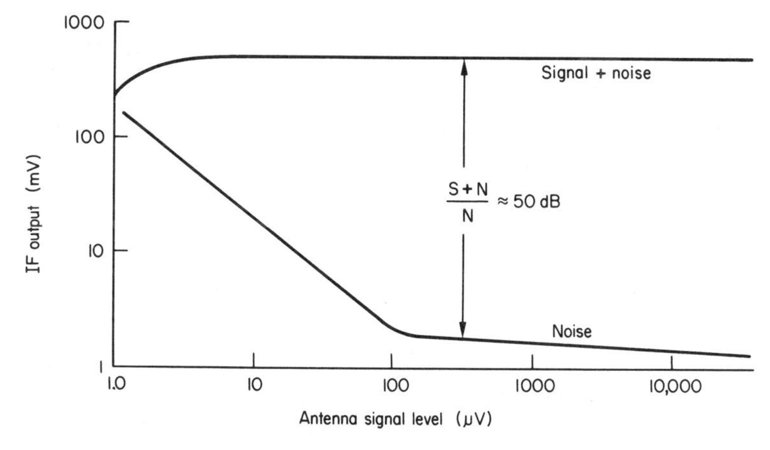

Figure 10-11 AGC of the IF amplifier will reduce the circuit gain to maintain a constant output to the detector and improve the S+N/N ratio.

If separate stages are gain-controlled, the stage closest to the detector should have its gain reduced first, and the gain reduction of the earlier stages should be delayed until the input signal is stronger. This will maintain a low noise level while preventing the higher-level stages from overloading. When delayed AGC is used in this manner, the stability of the AGC circuit should be considered in case two separately controlled amplifiers decide to fight it out and alternately reduce the gain of one and then the other.

AM and SSB receivers do not inherently enjoy the improved noise immunity provided by the limiter amplifiers, but some rejection of impulse noise (local lightning discharges and automobile ignition noise) can be obtained through the use of noise blanking circuits. As shown in Figure 10-12, short-duration, high-amplitude pulses will be shortened and broadened after passing through a narrow-bandwidth filter and so become less distinguishable (but just as bothersome). If an amplifier stage ahead of the narrow filtering were turned off for the duration of the short pulse, the loss of signal for the few microseconds would never be noticed when voice bandwidths are involved, and the pulse will have been eliminated.

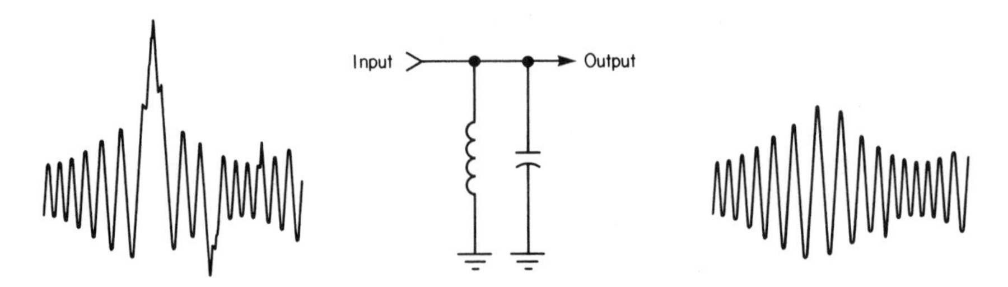

Figure 10-12 Impulse noise before and after a narrow-band filter.

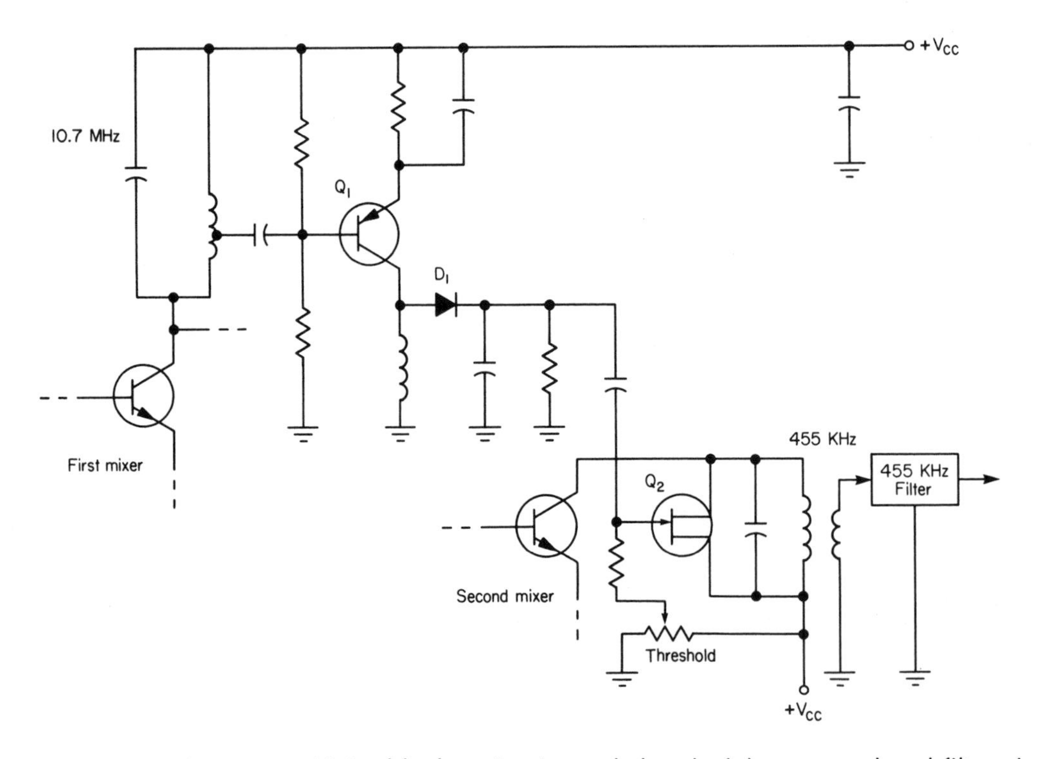

Figure 10-13 Noise blanker circuit used ahead of the narrow-band filters in a dual-conversion AM receiver.

A typical noise blanker for a dual-conversion AM receiver is shown in Figure 10-13. Transistor Q-1 amplifies the 10.7 MHz IF signal and any noise pulses. The signal is rectified by D-1 to form a positive pulse that is applied to the gate of FET Q-2. This FET is normally off and so does not interfere with the operation of the second mixer. But when a large-amplitude noise pulse appears, the FET turns into a low resistance that shunts the mixer load, greatly reducing its gain. The height of the pulse required to turn the FET on will be determined by the setting of the threshold potentiometer that sets the cutoff bias for the gate. Strong noise pulses will therefore be prevented from passing through the mixer stages into the narrow-bandwidth 455-kHz filter.

10-11 FM IF LIMITER AMPLIFIERS

The IF amplifiers for an FM receiver must amplify the weak signal from the mixer and remove any amplitude variations by acting as limiters. It is very important that any undesired signals be eliminated before the limiting action begins, since limiters are inherently nonlinear amplifiers. The most common amplifier for FM IF use is the IC limiter amplifier shown in Figure 10-14.

The emitter-coupled pairs are biased at relatively low current levels, so they cannot saturate. Limiting therefore takes place by alternately cutting off one transistor and then the other. For very weak signals, the first two pairs will operate as amplifiers and the last pair will perform the limiting. As the input signal increases, the limiting will occur in stages closer and closer to the input. The stages after the limiting one will not provide additional gain, since they will already be operating with maximum input signals. Therefore, no AGC is needed for the IF amplifiers (AGC may still be needed to keep the RF stages linear).

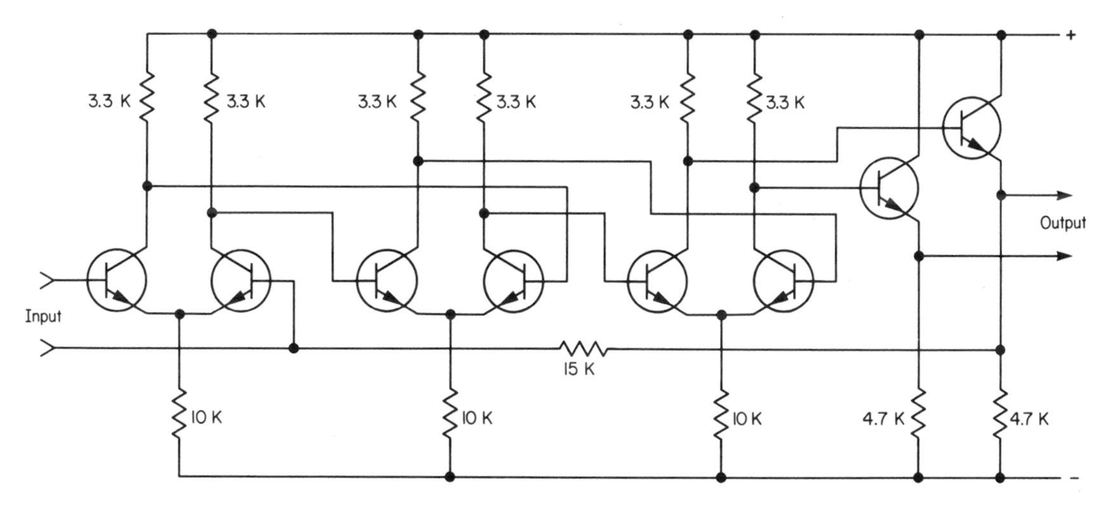

Figure 10-14 Circuit typical of many commercial limiter amplifiers used for the IF of FM receivers.

10-12 FM DEMODULATORS

A large number of FM demodulators are available; some of the more popular are described here. The Foster–Seeley discriminator and the ratio detector are shown in Figure 10-15. Both are high-quality detectors suitable for hi-fi stereo receivers but cannot be built in integrated-circuit form because of the special transformers. The operation of both detectors depends on the vector sum of two voltages at the point of rectification. One voltage is created across the inductor L and will always be 180° out of phase with the input signal. The other voltage is created across the tuned circuit and will be 90° out of phase with the input signal at the center frequency and will have a greater angle at higher frequencies and a smaller angle at lower frequencies. The audio output will be the difference between the absolute sum of these two different resultant voltages. The ratio detector goes one step further and includes a capacitor across the two outputs that tends to keep the sum more constant and so provide increased immunity to amplitude fluctuations at the input.

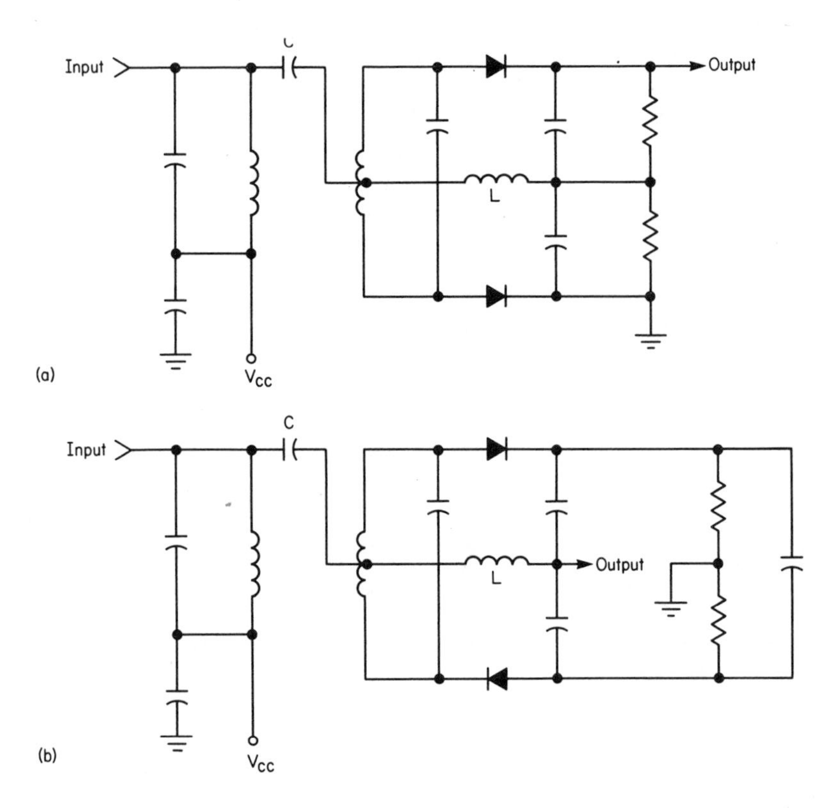

Figure 10-15 Foster–Seeley discriminator (a) and the ratio detector (b).

The phase-locked loop of Section 8-10 can also be used as an FM demodulator, as shown in Figure 10-16. The VCO will track the frequency variations of a signal applied to the reference input, and the demodulated audio will be available on the control voltage line, since it is the signal that is telling the VCO which frequency to go to. The linearity of this type of detector is the biggest problem, as it corresponds exactly to the linearity of the VCO's voltage-to-frequency characteristics. For FM broadcast reception, this would require better than 0.5% linearity over a 200-kHz range centered at 10.7 MHz.

For integrated-circuit fabrication, the quadrature detector of Figure 10-17 is popular. It still requires an external inductor but it is simpler than the

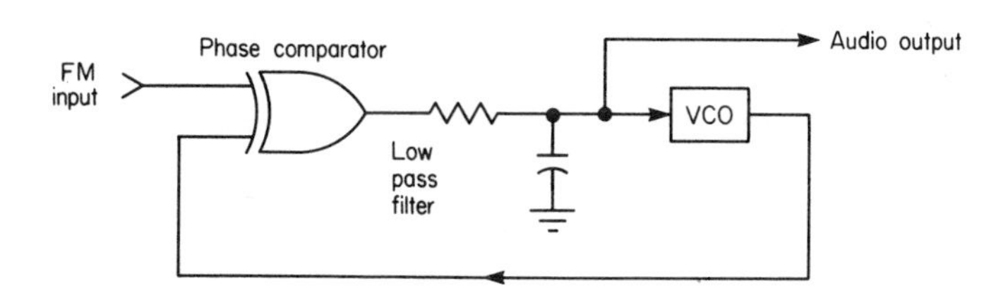

Figure 10-16 Block diagram of an FM demodulator using a phase-locked loop.

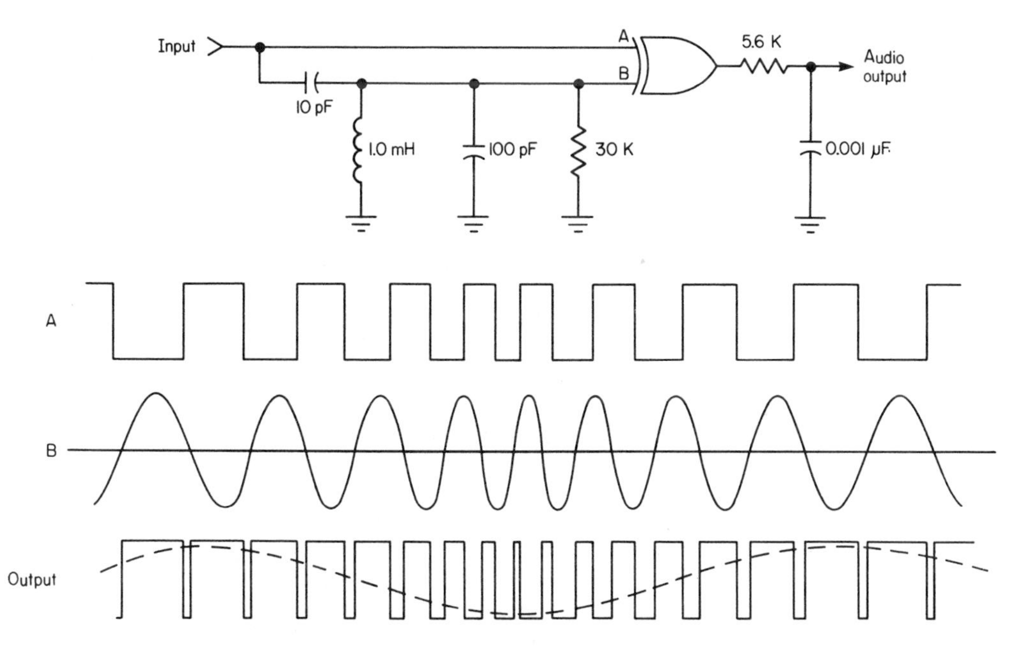

Figure 10-17 Quadrature FM detector, along with the two input waveforms to the exclusive-OR circuit and its output waveform. The audio component is shown as a dotted line on the output waveform.

transformers required for either the discriminator or ratio detector of Figure 10-15. The input to the detector could be a sine wave but is usually an FM square wave from a limiter amplifier similar to that of Figure 10-14. The detector itself consists of either an AND gate or an exclusive-OR gate, and a phase-shift network. In Figure 10-17, the phase-shift network consists of the two capacitors, C_1 and C_2, and the inductor, L. C_2 and L are parallel resonant at 500 kHz, and at the 455-kHz operating frequency of this detector the parallel circuit has an equivalent reactance of $+j16$ kΩ. Capacitor C_1 has a reactance of $-j30$ kΩ at the same frequency. The signal reaching input B will be a sine wave shifted almost 90° from the input at A (the parallel tuned circuit will filter out most of the square-wave harmonics). At frequencies below the center frequency, the phase shift will be increased, and at higher frequencies, the shift will be decreased. The two inputs and the output of the exclusive-OR circuit are shown in Figure 10-17. The audio component will be the average value of the output waveform and can be recovered with a low-pass filter. The internal construction of the logic gate was shown in Figure 8-20.

10-13 AM DETECTORS

At first appearance, the AM detector is the simplest of all detector circuits, requiring only a single diode and a small filter capacitor. But for some applications, such as television and weak-signal communications, significant improvements can be obtained by using a more sophisticated circuit.

The single-diode demodulator operates as a peak detector. The diode half-wave rectifies the modulated signal and the capacitor, if it is not too large, charges up to and follows the peaks, as shown in Figure 10-18. The average value of the capacitor voltage will be the recovered audio waveform. One of the

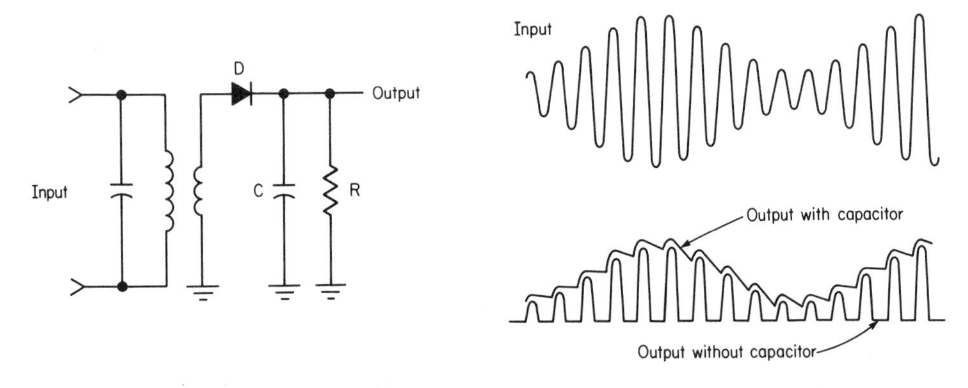

Figure 10-18 Single-diode AM detector; operates as a peak detector. The input and output waveforms show the operation of the circuit.

problems with this simple detector shows up when weak signals mixed with random noise are being received. Since the diode is a nonlinear device, intermodulation occurs between the noise and the desired signal. The signal-to-noise ratio after the detector will then be poorer than before the detector, and the deterioration will be worse as the input signal-to-noise ratio decreases. For AM receivers operating close to the transmitting station, the received signal usually has a high-enough signal-to-noise ratio that any deterioration is minor; but for long-range communications with very weak signals and low signal-to-noise ratios, the deterioration is significant. It is the intermodulation of the noise with the received signal that explains why the noise level at the output of a receiver will increase when a weak signal is tuned in.

The input and output signal-to-noise ratios of an ideal AM detector and a typical diode detector are compared in Figure 10-19. If the input signal is the same strength as the input noise (0 dB on the horizontal scale), the ideal detector will produce an output signal that is 3 dB stronger than the noise. The typical diode detector, however, will produce a desired output that is 5 dB weaker than the noise. This is a deterioration of 8 dB. For weaker input signals, the difference increases, and for stronger inputs, the difference drops and approaches 6 dB.

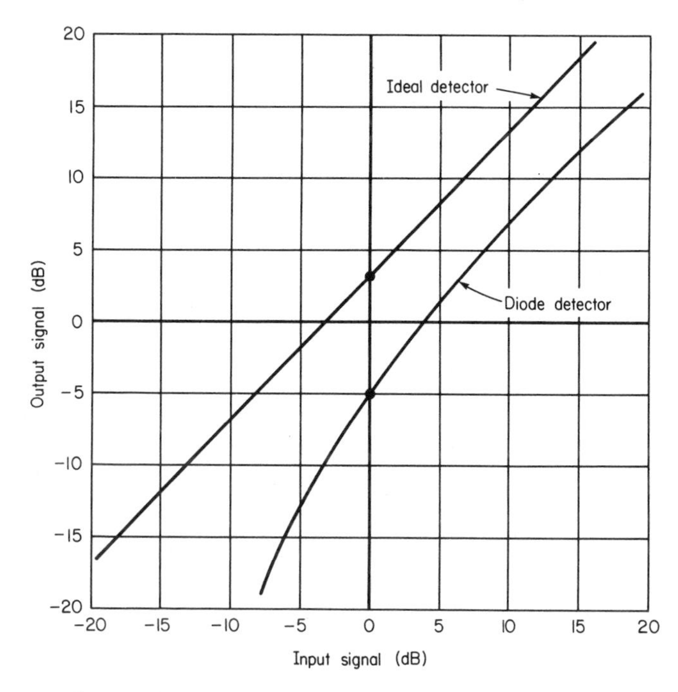

Figure 10-19 Relationship between input and output signal-to-noise ratios for an ideal AM detector and for a typical diode peak detector. The scales are calibrated for signal level relative to a 0-dB noise level.

The second problem with the peak detector circuit shows up when it is used to demodulate a television video signal. The vestigial sideband modulation used to conserve bandwidth means that video information higher than 1.0 MHz is transmitted in only one sideband.

As shown in Figure 10-20, a 2.0-MHz video signal and the video carrier produce an envelope that appears distorted. If a diode peak detector is used, and it frequently is, the demodulated waveform will also be distorted. The problem is eliminated if the amplitude of the modulated waveform is sampled periodically in step with the peaks of the unmodulated carrier rather than at the peaks of the modulated wave, which will be shifted in time. The correctly demodulated waveform is shown by the dotted line.

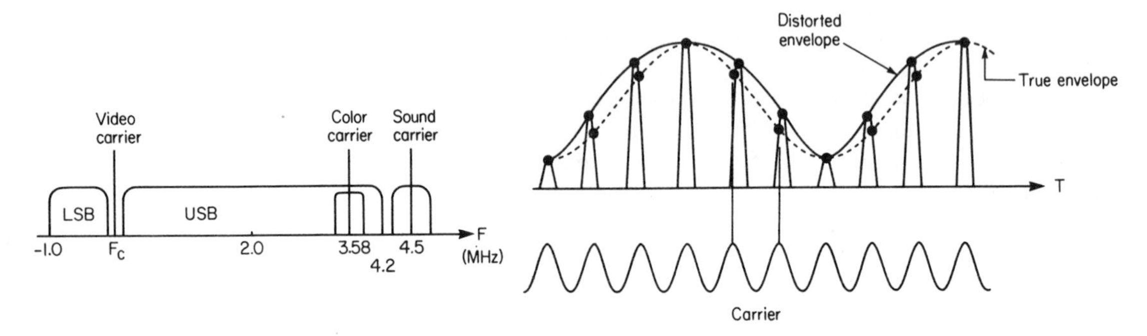

Figure 10-20 Television video signals above 1.0 MHz are transmitted in only one sideband, thereby causing the envelope of the modulated carrier to appear distorted.

10-13.1 Ideal AM Demodulator

The ideal AM demodulator is nothing more than a square-law mixer circuit, as shown in Figure 10-21. If the oscillator input to the mixer has the same frequency as the carrier of the modulated signal, the difference signal at the mixer output will be the demodulated audio free of envelope distortion and with a 3-dB better signal-to-noise ratio than was available at the mixer/demodulator input (assuming that the audio bandwidth is half the RF bandwidth). The 3-dB improvement is due to the multiplying or sampling action of the mixer, which demodulates only the noise component that is in-phase with the carrier; it is not due to the apparent change in bandwidth.

Any of the mixer circuits of Figure 10-6 can be used, although the circuits of (c), (e), and (f) are most common. The output circuit would have to be modified to provide a load for the audio instead of RF frequencies. The exclusive-OR circuit of Figure 8-20 can also be used, as it is equivalent to a double-balanced mixer. The problem of setting an oscillator to exactly the

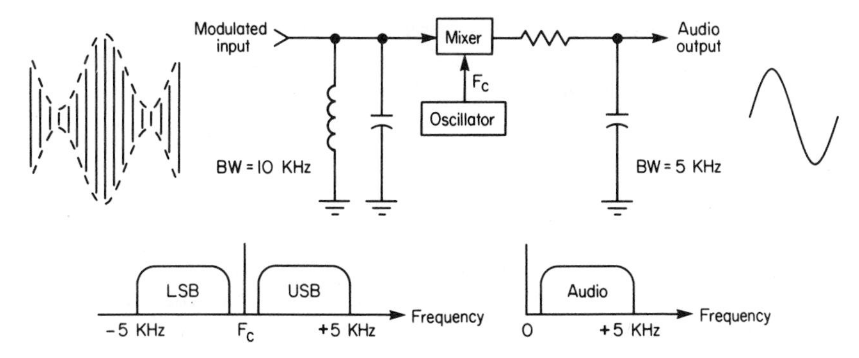

Figure 10-21 Ideal AM detector; a square-law mixer with an oscillator operating at the carrier frequency.

same frequency as the incoming carrier is greatly eased if a phase-locked loop is used to lock onto the carrier of the modulated signal. The local oscillator signal must also be more-or-less in phase with the carrier at the mixer. If it happened to be exactly 90° out of phase either way, a look at Figure 10-20 would show that the output would always be zero. Phase shifts of less than 90° cause a reduction in audio output level.

A diagram of a *synchronous detector* is shown in Figure 10-22. The phase comparator and voltage-controlled oscillator act as a very narrow band filter to extract the carrier from the incoming modulated wave. This drives alternate

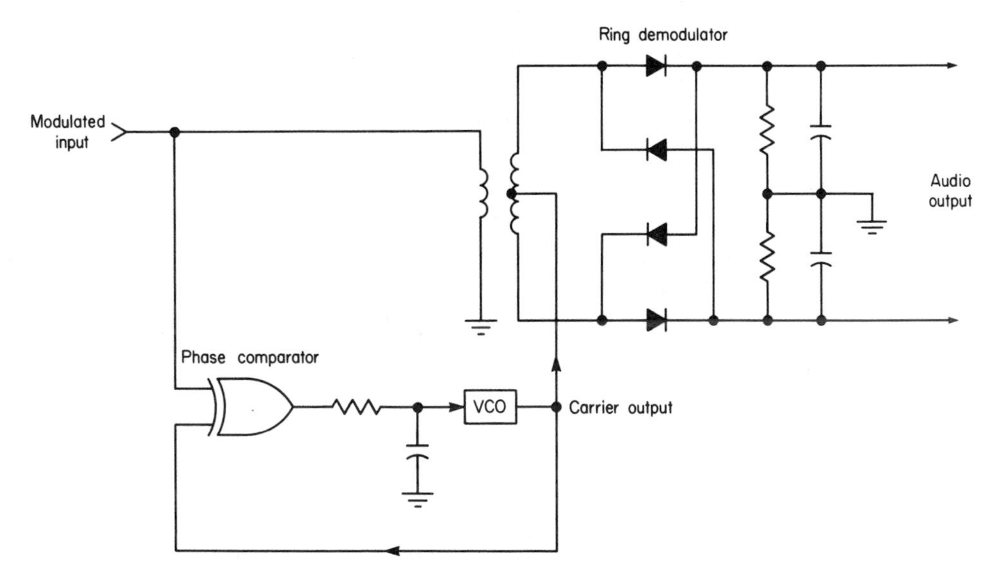

Figure 10-22 Ideal AM detector; a balanced ring demodulator and a phase-locked loop carrier oscillator.

pairs of diodes on in the balanced *ring demodulator* in step with the carrier. The sampled/full-wave-rectified output is filtered to remove any high-frequency components leaving the perfectly demodulated audio signal.

QUESTIONS

1 Explain why the mixer is the most critical section of a receiver.

2 What advantages are obtained by using an "up-converter" type of mixer whose IF output is at a higher frequency than the incoming RF signal? What are the disadvantages?

3 For a mixer with the characteristics shown in Figure 10-5, what is the maximum input power (dBm) that will keep all undesired mixing products 20 dB below the desired one?

4 Why are VHF receivers usually designed with much better sensitivity than MF broadcast band receivers?

5 The first stage of a receiver has a 3.5-dB noise figure and a 9.5-dB gain. The tuned circuits ahead of the first stage have a 0.5-dB insertion loss due to resistive losses. The second stage amplifier has a 4.7-dB noise figure. What is the overall noise figure of this receiver?

6 A receiver for 18.0 MHz has an IF at 4.5 MHz. What loaded Q will be needed for a single-tuned circuit to reduce the image signal to 45-dB below the desired signal level if (a) the local oscillator is on the lowside, or (b) the oscillator is on the high side of the 18-MHz signal?

7 Explain what would happen in a receiver if the local oscillator level were to decrease.

8 Describe three defects of the single-diode AM detector.

9 For a 40-MHz FM receiver-transmitter system, the modulation extends ± 8 kHz about the carrier. Both the transmitter and receiver are equipped with $\pm .003\%$ crystals. The receiver IF is 10.7 MHz. Calculate the required IF bandwidth necessary to accommodate the modulation and all tolerances if (a) the local oscillator is on the lowside, and (b) the oscillator is on the high side of the RF signal.

REFERENCES

Brown and Glazier. 1974. *Telecommunications*. 2nd ed. London: Chapman and Hall Science Paperbacks.

Kennedy, G. 1970. *Electronic communication systems*. New York: McGraw Hill.

News from Rohde and Schwarz, Issue 78, Munich, Germany: Rohde and Schwarz.

Rohde, U. L. Feb. 20, 1975. Eight ways to better radio receiver design. *Electronics*.

appendix a
charts

This Appendix contains four charts often used in high-frequency design work. The first is a reactance chart that can be used to determine the reactances of capacitors and inductors and also find approximate resonant frequencies.

The other three charts are Smith charts as used for impedance matching (see Chapter 5). The first Smith chart is marked with impedance coordinates and has a 50-Ω center. The second Smith chart is marked with admittance coordinates and has a 20-mmho center. The last chart is the "universal" admittance/impedance chart with 1.0 at its center.

HIGH-FREQUENCY REACTANCE CHART

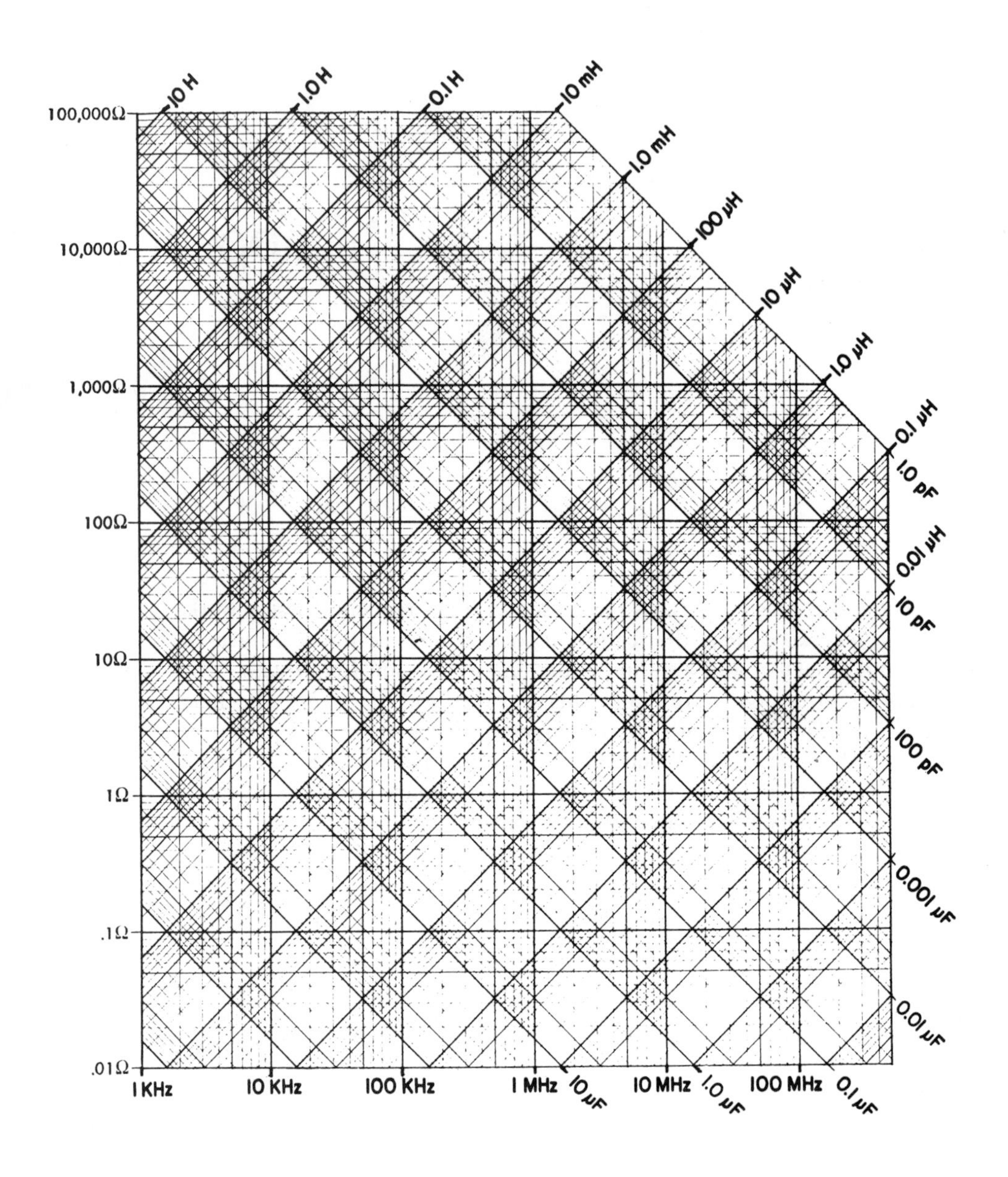

IMPEDANCE COORDINATES—50-OHM CHARACTERISTIC IMPEDANCE

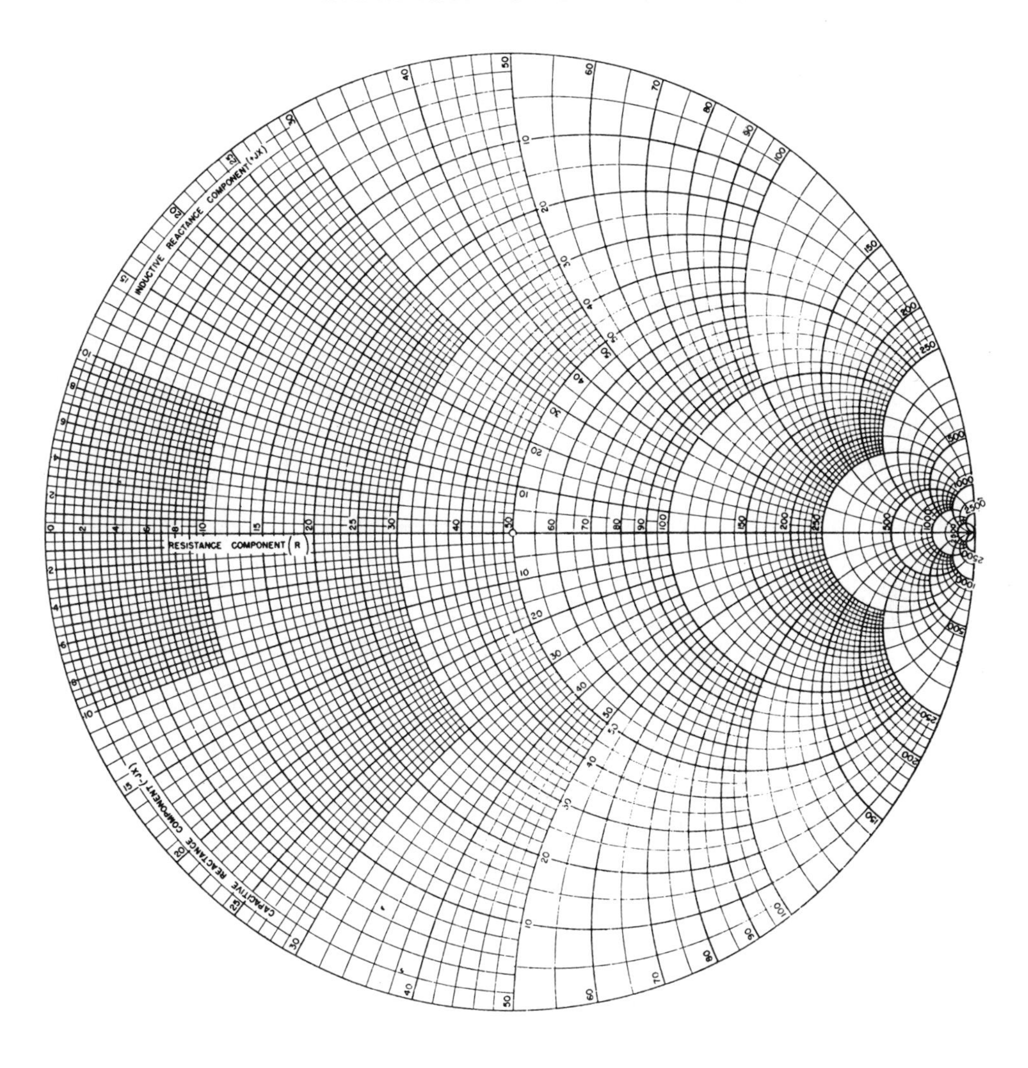

ADMITTANCE COORDINATES—
20-MILLIMHO CHARACTERISTIC ADMITTANCE

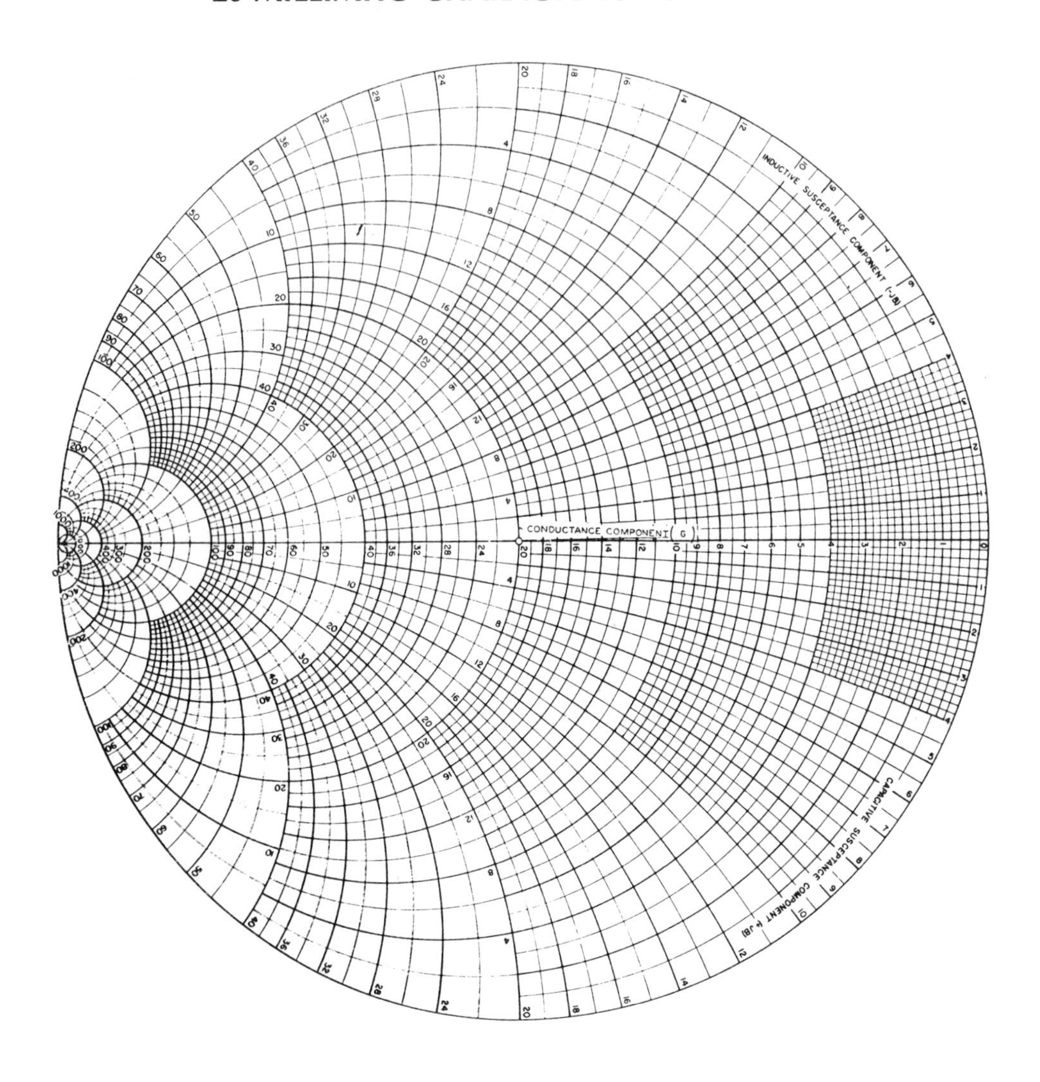

IMPEDANCE OR ADMITTANCE COORDINATES

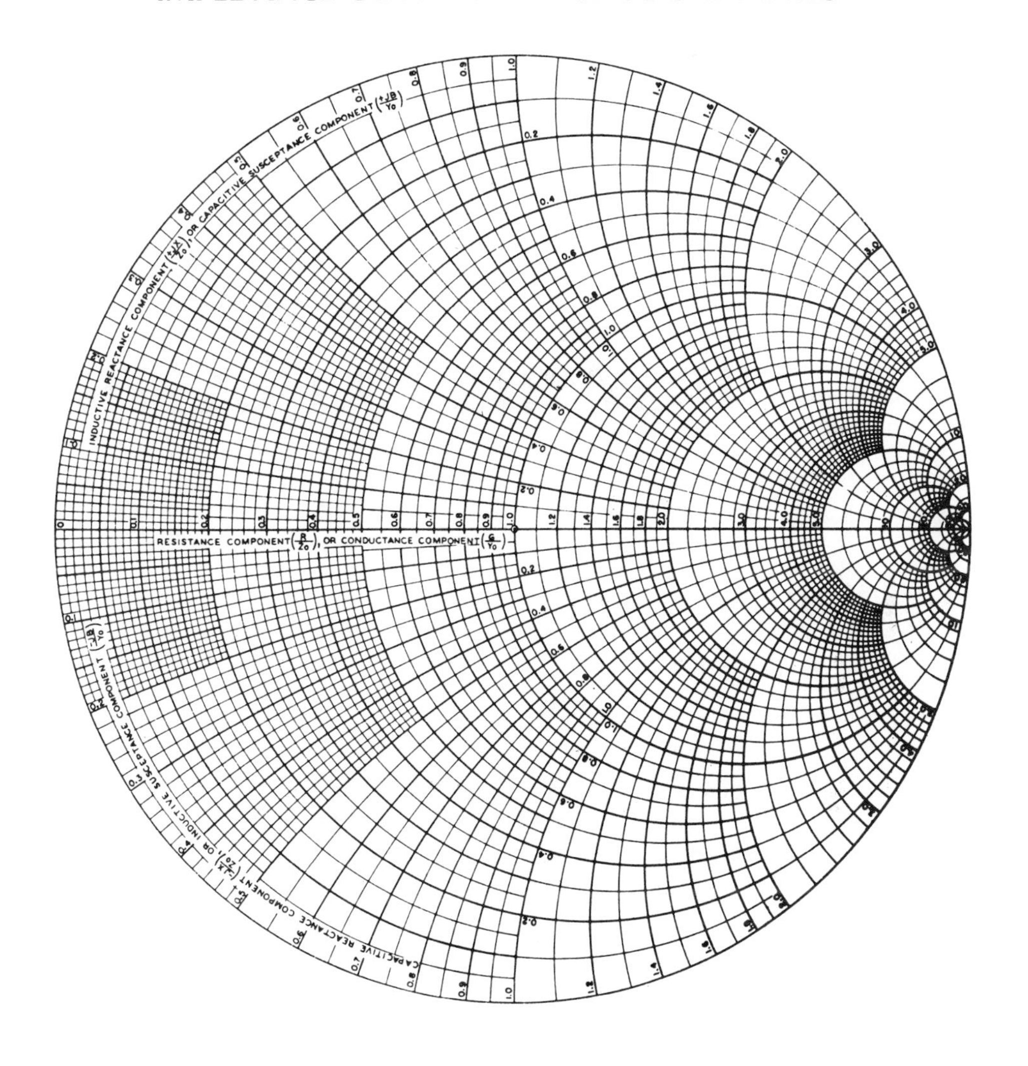

appendix b

answers to questions

CHAPTER 1

1 $e_n = 7.19\ \mu V$
2 S/N ratio $= 16.8$ dB
3 $NF = 8.11$ dB
4 Three sets of points are $e_{in} = 0.5, 2.0, 20.0$ mV;
fundamental $= 10.0, 40.0, 400$ mV;
second harmonic $= 0.375, 6.0, 600$ mV.
5 Second order intercept at $e_{in} = 13.33$ mV;
largest input signal 1.33 mV.
6 Third order intercept at $e_{in} = 163.3$ mV
7 Envelope delay $= 2.78\ \mu s$

CHAPTER 2

1 Resistance (dc) $= 13.5\ \Omega$
2 Resistance (ac) $= 71.5\ \Omega$

3 5.8 turns, 1.27 cm length, 0.637 cm radius.

4 247 turns, $b=0.40$ cm, $a=0.80$ cm.

5 46.3 turns of #26 enamel coated wire.

6 Resonant frequency (20°C)=11.397 MHz increase at 75°C=194.9 kHz.

7 X4V characteristic, +22 to −82% change for −55°C to +65°C temperature change.
Y5P characteristic, ±10% change for −30°C to +85°C temperature change.

CHAPTER 3

1 $L=3.94$ μH, $C=670$ pF.

2 bandwidth=227.7 kHz,
insertion loss=2.04 dB.

3 $L=1.48$ μH, $C=9400$ pF, turns ratio=2.73.

4 $L=1.82$ μH, $C=676$ pF, C coupling=15.9 pF.

CHAPTER 4

1 Maximum phase shift 450°;
final attenuation slope 30 dB/octave;
higher initial slopes possible with high-Q filters.

2 $L=280$ μH, $C=0.0124$ μF.

3 4 elements, $C_1=14.5$ pF, $L_2=12.6$ μH,
$C_3=35.0$ pF, $L_4=5.22$ μH.

4 $C_2=334$ pF, $L_1=L_3=2.57$ μH.

5 $C_1=C_3=1975$ pF, $L_1=L_3=0.424$ μH,
$C_2=110$ pF, $L_2=7.64$ μH.

6 $L_1=L_3=0.955$ μH, $C_2=943$ pF,
input impedance$=4.5+j36$ Ω.

7 On the graph the delay at 3.5 MHz is 0.141 μs.

CHAPTER 5

1 $C=1370$ pF, $L=2.91$ μH, bandwidth=3.87 MHz.

2 $L_1=0.759$ μH, $C_1=423$ pF, $L_2=1.11$ μH, $C_2=289$ pF.

3 $L_1=0.327\ \mu H$, $C=727$ pF, $L_2=0.545\ \mu H$.

4 $L=0.160\ \mu H$, $C_1=0.212\ \mu F$, $C_2=0.0801\ \mu F$.

5 $L=0.0424\ \mu H$, $C=84.9$ pF.

6 $C_1=1770$ pF, $L=0.212\ \mu H$, $C_2=424$ pF.

7 $Z_0=23.8\ \Omega$, width = 2.0 cm, length = 19.6 cm.

CHAPTER 6

1 $f_B=1.77$ MHz, $f_T=150$ MHz, current gain at 25 MHz = 6.0

2 Current gain = $2.59\angle -88.3°$
voltage gain = $45.3\ \angle -38.6°$

3 Power gain = 18.8 dB

4 $Z_{in}=20.7-j\,24.3$ with $R_L=560\ \Omega$,
$Z_{in}=21.9-j\,56.4$ with $R_L=0$.

6 MAG = 15.5 dB.

CHAPTER 7

1 $K_S=1.71$, stable amplifier.

2 $m=2.17$, $G_S=13.0$ mmho, $G_L=0.522$ mmho.

3 0.28 dB loss.

4 $Y_{in}=53.2+j4.85$ mmho,
$Y_{out}=3.25+j4.59$ mmho.

5 1.31 μH across input, 1.39 μH across output.

7 $Q_u=65.6$ minimum.

8 Feedback reduced 6 dB.

CHAPTER 8

2 $C_1=318$ pF, $C_2=8600$ pF, $L=0.825\ \mu H$.

3 $C=307$ pF, $L=0.825$ tapped at 1/27.

4 $C=465$ pF, $L=1.37\ \mu H$, C_1 and C_2 same as Colpitts.

5 Highest frequency 27.501,375 MHz, lowest frequency 27.498,625 MHz.

6 Reference crystal 20 kHz, dividers must operate ÷ 750, 751, 752 · · · 799.

CHAPTER 9

1 Drive power 1.7 watts,
amplifier gain 9.45 dB,
parallel input resistance 2.5 Ω,
reactance $-j$ 15 Ω.

2 $R_L = 24.8$ Ω.

3 Deviation $= \pm 1.333$ kHz, or $\pm 0.008\%$.

5 Load resistance $= 12.6$ Ω,
DC power $= 11.8$ watts.

6 Audio power $= 5.9$ watts,
secondary impedance $= 16.1$ Ω.

7 $BV_{CES} = 55$ volts minimum.

8 5 elements.

CHAPTER 10

3 Maximum input -7 dBm.

5 NF $= 4.41$ dB.

6 (a) $Q = 120$, (b) $Q = 210$.

9 (a) BW $= 20.2$ kHz, (b) BW $= 21.4$ kHz.

index

a

b

c

d

e

f

g

h

i

j

l

m

n

o

p

q

r

s

t

u

v

w

y

z